Wittgenstein's *Tractatus*

MIND ASSOCIATION OCCASIONAL SERIES

This series consists of carefully selected volumes of significant original papers on predefined themes, normally growing out of a conference supported by a Mind Association Major Conference Grant. The Association nominates an editor or editors for each collection, and may cooperate with other bodies in promoting conferences or other scholarly activities in connection with the preparation of particular volumes.

Director, Mind Association: M. Fricker
Publications Officer: M. A. Stewart

Recently published in the series:

Philosophical Perspectives on Depiction
Edited by **Catharine Abell** and **Katerina Bantinaki**

Emergence in Mind
Edited by **Cynthia Macdonald** and **Graham Macdonald**

Empiricism, Perceptual Knowledge, Normativity, and Realism
Essays on Wilfrid Sellars
Edited by **Willem A. deVries**

Spheres of Reason
New Essays in the Philosophy of Normativity
Edited by **Simon Robertson**

Dispositions and Causes
Edited by **Toby Handfield**

Images of Empiricism
Essays on Science and Stances, with a Reply from Bas C. van Fraassen
Edited by **Bradley Monton**

Philosophers on Music
Experience, Meaning, and Work
Edited by **Kathleen Stock**

Values and Virtues
Aristotelianism in Contemporary Ethics
Edited by **Timothy Chappell**

Identity and Modality
Edited by **Fraser MacBride**

Truthmakers
The Contemporary Debate
Edited by **Helen Beebee** and **Julian Dodd**

Wittgenstein's *Tractatus*

History and Interpretation

EDITED BY
Peter Sullivan and Michael Potter

OXFORD
UNIVERSITY PRESS

OXFORD
UNIVERSITY PRESS

Great Clarendon Street, Oxford, OX2 6DP,
United Kingdom

Oxford University Press is a department of the University of Oxford.
It furthers the University's objective of excellence in research, scholarship,
and education by publishing worldwide. Oxford is a registered trade mark of
Oxford University Press in the UK and in certain other countries

First Edition published in 2013

Impression: 1

British Library Cataloguing in Publication Data
Data available

Library of Congress Cataloging in Publication Data
Data available

ISBN 978–0–19–966578–5

Printed in Great Britain by
MPG Books Group, Bodmin and King's Lynn

Contents

Acknowledgements vi
List of contributors vii

1. Introduction 1
   Michael Potter and *Peter Sullivan*
2. Wittgenstein's pre-*Tractatus* manuscripts: a new appraisal 13
   Michael Potter
3. Why does Wittgenstein say that ethics and aesthetics are one and the same? 40
   Hanne Appelqvist
4. Kierkegaard and the *Tractatus* 59
   Genia Schönbaumsfeld
5. What is Frege's 'concept *horse* problem'? 76
   Ian Proops
6. *Tractatus* 5.4611: 'Signs for logical operations are punctuation marks' 97
   Peter Milne
7. Logical segmentation and generality in Wittgenstein's *Tractatus* 125
   Thomas Ricketts
8. Does the *Tractatus* contain a private language argument? 143
   William Child
9. Logic and solipsism 170
   James Levine
10. Was the author of the *Tractatus* a transcendental idealist? 239
    A. W. Moore
11. Idealism in Wittgenstein: a further reply to Moore 256
    Peter Sullivan

Index 271

Acknowledgements

Most of the chapters in this volume derive from papers presented at a conference held at the University of Stirling in September 2005. This conference was the main public meeting of a three-year AHRC-funded project on the interpretation of Wittgenstein's *Tractatus*. Additional support for the conference was provided by the Mind Association, the Aristotelian Society, the Scots Philosophical Club (now, the Scots Philosophical Association), and the Department of Philosophy at the University of Stirling. We gratefully acknowledge the support of all these bodies.

We should also like to thank those institutions and people who contributed to the success of the project more generally, first, Stirling's Philosophy Department and its Heads, Alan Millar and Antony Duff, for hosting the project, and secondly an impressively committed group of *Tractatus* scholars who made regular journeys to Stirling over three years to attend its series of workshops.

The volume has been quite some time in the making, and so we owe to our contributors thanks not only for the quality of their work but for their remarkable patience with the delays in its publication. Finally, we are extremely grateful to Peter Momtchiloff for the support and advice he has offered, and the forbearance he has shown, throughout this time.

PMS, MDP

List of contributors

HANNE APPELQVIST Academy of Finland Postdoctoral Fellow at the University of Helsinki

WILLIAM CHILD Lecturer in Philosophy at the University of Oxford and Fellow of University College

JAMES LEVINE Senior Lecturer in Philosophy at Trinity College, Dublin

PETER MILNE Professor of Philosophy at the University of Stirling

A. W. MOORE Professor of Philosophy at the University of Oxford and Fellow of St Hugh's College

MICHAEL POTTER Professor of Logic at the University of Cambridge and Fellow of Fitzwilliam College

IAN PROOPS Professor of Philosophy at the University of Texas, Austin

THOMAS RICKETTS Professor of Philosophy at the University of Pittsburgh

GENIA SCHÖNBAUMSFELD Senior Lecturer in Philosophy at the University of Southampton

PETER SULLIVAN Professor of Philosophy at the University of Stirling

1

Introduction

Michael Potter and Peter Sullivan

Wittgenstein wrote the *Tractatus* while he was serving, first as a volunteer soldier, then as an officer, in the Austrian Army during the First World War. Although he completed it in the summer of 1918, it did not appear in print for three more years. Austria's parlous economic state contributed to the delay, no doubt, but so did the book's impenetrability. Frege was too perplexed to make much headway with it; even Russell needed a week of explanations from the author. Yet when the book did appear, philosophers quite soon began to pay attention to it. Keynes said in 1924 that it dominated all fundamental discussions in Cambridge (Wittgenstein 2008: 151). With Schlick's encouragement the Vienna Circle treated it as their bible. By the time Wittgenstein returned to Cambridge in 1929, he was a noted figure not just there but abroad. His stance in the philosophy of mathematics, for instance, was judged sufficiently important to merit its own session at the Königsberg conference in 1930, alongside logicism, intuitionism, and formalism.

Throughout the 1930s, however, rumours emanated from Cambridge that Wittgenstein had rejected the views he expressed in the *Tractatus*. He did not publish his new views: just what they were could at first be gleaned only from samizdat copies of typescripts he had prepared for his Cambridge students. Only after his death did his literary executors publish the *Philosophical Investigations*. The next forty years were a period of gentle decline in interest in the *Tractatus*. At first, this was because so much effort had to be put into understanding and assimilating the lessons of Wittgenstein's late work. Then there was a flow of further material to take account of, as his executors oversaw the gradual publication of his *Nachlass*. Most of all, though, for those who were not interested in the minutiae of the texts there seemed little point in studying a book whose claims its author had himself rejected as mistaken.

More recently, however, there has been a resurgence of interest in the *Tractatus*. Why so? One simple answer would point to the process by which an uncomfortable and old-fashioned house is transformed by the passage of time into a treasured monument: the *Tractatus* is now far enough removed from us to assume the status of a classic text, a status that has encouraged a different, more scholarly and historical mode of engagement with it. But another, undoubtedly important reason has been the excitement and controversy generated by what has become known as the 'new' or 'resolute' reading. As the literature has grown it has, perhaps, become less and less clear what its proponents take a resolute reading of the book to be, but at its centre is the issue of what follows from taking seriously the book's claim that most of the sentences it contains are sheer nonsense. The controversy generated by the resolute reading was surely to some extent deliberate: one of the most striking and strikingly contestable claims made on its behalf has been its novelty, and it has often been presented by means of a contrast with another ('traditional' or 'irresolute') reading to which it is supposed to stand in opposition.

Controversies of this kind have a shelf-life. While polemical advocacy of schematically characterized general approaches can usefully serve to stimulate interest, satisfaction of the interest thus stimulated demands that the schematic outlines be filled in. As leading advocates and critics of the resolute reading have turned to this more substantial work, the sense of a simple or direct opposition between them has become hard to sustain. The more closely they have engaged with the *Tractatus* itself, the more constructively they can be seen to engage with each other. This is, in our view, a welcome development, and one to which we hope that this volume will contribute. Even a cursory glance at the chapters in it will make it clear that evaluating the resolute reading is not their primary aim. Some of the contributors would count themselves resolute, others not, but all have focused here principally on the part of the task of understanding the book that stands before the moment when the ladder is thrown away rather than on what commitments or understandings that moment may leave behind. This is not because the second issue is unimportant; it is rather because a rich or developed account of it can only be approached through the first.

The volume also contributes to the trend first mentioned, of engaging with the writings of the early analytic philosophers as texts in the history of philosophy, rather than as slightly outmoded contributions to contemporary debates. The historical orientation its title invokes is, by and large, one that sets the *Tractatus* against *its own* history, rather than against the context of subsequent developments. Included under that broad heading is a concern with the immediate circumstances of the composition of the book (Potter); with more distant or less familiar sources for some of its key ideas (Appelqvist, Schönbaumsfeld); with the exact form of Wittgenstein's reaction to the theories of his direct predecessors, Frege and Russell, in the areas of logic (Proops, Milne, Ricketts), mind (Child), and metaphysics (Levine); and finally with how the book stands towards the central commitments of the broadest tradition that embraces it, the tradition of critical or transcendental philosophy initiated by Kant (Moore, Sullivan). This primary orientation does not however preclude forward glances, to Wittgenstein's later thought or to the shape taken in contemporary philosophy by issues he addressed in the *Tractatus*. In this way and more generally the volume demonstrates that a historical approach is not an alternative to focusing on what is philosophically most fertile in the *Tractatus*, but a way of sharpening that focus.

Much is now known about the genesis of the *Tractatus*, but there is also much that can only be conjectured. The central piece of evidence, of course, is the hardback notebook, now in the Bodleian Library in Oxford, in which Wittgenstein assembled the complete draft of his book known as the *Prototractatus*. Yet controversy has remained over just when Wittgenstein compiled the various parts of this volume and from what sources. In the opening chapter of the volume Michael Potter attempts to clear up this uncertainty to the extent that the available evidence permits. Some readers may find this sort of detective work interesting in its own right. And there is of course a biographical interest in understanding the circumstances in which the *Tractatus* was written. In particular, what is now known about how Wittgenstein compiled the *Tractatus* shows him to have been extremely methodical, and the processes he used deliberate and careful. The main point of such detective work, however, is, or ought to be, philosophical. There are many claims in the *Tractatus* for which Wittgenstein does not supply any argument in the text, or for which the argument which he does supply is plainly inadequate. If we are to work out why he held these views, it is enormously valuable to know when he adopted them.

The reasons to be interested in the influences on Wittgenstein are similar. There is a biographical interest in finding out which of the works he read had an effect on his philosophy, but there is a philosophical purpose too. Wittgenstein often uses words and phrases taken from other writers but does not stop to explain what he means by them. He often echoes or alludes critically to the writings or theories of others, sometimes without even mentioning the source and typically without detailing his understanding of it. In such cases identifying and interpreting the source may be a useful step on the way to understanding Wittgenstein. In the 1930s Wittgenstein made a list (nowadays endlessly discussed) of authors whose work had influenced him: Boltzmann, Hertz, Schopenhauer, Frege, Russell, Loos, Kraus, Weininger, Spengler, Sraffa. Some of these authors (Spengler, Sraffa) plainly had their effect on Wittgenstein long after the *Tractatus* was finished. Of the others, it is obvious that Hertz, Schopenhauer, Frege, and Russell all influenced the book quite directly in various places. Nonetheless, even if it is obvious *that* they influenced the book, it is not always obvious *how* they did. There therefore remains considerable scope for work that explains which of the echoes of these authors that we can hear in the *Tractatus* are superficial and which are deep borrowings. There are also strong reasons, both internal to the text and from biographical sources, for thinking that this list is incomplete. Kant, Kierkegaard, and, perhaps most strikingly, Tolstoy are all omitted. In each case we know less than we should like to know about the exact character and chronology of Wittgenstein's engagement with them, but it is at least clear that they occupied his thoughts at relevant times during the composition of the *Tractatus*.

Hanne Appelqvist's chapter is concerned with the first of these omissions. Wittgenstein read the *Critique of Pure Reason* in the prisoner of war camp in 1919 when the *Tractatus* was complete, and he is generally supposed to have read Kant at some point before that, though we do not know quite when. (In his biography

McGuinness reports that Wittgenstein would refer to 'the great Kant' in conversation with friends he made in Olmütz in the autumn of 1916; see McGuinness 1988: 252–3.) Kant's influence on the *Tractatus* has been mentioned by many commentators, but usually on Wittgenstein's metaphysical rather than his ethical views. In her discussion of the latter, Appelqvist's approach contrasts notably with that of commentators who present the Kantian elements in Wittgenstein's thought about ethics as strongly coloured by Schopenhauer's mediation. Wittgenstein's concern, on her account, is with the necessary conditions of value, the underlying form of the ethical attitude. This form, she contends, is in the first instance aesthetic, and best elucidated through structural parallels with the analysis of judgements of taste that Kant develops in his third *Critique*.

A less structural and perhaps a more romantic perspective on Wittgenstein's ethics is offered in Genia Schönbaumsfeld's exploration of connections between the *Tractatus* and the thought of Kierkegaard, the second omission from Wittgenstein's list. Again, we know that Wittgenstein read him, but unfortunately we do not know exactly what or when. Russell mentioned Wittgenstein's reading of Kierkegaard in a letter he wrote just after their meeting in the Hague in December 1919 (Wittgenstein 2008: 112). His name crops up in the surviving wartime correspondence only in a letter to Wittgenstein of November 1917 from his sister, Hermine, mentioning that she had recently sent to Wittgenstein a number of Kierkegaard's books, but failing to mention which (Wittgenstein 1996: 48). Kierkegaard's possible influence on Wittgenstein is therefore tantalizingly uncertain. The issue is especially intriguing now that some resolute readers of the *Tractatus* have suggested that one way of understanding the book is to think of Wittgenstein as deliberately playing with the notion of authorship in something like the way that Kierkegaard does in his various pseudonymous works. Schönbaumsfeld considers and rejects this suggestion. She contends instead that Wittgenstein's real affinity with Kierkegaard is apparent in what Engelmann described as 'an ethical totalitarianism . . . , a single-minded and painful preservation of the purity of the uncompromising demands of ethics' (Engelmann 1970: 18). Schönbaumsfeld's elaboration of the quasi-religious conception of ethics that both philosophers endorsed leads, in the final section of her chapter, to a striking interpretation of Wittgenstein's famous claim, in a letter to Ficker, that the point of his book is an ethical one.

With Ian Proops's chapter we reach Wittgenstein's list of acknowledged influences, but we also, in a sense, reverse the trend of discussion. Whereas the two previous authors emphasize affinities that are too little recognized, Proops questions a claim of influence which, he holds, has been too easily accepted. His topic is a doctrine of Frege's whose working out in the *Tractatus* has for a long time been a preoccupation equally of 'resolute' and 'traditional' interpreters, the doctrine that issues in Frege's incongruous claim that the concept *horse* is not a concept. This claim arises from the distinction which Frege draws, in analysing the structure of a sentence, between saturated and unsaturated components. For instance, the sentence 'Marengo is a horse' decomposes into a name, 'Marengo', and a predicate, 'ξ is a horse'. The name is saturated (or complete); the predicate is unsaturated (or incomplete) because the 'ξ' in it is the marker of an argument place awaiting completion. Frege also held that this division among the components of a sentence mirrors an exactly corresponding division among the entities these components refer to. The name 'Marengo' refers to a certain object, namely Napoleon's horse. We should like to be able to say, similarly, that the predicate 'ξ is a horse' refers to a certain concept, namely the concept *horse*. Yet we cannot say this, since concepts are unsaturated whereas the phrase 'the concept *horse*' is not. The conclusion we reach is that, by what Frege called an 'oddity of language', we are unable to refer to the concept at all: our attempts to do so inevitably miss their intended target. Ever since Geach's seminal article of 1976 it has been common to present Frege's problem as an instance of and the inspiration for Wittgenstein's doctrine of unsayability in the *Tractatus*. Wittgenstein's doctrine, that is to say, is to be reached in two stages: the first stage is to recognize the seriousness of the difficulty and not dismiss it, as Frege did, by asking his readers to grant him a pinch of salt; the second stage is to diagnose the difficulty as lying not with the particularities of Frege's bipartite division between the saturated and unsaturated, but with any attempt to give contentful expression to the intimate relationship between language and world which would make that expression possible. While Proops does not deny that this is a possible route to distinctively Tractarian doctrines, he challenges Geach's claim that Wittgenstein simply 'takes over' and generalizes a commitment to inexpressibility that is already present in Frege. To this end he distinguishes various more specific sub-problems often lumped together as constituting Frege's 'problem of the concept *horse*', and severally evaluates Frege's responses to them. He concludes, first, that only one of these sub-problems is plausibly seen as an ancestor of Wittgenstein's view, and secondly that this sub-problem is the one for which there is least evidence that it was of real concern to Frege, or even clearly grasped by him. If they are accepted, Proops's contentions will call at least for substantial revision to one of the most widely accepted strands in our account of the origins and motivations of Wittgenstein's thought in the *Tractatus*.

Russell famously said in his Introduction that the *Tractatus* was concerned with the conditions for a logically perfect language. Ramsey—in a critical notice written before he and Wittgenstein had met—was the first to suggest that this might be a misinterpretation. Nonetheless, Wittgenstein does undoubtedly make a number of specific proposals as to what a formal language should be like. He tells us, for instance, that in a Tractarian formal language two objects will never have the same name, and that the use of different names will always indicate that different objects are referred to. He also tells us that every proposition can be obtained from the elementary propositions by a recursive process of applying the N-operator (a generalization of the Sheffer stroke which expresses the joint denial of the propositions to which it is applied) to classes of propositions to obtain further propositions (6.001). On the details, however, Wittgenstein is frustratingly vague. He tells us three ways in which a class of propositions to which the N-operator is to be applied may be specified, but he does not explain in detail how they work, and he seems to leave it open that there could be ways of specifying a class of propositions other than these three (5.501). As a result of this uncertainty it remains controversial, for instance, whether he would permit higher-order quantification or not. The chapters by Peter Milne and Thomas Ricketts aim to bring clarity to different aspects of this account.

Milne's focus is on what Russell and Wittgenstein called 'molecular' propositions, finite truth-functions, where the class of propositions to which a truth-operator is applied can be given by a list—the first of the three ways mentioned at 5.501. Supposing that an initial stock of elementary propositions is given, the range of propositions constructible in this way of course raises no difficulties (4.42). What is less clear, though, is how this way of constructing molecular propositions relates to the view of them encouraged by the alternative 'TF-notation' that Wittgenstein recommends at 4.442. In the TF-notation a truth-function of elementary propositions is, as it were, given 'neat'; in Milne's phrase, the notation 'washes out' all indication of a route by which the specified truth-function might have been determined by the successive application of truth-operators, leaving only an extensional record of the result. If this notation perspicuously captures the sense of a molecular proposition, it follows that truth-operators—Wittgenstein's own N-operator, or the more familiar logical constants of Fregean and Russellian notations—make no essential contribution to that sense. Milne finds in this consequence the core meaning of Wittgenstein's 'fundamental thought', that the logical constants are not representatives (4.0312). He develops a general argument to show that this thought is a consequence of basic commitments of a truth-conditional conception of meaning, commitments shared between Frege and Russell on the one hand, and by many contemporary theorists on the other, thereby revealing the strength of the compressed argument for his fundamental thought that Wittgenstein presents at 5.4–5.41.

If the logical constants make no essential contribution to the sense of a proposition, how are we to conceive of their meaning? Developing suggestions implicit in Dummett's work, Milne answers this question by distinguishing between an expression's contributing to the truth-conditions of a proposition and its contributing to the *determination* of those truth-conditions. He explains how Wittgenstein's contrast of 'operations' and 'functions' in the 5.4s shows a clearer understanding of this distinction than was attained by either Frege or Russell, and how in turn it might explain Wittgenstein's puzzling pronouncement that the logical constants are 'punctuation marks' (5.4611). Milne concludes, however, that this pronouncement overshoots the mark, in a way that displays Wittgenstein's primary concern with the sense a sentence expresses, and his relative lack of concern with the question—on contemporary conceptions, surely, the crucial question—of how a language-user grasps that sense.

Wittgenstein allows for generalization by means of the second and third methods mentioned at 5.501 of specifying a class of propositions to which a truth-operation can be applied. The second method specifies a class of propositions as comprising all values of a propositional function. This is quantificational generality, familiar to us from the work of Frege and Russell. But what generalizations did Wittgenstein take to be constructible by this method? In particular, which propositional functions did he recognize as suitable to determine the instances of a quantificational generalization? The third method specifies a class of propositions as those constructed in accordance with a formal law. This method, Wittgenstein says, was overlooked by Frege and Russell (4.1273). Why is it needed? And what additional expressive power does its recognition bring to the formal language Wittgenstein envisaged?

In answering these questions Ricketts builds on Milne's conclusions about molecular propositions. As Milne argues, and as Ricketts agrees in finding encapsulated in the TF-notation, the particular way in which a molecular proposition is formulated through the successive application of truth-operators is inessential to the sense expressed. But it must not follow from this that the iterative structure of truth-operations has no logical import. As is plain from the well-known debate between Fogelin and Geach, an understanding of how Wittgenstein's language can encompass even embedded first-level quantifiers depends on an account of how generalization interacts with that iterative structure.

Ricketts's account of this interaction, which determines his subsequent conclusions, can only be sketched here. Suppose, with Frege in §9 of *Begriffsschrift*, that the first key step towards a quantificational generalization is taken in articulating a proposition as function and argument by conceiving some component in it as replaceable by others, the invariant component being thus highlighted as a function to which the variable component appears as argument. Wittgenstein endorses this model at 3.318, holding that a proposition may be conceived 'as a function of the expressions contained in it'. But on Ricketts's account this endorsement of Frege's model is at the same time a restriction on it, since an 'expression' is explained as any component *essential to its sense* that a proposition may have in common with another (3.31). By Wittgenstein's fundamental thought, a truth-functionally complex predicative component of a proposition—for instance, the component '$(F\xi \vee G\xi)$' in the proposition '$\neg(Fa \vee Ga)$'—is *not* essential to its sense, and so is not an 'expression' in it. Such a component may therefore play the role of function in Wittgenstein's understanding of Frege's model, but it cannot play the role of argument. The first of these points explains how Wittgenstein's quantificational language extends to include all first-level quantifications. The second, according to which Wittgenstein recognizes no propositional functions taking arguments of arbitrary logical complexity, explains why *quantificational* generality cannot extend further than this. Higher-level generalization must therefore be accommodated—if at all—in a quite different fashion. In the remainder of his chapter Ricketts explains how, and how far, *formal* generalization, the third method from 5.501, makes good this deficit.

It will be clear even from this brief sketch that much in Ricketts's account depends on his understanding of Wittgenstein's notion of an 'expression'. So it is notable that Ramsey, who reached a very different conclusion about the scope of a Wittgensteinian quantificational language, also understood this notion very differently. Taking his lead from Wittgenstein's remarks about the symbolism of generality at 5.522, rather than from 3.318, Ramsey associated the notion with something that in Frege's model plays the role of function rather than that of argument (1931: 123–4). On first encountering it one might suppose this to be a minor exegetical difference. Ricketts's argument shows instead that resolving it will be crucial for understanding the logic of the *Tractatus*.

Wittgenstein's understanding of generality remains of central importance through the following two chapters, where William Child and James Levine explore consequences of his claim that 'the understanding of general propositions *palpably* depends on that of elementary propositions' (4.411). The view summarized here, and amplified in Wittgenstein's account of a quantified proposition as a truth-function of its instances, contrasts with Russell's theory that a true generalization 'everything is *F*' expresses a further fact, not reducible to or implied by the truth of all its formulable instances of the form '*a* is *F*'.

Russell makes use of this theory in explaining how I can have knowledge by description of the mind and the specific thoughts and experiences of another person, items with which it is, in some sense, impossible that I should be directly acquainted. For instance, I cannot have that direct acquaintance with the person Bismarck that Bismarck himself has, and that he will express by his use of the first person. I can nonetheless grasp a proposition ordinarily expressed by the sentence 'Bismarck was an astute diplomatist'. I will do this by framing, by reference to things and concepts with which I am directly acquainted, a description uniquely satisfied by Bismarck, and forming thereby the proposition that whatever uniquely satisfies this description was an astute diplomatist. Assuming that it is true, this general proposition, which I can understand, will be true in virtue of the truth of a singular instance in which Bismarck is directly named, and which therefore I, who am not acquainted with Bismarck, cannot understand. In this way my general thoughts can reach out to things that my singular thoughts cannot.

Wittgenstein's understanding of generality apparently commits Wittgenstein to reject this account. Cora Diamond has argued that Wittgenstein clearly recognized this commitment. Another's inner self, or the thoughts and experiences of another self, conceived as private objects knowable only to their possessor, cannot play any role in my language—which is to say, they cannot figure in the truth-conditions of a general proposition framed in my language. Diamond proposes that Wittgenstein's drawing of this consequence amounts to a Tractarian 'private language argument'. William Child's chapter critically examines this proposal. His direct counter-argument holds that Diamond's case depends on an epistemically constrained understanding of key notions—for instance, of what it is for an expression to have a 'significant use', or what it is for one proposition to be a 'ground' of another—which is in keeping with Wittgenstein's later philosophy but which has no basis in the text of the *Tractatus*. But Child's discussion extends far beyond points of local criticism, to consider what sense Diamond's proposal can make of the later developments in Wittgenstein's thought about the 'inner', and whether in general the atomistic analyses envisaged in the

Tractatus should be understood as amounting to a form of realism or anti-realism about the objects of ordinary discourse.

Russell's account of how I might have descriptive knowledge of the thoughts and experiences of another is an instance of a more general view. According to Russell, my irreducible grasp of generality, of what it is for a property to hold in some or all cases, enables me to frame the proposition that there is something with which I am not acquainted. I can understand this general proposition, and have good reason to believe that it is true, even though it would involve a kind of contradiction to cite some singular instance in virtue of which it is true: for, were I to form a singular proposition, 'I am not acquainted with *a*', saying of some particular thing that I am not acquainted with it, my understanding the proposition would be enough to show that it is not true. For Russell, this general proposition is a denial of solipsism. And it seems that an argument similar to that given by Diamond in the special case of another's private objects will make it impossible for Wittgenstein to join Russell in that denial.

It will be useful here to revert to Child's articulation of a simpler argument than Diamond's to that effect. It proceeds, he says, in three steps: first, a general proposition that I can understand cannot have as one of its instances a singular proposition that I cannot understand; second, I cannot name an object, nor therefore understand a singular proposition concerning that object, unless I 'know' the object; and third, to 'know' an object, in the sense invoked at the argument's second step, requires that I be 'acquainted' with the object—in Russell's sense of acquaintance. Child accepts that Wittgenstein's understanding of general propositions and of the elementary propositions that (ultimately) instantiate them commit him to the first two steps of this argument. But he denies that Wittgenstein would accept the third step: Wittgenstein's account of what is involved in 'knowing' an object, he contends, makes it altogether unlike Russell's notion of acquaintance.

In his discussion of the more general issue of Russell's denial of solipsism, James Levine holds that Wittgenstein is committed to all three of these steps. Superficially, then, Child and Levine are directly opposed on the question whether to 'know' an object involves, or is, to be 'acquainted' with it.[1] However, this superficial diagnosis will not survive a closer reading. In his endorsement of the third step Levine emphasizes structural semantic principles shared by Russell and Wittgenstein which collectively serve, as it were, implicitly to define the role of 'acquaintance' in his discussion; he is by no means insensitive to the points raised by Child, which distance Wittgenstein from the epistemological atomism of Russell's own understanding of acquaintance. So focusing on this one apparent disagreement between Child and Levine would be an excellent way of arriving at a deeper understanding of what Wittgenstein's thought here owes to Russell.

[1] The two chapters therefore offer contrasting accounts of Wittgenstein's comments on the translation of 2.0123, where 'to be acquainted with' is an inviting rendering of the first occurrence of *kennen*, but would be grammatically discrepant, or at least strained, as a rendering of its second occurrence.

Let us for the present accept Levine's argument. The conclusion it yields is, as we put it, that Wittgenstein cannot endorse Russell's denial of solipsism. What then *is* Wittgenstein's attitude to the position? Does he endorse instead the statement that Russell is concerned to deny, that there is nothing beyond my acquaintance? Or, if there are problems for Wittgenstein in actually endorsing this statement, is it, in some sense, his unspoken view? What in general might it be to harbour an unspoken view that one cannot explicitly endorse? And in any case, is there reason to suppose that the position *Russell* labelled 'solipsism' is the view that Wittgenstein considered under that name in the *Tractatus*? Levine's chapter offers answers to all of these questions, bringing creative order to their discussion through a structural parallel with Berkeley's Master Argument for idealism. In a final section Levine then argues that the dispute between Nagel and Davidson over the latter's alleged 'idealism' manifests again the same structure.

In that dispute 'idealism' has meant, very roughly, 'cutting the world down to size', or accepting that reality is restricted by the limits of one's understanding. Whether or not Wittgenstein addressed solipsism in the specifically Russellian form that is Levine's primary concern, it is undeniable that he is concerned in the *Tractatus* with this more general view. In the final two chapters of the volume Adrian Moore and Peter Sullivan continue a debate begun in previous papers about the nature of that concern.

It seems that the issue ought to be less complicated than it appears to be in these two chapters. Moore and Sullivan agree that Wittgenstein was strongly drawn towards a view of the broad kind just outlined. They also agree that this temptation took a more specifically Kantian form. The idealism in question would hold that reality is unified by forms of representation, whose universality is demanded by a requirement to secure the unity of the representing subject. It would thus hold that 'the subject is . . . a limit of the world' (5.632): that any way that things might be must be such as can be integrated into the subject's unifying perspective on how things are. The question is whether Wittgenstein endorsed this view. So can we not just look to the text and see? Indeed, have we not just done so, and seen that, in the remark quoted, he did?

One kind of reason why the issue is not so quickly resolved is not at all specific to the interpretation of the *Tractatus*. In the remark just quoted Wittgenstein says that the subject is a limit of the world. Ten lines earlier he says, 'The thinking, representing subject does not exist' (5.631). The apparent clash between these statements invites exegetical manoeuvres of a kind familiar to anyone working in any area of the history of philosophy. (In this familiar vein one might ask, for instance: whether 'the subject' denied existence in one remark is different from 'the subject' apparently affirmed in the other; whether 'existence' has some narrow meaning in the first remark, so that something might in that narrow sense fail to 'exist' without altogether failing to be; whether the second remark does not *affirm* the being of a subject, but only *explains* what is to be understood by 'the subject' when it is denied in the first remark that there is any such thing; and so on.) But a second kind of reason *is* specific to the *Tractatus*. Whatever we find in the text, when ordinary exegetical uncertainties are resolved, must be reconciled with what its author tells us we should expect to find: sheer nonsense.

This complexity explains why there is room for Moore and Sullivan to disagree over whether Wittgenstein is, in any sense, committed to an idealism of the kind just sketched. On the face of it, it does not explain why their disagreement should replicate itself at the 'meta' level, leaving them unable to agree even as to whether they are in disagreement—a turn of events that would surely tax any reader's patience. Here we offer a simple diagnosis of that to dispel any air of mystery. Moore and Sullivan do disagree, in the ordinary and familiar way in which interpreters of Locke, or Hume, might disagree: they resolve a multitude of first-level exegetical uncertainties of the kind just illustrated in contrary directions, Moore finding in 'the body' of the text commitment to an idealism which, on Sullivan's account, is present only as something to be diagnosed and dispelled.[2] But this ordinary, first-level disagreement is then caught up in the distinctively Wittgensteinian 'second-level' issue of how what we thus find in the text is to be reconceived in the light of the book's final self-renunciation.

We remarked at the start of this introduction that, while 'second-level' issues are unavoidably present in the interpretation of the *Tractatus*, it is not the *primary* concern of any of the chapters gathered here to address them. Speaking more clearly for ourselves, we would say that such issues *cannot* be an interpreter's primary concern. These abstracted 'second-level' issues have by now a familiar form. Without investigations of the kind pursued by our contributors that is all they will have: they will lack substance.

References

Engelmann, Paul (1970) *Ludwig Wittgenstein: Briefe und Begegnungen*. Vienna-Munich: Oldenbourg.

Geach, Peter (1976) 'Saying and Showing in Frege and Wittgenstein', in J. Hintikka *et al.*, eds, *Essays in Honour of G. H. von Wright, Acta Philosophica Fennica* 28, 54–70.

McGuiness, Brian (1988) *Wittgenstein: A Life—Young Ludwig (1889–1921)*. London: Duckworth.

Moore, A. W. (1997) *Points of View*. Oxford: Oxford University Press.

Ramsey, F. P. (1931) *The Foundations of Mathematics and Other Logical Essays*, ed. R. B. Braithwaite. London: Routledge and Kegan Paul Ltd.

Sullivan, Peter (1996) 'The "Truth" in Solipsism, and Wittgenstein's Rejection of the A Priori', *European Journal of Philosophy* 4, 195–219.

[2] To give one instance of this, Moore understands Wittgenstein as introducing Schopenhauer's analogy of the eye and the visual field as an encapsulation of his (1997: 154) own vision, while Sullivan (1996: 202) holds that the analogy is appealed to by Wittgenstein's dialectical interlocutor.

Wittgenstein, Ludwig (1922) *Logisch–Philosophische Abhandlung/Tractatus Logico–Philosophicus* (TLP). Trans. C. K. Ogden, London: Kegan Paul Ltd, 2nd edition 1933. Trans. D. F. Pears and B. F. McGuinness, London: Routledge and Kegan Paul Ltd, 1961, 2nd edition 1974.

—— (1996) *Wittgenstein Familienbriefe*, ed. B. F. McGuinness, M. Ascher, and O. Pfersmann. Vienna: Verlag Hölder-Pichler-Tempsky.

—— (2008) *Wittgenstein in Cambridge: Letters and Documents 1911–1951*, ed. B. F. McGuinness. Oxford: Blackwell.

2

Wittgenstein's pre-*Tractatus* manuscripts: a new appraisal

Michael Potter

Although he began the work that led to the *Tractatus* when he arrived in Cambridge in 1911, Wittgenstein composed the book itself while he was serving in the Austrian army during the First World War. When he was taken prisoner by the Italians at the armistice in November 1918, he had among his belongings a typescript differing at most in a few small details from the version eventually published. When and how did the composition take place? Did Wittgenstein, as the popular myth has it, write the book at the front with the sound of gunfire in his ears? The available evidence is tantalizingly incomplete, but it does at least allow some plausible inferences to be drawn about his working methods and the circumstances of the book's composition. Much of the narrative is well known from the pioneering work of McGuinness (1989), but more recently Geschkowski (2001), Bazzocchi (2005), and Kang (2005) have questioned various of his conclusions. What I attempt here is a new assessment of all the evidence currently available.

1 The *Notes on Logic*

The letters Wittgenstein wrote to Russell during his time at Cambridge between 1911 and 1913 give us glimpses of his philosophical thoughts during this period. However, the earliest text to have survived that could be described as a philosophical work by Wittgenstein is the *Notes on Logic*. I have written at length elsewhere (Potter 2009: Appendix A) about the circumstances in which Wittgenstein compiled these notes in October 1913, just before leaving to spend a year in Norway, as a kind of summary of the work he had done under Russell's supervision at Cambridge. I shall not repeat that discussion here, but shall only summarize those of its conclusions that are directly relevant.

One of the most striking features of the notes is that they exhibit the rather idiosyncratic method by which Wittgenstein wrote philosophy throughout his working life. This method had at its centre the working notebooks, *Tagebücher*, in which he wrote down his philosophical thoughts day by day. The *Notes on Logic* were largely composed, I believe, by extracting remarks from a number of these working notebooks which he had used while at Cambridge. He dictated these remarks to a shorthand typist in Birmingham, and part of what has survived is Russell's translation of the typescript thus produced. The remainder of the *Notes* is a record of discussions Wittgenstein had with Russell in Cambridge in the next few days.

When he departed for Norway shortly afterwards, Wittgenstein left some of the working notebooks from which the Birmingham typescript had been compiled in Cambridge, where they remained until he asked Russell to destroy them after the War. Yet many of the remarks in the Birmingham *Notes* are repeated in the *Tractatus*. Since Wittgenstein did not have access to his Cambridge notebooks, from which these remarks must have been drawn, while he was compiling the *Tractatus*, my assumption is that he must instead have had a copy of the Birmingham typescript itself.

2 The notes dictated to Moore

A few letters survive to Russell written by Wittgenstein from Norway during the autumn of 1913, but early in 1914 there was a rift between the two men and the correspondence was broken off for a while. Whatever the immediate cause of the rift may have been—the letter which began it has not survived—the available evidence makes it plain that it was Wittgenstein who ensured that it lasted despite Russell's best endeavours to engineer a reconciliation.

The next evidence we have of Wittgenstein's philosophical development is the notes he dictated to Moore during the visit the latter made to Norway in April 1914. These notes are preserved in a series of three soft-covered jotters of the sort commonly used in schools. They present fewer puzzles for the historian than the *Notes on Logic* (although just as many for the philosopher).

Although it is now lost, we know that Wittgenstein kept a notebook during his period in Norway: I shall refer to it as NB0. He had this 'large writing book' with him when he joined up for military service in August 1914, but immediately started a new notebook, in the first diary entry of which (9 August 1914) he reported that he was sending the Norwegian notebook, via a helpful army lieutenant, back to Vienna for safe keeping. Wittgenstein had shown this notebook to Moore during the latter's visit to Norway, and many of the remarks in the notes dictated to him no doubt reflect its contents. The evidence of the *Prototractatus* suggests, however, that the match is not especially close: although some sentences in Moore's dictations are similar to sentences in the *Prototractatus*, they are not direct translations. In this respect the notes resemble the Cambridge part of the *Notes on Logic*, which, I have argued, was also written or dictated in English. In neither of these cases does the evidence compel us to the view that what we now have are line-by-line translations by Wittgenstein of German sentences from a notebook open in front of him as he spoke.

3 Wittgenstein's 'dissertation'

We have noted how Wittgenstein drew back from Russell at the beginning of 1914—pulled away from him in a manner that seems to have been deliberate. Shortly after Moore's visit Wittgenstein did something very similar with him. The occasion (or perhaps the excuse) was a task Wittgenstein had evidently asked Moore to carry out on his behalf. When he returned to Cambridge, Moore had in his possession a piece of Wittgenstein's work and was to investigate the possibility of submitting this work as a dissertation for the degree of BA by research. Moore consulted Wittgenstein's tutor at Trinity, a man called Fletcher. At that time, and indeed until very recently, it was the practice in Cambridge that dissertations were submitted through the candidate's college tutor. Fletcher was therefore the person who had the responsibility of ensuring before Wittgenstein's dissertation was submitted that it complied minimally with the university's regulations. What he evidently told Moore (which Moore could in any case easily have found out for himself) was that those regulations required the candidate to add a signed statement of the extent to which the work was original. What engendered the rift between Wittgenstein and Moore was evidently the former's outrage at the latter's letter explaining this requirement.

A question arises, therefore, as to which piece of work it was that Wittgenstein wanted to submit for a Cambridge degree. Some writers have been inclined to think that it was a distinct work now lost, in effect a very early draft of the *Tractatus* itself. If such a work existed, the explanation for the fact that it has not survived would presumably have to be either that Moore returned it to Wittgenstein along with the letter which triggered their falling out or that he never retrieved it from W. E. Johnson, the Cambridge philosophical logician to whom he lent it on the same day he spoke to Fletcher about the matter. (We can be fairly sure at least that Moore, an almost obsessive hoarder of paper, would not have thrown it out, however strong his feelings about Wittgenstein's treatment of him.)

There is reason to doubt, however, that what Moore had in his possession could have been a distinct work. If it were, it would presumably have been written in German—throughout his life Wittgenstein was willing to talk about philosophy in English, but hardly ever to write in it—and it seems improbable that Moore would have discussed with Fletcher the question of submitting it for a degree before he addressed the more significant issue of how it was to be translated into English. Wittgenstein himself referred to his putative dissertation as 'Logik', but the spelling should not be taken to imply that the work he was referring to was in German as his English spelling was frequently erratic in this kind of way.

One further piece of evidence which might be taken to suggest that Wittgenstein did write a separate work on logic while he was in Norway is an entry Pinsent wrote in his diary in April 1914, just after Moore returned from Norway:

> About a month ago Ludwig wrote to me that he had settled Types and in fact put the whole of Logic in order and was writing a book on the subject. (Von Wright 1990: 90)

On the other hand, we need not read this as much more than the expression of an intention. (I myself started telling people I was writing a book about the *Tractatus* almost a year before I began to write it down.) Moreover, as we shall see later, the list of manuscripts that Wittgenstein made in January 1917 does not include anything that seems easy to identify as a now-lost Norwegian *Tractatus*. It seems most likely, therefore, that whatever he was writing in Norway was contained in the large Norwegian notebook NB0 rather than being a distinct manuscript.

If Wittgenstein had not yet written a complete book, the work that Moore referred to in his diary as a 'dissertation' can only have been either the *Notes on Logic* or the dictated notes which Moore had just brought back from Norway himself. Which of these is the more likely depends to some extent on the circumstances in which what is now known as the Costello version of the *Notes on Logic* was compiled. This is a rearrangement of the *Notes on Logic* in which the remarks are grouped according to topic and repetitions deleted. The assignment of the remarks to the various groups was done by Russell in February 1914 in preparation for lectures he was to give at Harvard, in which he intended to discuss Wittgenstein's ideas extensively. Russell told Ottoline Morrell in a letter that he had finished this editorial work, but he did not specify what he meant precisely. On the same day he met Moore to discuss Wittgenstein's views and Moore took away with him something of Wittgenstein's to read. If the editing work was left unfinished by Russell and completed later at Harvard by Costello (who was Russell's teaching assistant for his course there), then what Moore took away with him will most likely have been a copy of the second-stage typescript, which he will have had to give back to Russell before he left for America the following week. In that case the 'dissertation' was simply the notes dictated to Moore in Norway by Wittgenstein.

I am inclined to regard this as unlikely. As we noted earlier, the lack of any close correspondence between Moore's notes and the *Prototractatus* suggests that when Wittgenstein was dictating—in effect lecturing—to Moore, he did not have his notebook open in front of him. Wittgenstein (1995: 91) later described them as Moore's 'notes' (*Notizen*), and if they are indeed to be thought of more as lecture notes than straight dictation, that would certainly explain Wittgenstein's sensitivity about having to state in a signed declaration 'the extent to which he had availed himself of the work of others'. By the same token, though, it also renders rather implausible the notion that he might later have referred to the notes as something that he, Wittgenstein, had written. Moore himself also calls them 'my notes of Wittg[enstein]' in his diary, whereas elsewhere he twice refers to the 'dissertation', and it is hard to read these as references to the same item. (This is not quite conclusive, though, since the reference to 'notes' dates from July 1914, when the possibility that Wittgenstein might submit a dissertation had vanished.)

On the other hypothesis, according to which Russell completed all the editing of the Costello version himself in February, his secretary would have made a carbon copy of it, which Russell could have lent to Moore. In that case this copy of the Costello version could well be the 'dissertation' which Wittgenstein wanted to submit for a Cambridge degree. This hypothesis also supplies a ready explanation, of course, for Wittgenstein's sensitivity on the subject of the signed statement which he refused to submit. The Costello version is indeed not entirely Wittgenstein's own work: by translating and rearranging Wittgenstein's notes Russell had plainly gone well beyond the help which a diligent supervisor might reasonably be expected to offer to a dissertation candidate. Wittgenstein's explosion would then have been partly a way of avoiding the embarrassment that would have been involved in submitting a statement that stated honestly the contribution Russell had made.

4 The wartime notebooks

Three of Wittgenstein's wartime notebooks survive. The first two of these, which I shall call NB1 and NB2, form a pair, representing a continuous period of work from August 1914 to June 1915, during which Wittgenstein served first as a searchlight operator on a boat and then as a mechanic at an engineering works. He began to use NB1 in August 1914, a couple of days after he had volunteered, and filled it to the end before going on to NB2. At first the personal diary entries are written in clear text, but a week later he began to use a code. It is not elaborate ('a' for 'z', 'b' for 'y', etc.) and its only purpose is presumably to prevent other soldiers who happened upon the notebook from reading his private thoughts. Soon a firm practice was established: philosophical remarks he wrote on the right-hand page of a two-page spread in clear text, personal diary entries on the left-hand page in code. From late August onwards Wittgenstein worked with some regularity, writing philosophical remarks almost every day, except between January and April 1915, when his coded diary entries indicate (without really specifying the reason) that he found it hard to work.

The last entries in NB2 date from June 1915: we shall discuss shortly why Wittgenstein broke off from entering remarks in it at this point. A few weeks later Wittgenstein was injured in an incident at the depot. (A cannon exploded while it was being repaired.) After spending the second half of July in hospital Wittgenstein went home to Vienna on leave. This was the first substantial period of leave he had enjoyed since the beginning of the war. Although not yet at this stage a front-line soldier, he had been on active service almost continuously for a year. His experience was not in this respect atypical: the belief that the war would 'be over by Christmas' was widespread in Austria as much as in Britain, with the consequence that soldiers were not at first granted appropriate periods of leave. Later, practices changed: Wittgenstein was granted leave more frequently in the second half of the war than in the first.

After his leave Wittgenstein received a new posting in September to an engineering train at Sokal in what is now western Ukraine. No notebook survives for this period.

The only other notebook we have, which I shall call NB3, begins in April 1916, at the point when Wittgenstein was posted to the Eastern front, no longer an engineer but now an observer in an artillery battery.

One further hardback volume from this period survives. Unlike the others, though, Wittgenstein did not use it as a diary, but instead compiled in it the early version of the *Tractatus* now known as the *Prototractatus*. As is well known, Wittgenstein used the decimal numbering system in this notebook as a way of registering the intended order of the remarks in the final book, which did not coincide with the order in which he entered them in the volume. It will therefore be useful to have a name for the volume, to distinguish it from the book which it gradually came to contain. I shall follow McGuinness in calling it *Bodleianus* after the library where it now resides.

5 Hermine's list

In order to understand the relationship between the various notebooks and the book that Wittgenstein derived from them, we need to examine in some detail a list of manuscripts and typescripts 'dictated by Ludwig January 1917' to his sister Hermine:

(1)	Large chancellery book with Trenkler	Handwritten, exists also in a corrected typescript in Olmütz
(2)	2 quarto books with Trenkler	
(3)	1 quarto book	only handwritten (part of it is included in the typescript)
(4)	1 octavo book	literally every proposition in series order without any correction
(5)	Large chancellery book containing the working over of (1) and (2) for publication	

Russell gets (3), (4), and (5), and (1) and (2) in typescript and the gold clock.
Pinsent gets (1) in manuscript.
Typescript with Trenkler to be destroyed.

These are Wittgenstein's instructions, dictated just before he left to rejoin his regiment at the end of his winter leave, on how to dispose of his works in the event of his death in action. The following June Hermine made a copy of these instructions to add to her will and at that point sent the original list back to Wittgenstein with further queries about various items on it. The list must, fairly plainly, include all of Wittgenstein's manuscripts that his sister might come across on his death. The question we need to settle is, which is which?

Item (1) on the list is surely Wittgenstein's Norwegian notebook NB0, described here as a 'large chancellery book' and in his diary as a 'large writing book'. If any of the notebooks was to be given to Pinsent as a memento in the event of his death, this would be the most appropriate, since it would naturally trigger memories of the holiday they had spent in Norway before the war. Moreover, if it contained uncoded personal diary entries as well as purely philosophical remarks, that would also make Pinsent a more appropriate recipient than Russell.

It is fairly clear, too, that item (2) refers to the first two surviving wartime notebooks, NB1 and NB2: they certainly existed in January 1917, and so must occur somewhere on the list; they are indeed quarto-sized notebooks; and it would have been very natural for Wittgenstein to list his notebooks in chronological order.

I have already related how Wittgenstein dispatched NB0 back to Trenkler, the family's business manager, for safekeeping at the beginning of the war. Later, as the list indicates, he must have deposited NB1 and NB2 with Trenkler as well. (Exactly when he did this will be discussed shortly.) The list also refers to a typescript containing a corrected version of all these notebooks. The wording of Hermine's list does not require us to suppose that the typescript contained every remark from the notebooks: as we shall see later, there is some reason to think that it did not. It seems most likely that this typescript was prepared in Vienna, since that is where the notebooks were deposited, and that it was produced by dictation just as the Birmingham typescript had been.

6 Geschkowski

The identity of items (3) and (4) on the list is harder to determine. According to McGuinness, item (3) is the last of the three surviving wartime notebooks, NB3, and item (4) is a missing notebook containing all the remarks from the other notebooks copied out in order; according to Geschkowski (2001), on the other hand, item (3) is a missing notebook, which I shall call NB2a, while item (4) is NB3.

The simple criterion of size, which we might have hoped to decide between these two hypotheses, unfortunately does not do so with certainty. The surviving NB3 is indeed somewhat smaller ($8\frac{7}{8} \times 6\frac{1}{8}$ inches) than NB1 ($9\frac{5}{8}x8$ inches) and NB2 ($9 \times 7\frac{7}{8}$ inches). So one *might* describe NB3 as octavo, NB1 and NB2 as quarto. But I do not imagine that Wittgenstein measured the volumes before issuing his instructions, and he might at a pinch have described NB3 as a quarto notebook too if he was contrasting it with a clearly smaller missing octavo book.

On any view, of course, one of the notebooks Wittgenstein is referring to here is now lost. An advantage of McGuinness's view is that it provides an obvious explanation for the loss: if item (4) simply contained all the remarks from the other notebooks in uncorrected form, it would have served no purpose once the *Tractatus* was complete and might naturally have been discarded. By the same token, however, it is not entirely clear why in that case it was to be conveyed to Russell (or indeed what purpose there was in laboriously copying everything into another notebook at all).

The most striking consequence of McGuinness's view, of course, is that according to it there can have been no notebook to cover the period from June 1915 to April 1916. Wittgenstein must therefore have stopped writing his journal completely for nine months. That is not to say, of course, that Wittgenstein stopped *working*: on McGuinness's account he made significant progress with the composition of the *Prototractatus* instead. It is notable, though, that he carried on writing in NB3 during the period from September to December 1916, when on any version of events he was surely working on a compilation of some sort. Moreover, the method of working by which he noted down remarks day by day seems to have been very firmly entrenched, and it might therefore be thought rather unlikely that he stopped altogether for so long.

It is worth noting, though, that on either hypothesis, McGuinness's or Geschkowski's, the timing is curious. Wittgenstein filled NB1 to the end, but stopped using NB2 with a number of pages still unused. The last entries, both open and coded, are dated 22 June 1915. Why stop then? The accident in which he was injured, which might otherwise have provided an explanation, did not happen until about three weeks later. The last few entries provide no real clue. The philosophical remarks of June 1915, in particular, represent an enormously profuse passage of work which stops abruptly, seemingly in mid-flow. The few coded remarks from May and June certainly allude to difficulties in Wittgenstein's circumstances at the depot—'near to *tears*', 'broken and ill', 'surrounded by coarseness'—but also to the 'grace of work'. The last entry reads, 'Worked very much! Despite the *repulsive* surroundings!' (Wittgenstein 1991: 65).

If, on the other hand, Wittgenstein immediately started another notebook, why did he not wait until he had used up the previous one? The most likely explanation, I think, is that an opportunity arose at short notice for him to have the earlier notebooks NB1 and NB2 conveyed safely to Vienna and he decided to take that opportunity while he could. (His diary refers to the possibility that he might shortly be posted elsewhere.) Perhaps it is not a coincidence that letters from his family (in Wittgenstein 2004) refer to a visit paid to his mother in the first week of July by a soldier ('a particularly nice person', she thought) who brought her 'detailed news' about her son. She gave the soldier a letter to take back with him and deliver to Wittgenstein. It is surely plausible that in addition to news from her son the soldier was also bringing NB1 and NB2 to deposit with Trenkler.

Another consideration we should bear in mind is that Wittgenstein seems originally to have described the notebook typescript to his sister as containing only items (1) and (2); that it in fact included part of item (3) was an afterthought. On McGuinness's view it is perhaps a little surprising that Wittgenstein had forgotten this: between March and August 1916 he was continuously in the field, so the earliest he could plausibly have had a typescript made from NB3 was, when he dictated his instructions to his sister, only five months in the past, and at that point the notebook was already half-full. On Geschkowski's account, on the other hand, the typescript must have been dictated the previous August (the only time in the relevant period when Wittgenstein was on leave in Vienna): it therefore contained only perhaps the first eighth of the notebook in question (on this view, the missing NB2a), and had in any case been typed more than sixteen months ago.

The composition of the typescript in Olmütz creates a further difficulty for McGuinness's view. In NB1 and NB2 it is noticeable that although there are marks in the margins at various points, they do not correspond to the selections from these notebooks that occur in the *Prototractatus*: on the contrary, in all but a very few cases the marks are next to remarks which do *not* end up in the *Prototractatus*. The obvious explanation for this is that as a first stage in the selection process Wittgenstein marked those remarks in the notebooks which he did *not* want to include in the typescript. Then, once the typescript had been produced, he worked by and large with it rather than with the notebooks from which it had been extracted. The one significant exception is the following remark from NB2:

> The reality that corresponds to the sense of the proposition can surely be nothing but its component parts, since we are surely *ignorant* of *everything* else. (Wittgenstein 1979: 31)

Wittgenstein put a mark against this remark which, I am assuming, meant that it was not to be included in the notebook typescript. It nonetheless occurs word for word in *Bodleianus* on page 43, but (as with a handful of others) without a number. The reason for the remark's survival into *Bodleianus* after it had been marked for deletion might be nothing more mysterious than a last minute reprieve by Wittgenstein while he was dictating the typescript (or, of course, a mistake). However, it is perhaps also worth noting that the remark occurs in the midst of a passage of work (pages 40(4) to 43(8)) in which almost every sentence in *Bodleianus* derives from NB2, so it might be that here Wittgenstein was temporarily choosing to work directly from NB2 itself rather than from the typescript. (Another passage in which this might be the case occurs from page 71(4) to 74(2).) If so, Wittgenstein would presumably have had to return to the typescript at a later stage and tick off those remarks from NB2 which he had now used in *Bodleianus*.

These passages of work do not, however, affect my general contention, which is that Wittgenstein most of the time used the notebook typescript as an intermediate step between the notebooks from which it was copied and *Bodleianus*. If item (3) were NB3, as McGuinness claims, we would therefore expect there to be a difference in how parts of NB3 get transcribed: where in the *Prototractatus* Wittgenstein exploits remarks from the early part of NB3, which on this hypothesis is included in the notebook typescript, he might be expected (at least some of the time) to have copied them across from the typescript rather than from the notebook itself. But in fact there are *throughout* NB3 marginal marks which correspond closely to their use in *Bodleianus* (principally in the passage from page 81(4) to 86(6)). These marks cannot indicate the selection of remarks to *include* in the notebook typescript, since the last of them is next to the entry for 7 January 1917: if the typescript had been produced then, it would hardly have been in Olmütz three days later. This suggests that in transferring remarks Wittgenstein was throughout using NB3 itself and not a typescript extracted from it.

Geschkowski's view explains this better. It depends on reading the remark alongside item (4) as an *instruction* to Russell rather than a description of the book. (McGuinness reads the German as *jeder* (nominative, and hence a description), Geschkowski as *jeden* (accusative, and hence an instruction.) Unfortunately, in Hermine's script the two are hard to distinguish.) Wittgenstein is expressing his wish that in the event of his death Russell should publish not only the *Prototractatus* such as it was at that stage but also literally every proposition in NB3, in the order in which they appear there, without correction.

Is it plausible, though, that Wittgenstein should have considered one of his notebooks suitable for publication, unedited and unrevised? As we shall see, NB3 marks a significant transition in Wittgenstein's thought: it contains the germ of what became the remarks on mysticism, ethics, and religion in the 6.4s and 6.5s of the *Tractatus*. Eventually Wittgenstein transcribed many remarks from NB3 into *Bodleianus*, but on Geschkowski's hypothesis he had by the time he departed again for the front in January 1917 not yet got as far as doing this. If that is right, then it seems quite natural that he might want NB3 to be published as well, since otherwise this major shift in his thought would have been lost.

7 Kang

The identity of item (5) on the list is also disputed. McGuinness and Geschkowski both think it is *Bodleianus*; Jinho Kang (2005), on the other hand, thinks it is a missing notebook containing an intermediate stage in the development of the book, used by Wittgenstein in compiling the *Prototractatus*, which on this view Wittgenstein did not begin to write until some time later in 1917. Kang's view falls to be considered now.

In his work on the genesis of the *Prototractatus* Geschkowski deliberately adopted an approach which focused wholly on form: he analysed the structure of the various notebooks, the marks in the margins, and the cross-references between them, but he did not consider their content. Kang has adopted the opposite approach: he has focused on the philosophical content of the remarks, and argued on this basis for a radically different conclusion from other commentators. His view is that Wittgenstein did not start compiling *Bodleianus* until some time after the end of 1916. According to Kang the volume on Hermine's list of January 1917—the one which was to be sent to Russell for publication—is not *Bodleianus* but a precursor of it. At some later point, on his view, Wittgenstein became dissatisfied with his precursor and began all over again in *Bodleianus*.

Kang's argument for this radical view is that Wittgenstein cannot have been in a position to begin compiling *Bodleianus* during 1916. For there are various places in his 1916 notebook where Wittgenstein asks questions which are answered in the early pages of *Bodleianus*. Why would he have been asking these questions if he already knew the answers to them?

There is no doubt that Kang raises a significant issue. Should we be persuaded, therefore, to adopt his radical view? I do not think so. The overwhelming difficulty with Kang's proposal is one that arises from his lack of engagement with the structure of *Bodleianus*. As Geschkowski has shown, and as I have briefly mentioned here, the layers of *Bodleianus* show evidence of a series of distinct stages in its construction, when Wittgenstein drew on his various earlier notebooks in slightly different ways. Yet we know from Hermine's list that by January 1917 Wittgenstein already had a volume ready for publication containing at the very least a reworking of his notebooks from the period between October 1913 and June 1915. If this volume was not *Bodleianus*, it is incomprehensible how *Bodleianus* could exhibit the form that it does. The ordering of remarks in the part of the *Prototractatus* which draws on NB1 and NB2 does not seem to match Kang's hypothesis very well. There is certainly a degree of rewording, consistent with the idea that they are transcribed from a revised typescript. But if he had already compiled them into an intermediate notebook, I would have expected them to be ordered into themes to a significant extent, in much the way that large parts of pages 3–28 are.

This consideration is not, in the end, overwhelming. What I find much more implausible about Kang's view is a more philosophical consideration. I have already mentioned that in the summer of 1916 a fundamental shift occurred in Wittgenstein's view of the scope of his project, reflected in NB3. Yet this shift is not reflected at all in the structure Wittgenstein originally adopted for the *Tractatus*: it is clear from the *Prototractatus* that he originally intended the work to end with proposition 6, which states the general form of the truth-function. It seems inconceivable to me that he would have *started* mapping out a book with this structure after his work had 'opened out from the foundations of logic to the nature of the world' in the summer of 1916.

If these considerations are enough to cast doubt on Kang's hypothesis that item (5) on Hermine's list is not *Bodleianus*, it follows that we must look for other explanations of the questions Wittgenstein asked in his 1916 notebook. There are two responses that can be made to Kang, one general and one specific.

The general response is to note that Wittgenstein's working method was not linear. He therefore often addresses in his notebooks questions to which the reader might suppose that he already knew the answers. In this connection we might compare the hesitant remarks on 15 May 1915 with the treatment of facts and complexes in the *Notes on Logic* more than a year earlier: in 1915 Wittgenstein remarks, as if the thought is wholly new to him, that 'the proposition dealing with a complex will not be nonsensical if the complex does not exist, but simply false'. And it is similarly surprising to read Wittgenstein coming to the discovery on 12 September 1916 that 'thinking is a kind of language'.

The more specific response is that Kang's examples are related to the N-operator, which is unquestionably one of the more puzzling elements in the genesis of the *Prototractatus*. The N-operator is not mentioned at all in any of the surviving pre-*Prototractatus* material. It is very difficult to work out, therefore, when Wittgenstein adopted it. It does occur several times in the early pages of *Bodleianus*. Yet it is notable that almost all its occurrences show signs of alterations in the manuscript. It may well be, therefore, that the remarks in *Bodleianus* which Kang has highlighted were not in their final form when Wittgenstein asked his puzzling questions about them.

8 The 1915 *Tractatus*

In what follows I shall assume, hesitantly, that Geschkowski's identification of the various items in Hermine's list is correct. What conjectures does that allow us to make about the composition of the *Prototractatus*?

Many of the remarks in the *Prototractatus* are drawn from the sources we have already mentioned—the *Notes on Logic* and the surviving wartime notebooks—and many others are presumably taken from other notebooks now missing. Wittgenstein's working method, at least for a large part of the book's composition, was thus to search his notebooks, either directly or via typescripts compiled from them, for those remarks that he regarded as 'good', before transferring these to *Bodleianus* and allocating them numbers that positioned them in the gradually developing book. Although the order in which Wittgenstein entered these various remarks in *Bodleianus* is of course different from that of their intended order in the finished book, there frequently occur in *Bodleianus* what I shall call *runs*, that is to say sequences of consecutive remarks whose order in *Bodleianus* coincides with their intended order in the *Prototractatus*. The presence of runs is in itself hardly surprising: it would be natural for Wittgenstein to focus from time to time on a particular topic and search in his notebooks for relevant remarks that express a coherent progression of thought. When we examine *Bodleianus*, though, we notice that in both the respects just mentioned the section from page 4 to page 28 is distinctive. Very few of the remarks in this core section of *Bodleianus* are taken from the surviving pre-*Prototractatus* material, and none is a direct quotation of a length that requires us to suppose Wittgenstein had the original to hand when he entered it in *Bodleianus*. Moreover, there is a much greater degree of regularity in the core than in the later parts of *Bodleianus*: large parts of the core consist of runs, some of them rather lengthy.

These two features suggest rather strongly that when Wittgenstein compiled pages 4 to 28 in *Bodleianus*, he was relying on an earlier draft which had already done a significant amount of the work of getting his remarks into the right order. If this is right, then there are two questions for us to answer: when did Wittgenstein write the early draft, and when did he abandon this draft and start afresh by writing down the framework on page 3 of *Bodleianus*. One clue to this issue of dating is contained in a letter which Wittgenstein wrote to Russell in October 1915.

> I have recently done a great deal of work and, I think, quite successfully. I'm now in the process of summarizing it all and writing it down in the form of a treatise. Now: whatever happens I won't publish anything until you have seen it. But, of course, that can't happen until after the war. But who knows whether I shall survive until then? If I don't survive, get my people to send you all my manuscripts: among them you'll find the final summary written in pencil on loose sheets of paper. It will perhaps cost you some trouble to understand it all, but don't let yourself be put off by that. (Wittgenstein 1995: 103)

One possibility, therefore, is that the 'final summary' was the first draft on which pages 4 to 28 of *Bodleianus* are based. If this is right, then we can date the start of *Bodleianus* after October 1915. At the other end of the scale, by January 1917 *Bodleianus* had been filled at least as far as page 64 and perhaps further. So Wittgenstein could hardly have started work on *Bodleianus* any later than the early autumn of 1916, when he began his officer training in Olmütz.

9 Typescripts

I have not yet discussed the 'typescript with Trenkler' which was to be destroyed. One possibility is that this was simply a carbon copy of the notebook typescript in Olmütz. If so, there was indeed no reason to preserve it, since it would not have contained the corrections that the copy in Olmütz now had. On the other hand, it is also possible that the typescript with Trenkler was Wittgenstein's copy of the Birmingham typescript. In this case the reason to destroy it would have been that Russell already had a copy of it.

Around this time also, if my earlier conjectures are correct, Wittgenstein had a typescript prepared containing at least what he regarded as the philosophically interesting remarks in NB0, NB1, NB2, and that part of NB2a then existing. The first three of these could have been done without him present, of course, by sending instructions to Trenkler, the Wittgenstein family factor in Vienna, into whose care he had entrusted the notebooks. But for NB2a he would probably have done this by dictation, as he had done in Birmingham in 1913, so that he did not have to part with his current notebook. If he did not have this done in Sokal, it must have been when he was in Vienna on leave in August 1915, shortly before the time when, as we have just seen, he told Russell that he had been writing out a final summary of his work on loose sheets.

10 *Bodleianus* begins

I have suggested that at some point now hard to determine, but most likely around the end of 1915, Wittgenstein abandoned the attempt to compose his book entirely on loose sheets and began instead to use a hardback notebook for the purpose. He first of all laid out the structure of the book by means of the principal propositions, beginning with

> 1 The world is everything that is the case.

and ending with

> 6 The general form of the truth function is:

$$|N(\overline{p_0}),\ \overline{\alpha},\ N(\overline{\alpha})|$$

After this Wittgenstein began to fill in the details within this structure on pages 4–28. It is natural to suppose that this part of the book is very largely based on the loose leaf summary. This certainly fits with the manner in which the material appears in the

Prototractatus, which strongly suggests that a previously worked out text is being relied on.

It may well be that this is the point at which Wittgenstein used the famous decimal numbering system for the first time. Eventually, of course, this system served a stylistic goal—he told von Ficker the numbers were essential 'because they alone give the book surveyability and clarity and without them it would be an incomprehensible mess' (Wittgenstein 1969: no. 26)—but at this stage the reason for it may well have been only that the revisions he was making to the loose leaf summary had got out of hand. If he had been working in Vienna (or Cambridge), he could have relied on fairly frequent visits to typists to restore order to the document, but in Sokal we may conjecture that secretarial support was not easy to find. Wittgenstein had therefore to devise a means of inserting and rearranging propositions without making the document a complete mess, and the decimal numbering system (adapted, no doubt, from that in *Principia Mathematica*) was well fitted to this aim.

11 Bazzochi

At this point, though, we need to consider an argument offered by Luciano Bazzocchi to the effect that Wittgenstein started compiling *Bodleianus* very much earlier than I have been suggesting. Bazzocchi contends, in fact, that by June 1915 *Bodleianus* was already compiled at least as far as page 12. If he is right, it follows, of course, that the 'loose leaf summary' of October 1915 cannot be the early draft which Wittgenstein used in compiling the core part of *Bodleianus*.

Bazzocchi's argument relies on a diary entry for 18 June 1915 in which Wittgenstein makes, out of the blue, the following observation: 'Not: A proposition follows from another one, but: The truth of a proposition follows from the truth of the other'. In the context of his diary it is a puzzling remark, because it does not connect with what comes before or after it. But in a delightful piece of detective work Bazzocchi has discovered the target of Wittgenstein's otherwise puzzling remark. For on page 12 of *Bodleianus* Wittgenstein wrote this.

> A proposition follows from another one if all the truth-grounds of the first are truth-grounds of the second.

What is visible on the page, however, is that at some point Wittgenstein amended this remark by inserting 'the truth of' in just the way recommended by his diary entry.

It can hardly be doubted that Bazzocchi has correctly identified this remark in *Bodleianus* (which became 5.041 in the *Tractatus*) as the target of Wittgenstein's diary entry. What does this tell us, though, about the timing of the composition of *Bodleianus*? According to Bazzocchi, what it shows is that Wittgenstein must already have got as far as page 12 by the time of the diary entry; but this conclusion seems to me to be unwarranted. It is fairly plain, at any rate, that Wittgenstein cannot have had *Bodleianus* at his elbow in June 1915. For if he had, he would presumably just have made the amendment to 5.041 direct, rather than noting it in his diary. So on Bazzocchi's hypothesis Wittgenstein must have started to compile *Bodleianus* at some earlier stage and then presumably for safety's sake left it in, or sent it to, Vienna. It is far from clear when this might have happened. He seems to have been on military service continuously from September 1914 to July 1915 without any opportunity to return to Vienna to work on his book. If he started compiling *Bodleianus* while he was at the artillery workshop, why did he stop work before 18 June and send the volume back to Vienna?

It is difficult, therefore, to identify a period during which Wittgenstein might plausibly have worked on *Bodleianus* between the beginning of the war and the point in June 1915 when Bazzocchi supposes him to have wished to correct it. Moreover, the earlier we place the commencement of *Bodleianus* the more the sorts of worries posed by Kang multiply. If a significant part of *Bodleianus* was already filled in when Wittgenstein was still writing his two 1914–15 notebooks, many of the concerns voiced in them become very difficult to comprehend. The impression in these notebooks that Wittgenstein's system has not yet reached its settled form extends even to terminology: as late as 15 April 1915 he was still using the word *Sachverhalt* in a context that suggests he meant by it what in *Bodleianus* he always called a *Sachlage*.

Notice, too, that according to Bazzocchi's theory *Bodleianus* had already been filled some way by the time Wittgenstein wrote to Russell in October 1915. Why, in that case, did he tell Russell about the loose-leaf summary but remain silent about *Bodleianus*? If the loose-leaf summary was at this stage simply a copy of *Bodleianus* (but presumably in the correct order), what was the point of having *Bodleianus* at all?

The alternative hypothesis is that 5.041 occurred in its original form elsewhere—perhaps in Wittgenstein's Norwegian notebook, which, as we have already noted, he sent back to Vienna for safekeeping early in the war. When he realized the error, this notebook was not accessible, so all he could do was make a note of it for later correction. All that need be the case on this hypothesis is that at whatever point he transferred the remark into *Bodleianus*, he had temporarily forgotten the correction. At some later point when *Bodleianus* was to hand he either remembered the point or reread his diary entry of June 1915 and took the opportunity to make the correction.

That Wittgenstein might have had such a lapse of memory is surely made more plausible by the triviality of the point at issue. What, we might wonder, is wrong with the locution that one proposition follows from another? Wittgenstein's reasons for preferring to say that the truth of a proposition follows from the truth of another are obscure. Indeed, even for Wittgenstein little seems to hinge on the difference: although he took the trouble to correct 5.04 and 5.041, he left the succeeding remarks 5.042–5.07 in the run unchanged.

12 The 1916 *Tractatus*

As we have noted, Wittgenstein's circumstances changed markedly in April 1916. In the first months of the war, he had been close to the front line, as a searchlight operator, but from then on his role had been that of a mechanic in vehicle maintenance depots some way behind the lines. Now, though, he went to the front as an observer with an artillery battery. For the remainder of the war he lived the typical life of a front-line soldier, a series of postings to the Eastern and Italian fronts interspersed with periods of leave. This change in Wittgenstein's life, although undoubtedly something he had actively sought, created for him an immediate practical difficulty over how to continue work on the *Prototractatus*. There were obvious and narrow limits on what he would wish, or (at least until he became an officer) be allowed, to lug round with him in his rucksack.

When he left Sokal for the front, therefore, Wittgenstein started a new notebook, NB3, and continued to do philosophical work, but sent his other working papers back to Vienna for safe keeping and could therefore make no further progress with the composition of the *Prototractatus*. He laments this in a diary entry for May 1916: 'Would now have plenty time and peace for work. But it's no use. My stuff is a long way away' (Wittgenstein 1991: 71).

There are at various places in *Bodleianus* (on pages 28, 52, 60, 64, 70, 71, 78, 103, 106, 107, and 108) horizontal lines at the side of the page. We cannot be sure what these lines signify, or even that they all signify the same thing, but the circumstances of composition just described make it natural to conjecture that they mark breaks forced on Wittgenstein by the rhythm of his army service. It is my conjecture that Wittgenstein's working practices alternated from now on between two modes corresponding to two very different ways of living. While on active service he typically carried around with him one notebook in which to record his daily thoughts, but left the *Bodleianus* and his other notebooks in safe keeping either with his family in Vienna or with the new friends he was soon to make in Olmütz. Work on editing the remarks in his notebooks and transferring them into *Bodleianus* would from now on be confined to periods of leave.

I cannot prove this conjecture with certainty, though: its plausibility must rest largely on the degree to which it fits with the other available evidence. If it is correct, however, McGuinness is wrong to suppose that 'since it is written in large part in field-pencil, [the *Prototractatus*] was composed actually at the front' (1990: 265). As early as December 1914 (1995: 91) Wittgenstein was sufficiently convinced of the importance of his work that he had issued instructions (which he revised and renewed in January 1917, and no doubt on other occasions too) to ensure that his work would be sent to Russell in the event of his death in action. The reconstruction I shall offer supposes that Wittgenstein was careful never to consign *Bodleianus* itself to a war zone.

The first of the break points in *Bodleianus* occurs on page 28. Up to this point there have been, as I have noted already, a few propositions which bear some similarity to sentences that occur in the surviving manuscripts, but none is close enough in wording that it requires us to suppose it was copied direct. Directly after the break point on page 28, however, there is a whole series of propositions, pages 28(3)–34(3), which are copied directly from MSS 1, 3, and 4 of the *Notes on Logic*. After that comes a long section, 34(3)–64(3), in which many of the propositions are versions of remarks in NB1 and NB2. If the suppositions I made earlier about the identity of the items on Hermine's list are correct, this section of the *Prototractatus* will have been formed by extracting remarks not from the notebooks themselves but from the notebook typescript, containing the Norwegian notebook, NB1, NB2, and the first part of NB2a, which was typed in August 1915.

At the beginning of September 1916, after a period of leave in Vienna, Wittgenstein went to officer training school at Olmütz. It is natural to conjecture that he used this period, his first opportunity since March, to continue work on the *Prototractatus*. I am inclined to think that he compiled everything up to at least the break point on page 64 by the end of November 1916. This period thus covers break points on pages 52 and 60, which on this conjecture represent relatively minor pauses in composition. During this period it is noticeable that Wittgenstein was writing philosophical remarks in NB3 only sporadically (and coded remarks not at all). It is possible to posit ways in which some at least of the gaps in NB3 (the second half of August in Vienna, most of September, a fortnight at the end of October, ten days in November in Olmütz) might represent periods of work on the *Prototractatus*. We should bear in mind, however, that all these passages of work in *Bodleianus* contain remarks which I am supposing he extracted from the notebook typescript. Since he evidently left that in Olmütz on his return to Vienna, it seems likely that during December, when he was back with his family in Vienna, he did not work on the *Prototractatus*. Since there are hardly any remarks in NB3 for this period either, perhaps he simply took a rest. Another possibility, though, is that he compiled the passage on pages 64(4)–70(5) during this period too. These represent a new phase in the composition of the book, since now propositions with numbers greater than 6 are introduced for the first time. However, some of the material introduced is recognizable from the notes dictated to Moore in Norway, which suggests that it may be drawn from the Norwegian notebook.

A letter from his friend Pinsent (probably the last Wittgenstein ever received from him) is worth mentioning. 'I am delighted,' he says in response to a letter Wittgenstein had written on 18 November 1916, 'that you have completed your work' (Wittgenstein 1991: 103). (The version of the letter that has survived has been translated into German to get past the censor, so we cannot be quite certain that the word Pinsent used originally was 'work'.) If, as I am conjecturing, *Bodleianus* was by then filled as far as page 64, it would indeed have been describable as a complete work—perhaps he did even temporarily regard it as finished—containing, as he described it in his instructions to his sister, 'the working over of (1) and (2) for publication'. (Of course, this shows at most that the schedule of composition I am conjecturing is consistent with the available evidence, not that it is entailed by it. One of the oddities of Wittgenstein's method of composition was that the *Prototractatus* might reasonably be described as a complete work ready for publication at almost any point in its composition from now on.)

One of the friends Wittgenstein made in Olmütz, a lawyer called Heini Groag, recollected much later to McGuinness (1990: 260n.) that Wittgenstein had lent him a manuscript containing a work with numbered remarks. McGuinness's notes of the conversation placed the loan the following winter, but, as McGuinness himself has pointed out, that is hardly consistent with the evidence: Groag was by then a soldier fighting on the Italian front and his periods of leave that winter did not coincide with Wittgenstein's. I assume, therefore, that McGuinness is right to trust his memory, rather than his notes, and place the loan in the winter of 1916/17. The fact of the loan does not itself affect the narrative significantly, but it does add a little to the impression that perhaps by the end of his time at Olmütz Wittgenstein had a work that he, at least for a time, regarded as complete.

13 The *Prototractatus*

How did the 1916 *Tractatus* (roughly, I am supposing, *Bodleianus* as far as the line on page 70) become the *Prototractatus* (to page 103)? The last entry in NB3 occurs just before Wittgenstein set off for the front again on 10 January 1917. He will no doubt have started a fresh notebook at this point, just as he had when he was first sent to the front the previous April. 'I can work again, thank God,' he reported to Engelmann (1967: 3) shortly afterwards. However, it is my supposition that he left his other manuscripts and typescripts behind, variously in Vienna or Olmütz, so that 'work' here meant only entering fresh remarks in his new notebook, not additions to *Bodleianus*.

He was granted a month's leave some time in June and on 15 June he invited Frege to come to Vienna to meet him. (Frege declined.) It may have been at this stage that Wittgenstein added to *Bodleianus* the renowned proposition 7. This would explain the fact that a slight variant of it—'Whereof one cannot talk thereof one must be silent'—occurs in his hand as a sort of doodle on the letter from his sister mentioned earlier, which was written on 7 June 1917.

Wittgenstein left Vienna again on 2 July, and the record of his military service shows that by 9 July he was back at the front. From then until the disintegration of the Eastern front in October as a result of the Russian revolution Wittgenstein was involved in the fighting in Bukovina. For this period at least I am supposing that his work would have been restricted much as it had been the previous summer. The formal ceasefire—the effective end of the war on the Eastern front—came on 29 November 1917, and Wittgenstein arrived back in Vienna on leave some time after 9 December. He was back with his regiment early in January, and by 8 February he had been sent on a flying course. (The previous August he had been approached by an officer he had met in Sokal about the possibility that he might be transferred to the research section of the Austrian air ministry, based in the outskirts of Vienna. Perhaps the flying course was connected somehow with this proposal.) At the end of the course there seems to have been another short period of leave with his family before Wittgenstein was posted to the Italian front some time in early March.

As we have seen, it is likely that by July 1917 Wittgenstein had filled *Bodleianus* at least as far as page 71. Then he made another pass—pages 71(4)–78(10)—through the notebook typescript, fitting material into the newly extended structure that going beyond proposition 6 allows for. Pages 79(1)–81(3) may be from the part of NB2a that was not included in the typescript. After that there is a section (81(4)–86(6)) in which for the first time Wittgenstein mines NB3 for material: the nowadays notorious proposition 6.54 also occurs here. Then there is a long passage of work from page 87 to page 103 which may draw on the missing 1917 notebook as well as on passages from earlier notebooks.

We cannot now determine with any certainty when Wittgenstein carried out these various stages of composition. For reasons we shall come to shortly, however, we may deduce that when the compilation of *Bodleianus* had got as far as page 103 Wittgenstein had a typescript made of it. Engelmann's later recollection was that Wittgenstein 'dictated his manuscript to a typist in Vienna before his departure for the Italian front' (McGuinness 2002: 263 n.). This might, it is true, be a reference to the later production of a revised typescript in August 1918, before Wittgenstein left a second time for the Italian front, but it fits with the narrative I have been offering to suppose that it is the earlier production of a *Prototractatus* typescript that Engelmann was here describing. At any rate Engelmann was in Vienna at the beginning of March 1918, working on the interior design of one of the Wittgenstein family's houses, and he met Wittgenstein there. So if this is when Wittgenstein dictated the typescript, Engelmann was at least in a position to know.

Whatever the exact date of its production, though, the typescript will have contained the whole of what is now known as the *Prototractatus*. It was, that is to say, a copy of pages 3 to 103 of *Bodleianus*, the propositions rearranged now (one assumes) into the order that their decimal numbers imply.

14 The *Tractatus*

The next phase was evidently for Wittgenstein to revise the *Prototractatus* typescript and in particular to remodel the numbering so that now it served more accurately the purpose he advertised for it of indicating 'the logical importance of the propositions, the importance laid on them in [his] exposition'.

This phase of work does not seem to have involved adding propositions to the *Tractatus* but only revisions to the typescript. By the time Wittgenstein came to add further propositions, that is to say, the renumbering had already taken place. A natural explanation for this might be that when he did the renumbering Wittgenstein did not have his other notebooks available. He was on active service in or around Asiago from

March until early July 1918 apart from a fortnight attending a course in Trent in May and a short period in hospital in Bolzano. My conjecture is therefore that he may have left all his notebooks in safe keeping, either in Vienna or in Olmütz, and taken only the *Prototractatus* typescript with him to Asiago.

Eventually, though, he did return to the business of adding further propositions to the work. Curiously, though, he did this in *Bodleianus* rather than in the renumbered typescript, even though working in this manner was by now misleading because *Bodleianus* no longer represented the most recent revision of the book and so the numbers which he gave to the new propositions by reference to the typescript no longer indicated their correct positions in *Bodleianus*. These further propositions (which occur from page 103 onwards in *Bodleianus*) are generally treated as not belonging to the *Prototractatus*, and I shall share that practice here.

The work of revision I am now describing, which turned the *Prototractatus* into the *Tractatus*, was probably done in the summer of 1918 when Wittgenstein was granted a long period of leave beginning on 5 July. Wittgenstein's practice seems to have been to spend part of each period of leave with his family, whether in Vienna or at the Hochreit (their summer home), and part visiting the friends he had made in Olmütz. Paul Engelmann later recalled an incident that occurred during one such visit to Olmütz. (It is impossible now to date the visit in question with any precision.)

> On the day after his arrival [in Olmütz] we went out together in the morning to be able to talk undisturbed. In a remote, and at the time of day deserted, avenue of the city park we sat down on a bench, and there, quite contrary to my custom, I read the poem to him. (Engelmann 1967: 76)

Wittgenstein wrote to Engelmann—the letter is undated, but the wording suggests it was quite shortly after this visit to Olmütz—asking for a copy of the poem and the return of his manuscripts:

> Please tell Mr. Groag to be kind enough to send me my manuscripts, in spite of the danger. If they do get lost, it was God's will. Furthermore, please send me a copy of your poem. I shall not misuse it and will read the poem only to my sister Mining. . . . I was very glad to have seen you and talked to you. It has *refreshed* me. (Engelmann 1967: 7–9)

I mention this incident in order to observe that even if *Bodleianus* was one of the manuscripts Wittgenstein had left with Groag and now wanted back, the risk to it that Wittgenstein was willing to take need only have been that of an unreliable postal service and not that of sending it to a war zone. (Indeed the reference to his sister suggests that he was probably not at the front when he made the request.)

At any rate Wittgenstein spent part of the summer of 1918 at the Hochreit and part, it seems, with his uncle in Hallein near Salzburg. Engelmann, who was an invited guest of the family at the Hochreit that summer, later recalled that Wittgenstein had been working from seven notebooks when he compiled the *Tractatus*. If the conjectures I have been making here are correct, these will have been: the Norwegian notebook

NB0, two notebooks NB1 and NB2 from 1914–15, the missing NB2a, NB3 from 1916, a missing 1917 notebook NB4, and *Bodleianus* itself.

In August 1918 Wittgenstein had a new typescript prepared (along with a carbon copy) combining the worked-over typescript with the new propositions he had entered in *Bodleianus* from page 103 onwards. Both copies of this new typescript have survived: one ended up much later in the hands of Wittgenstein's friend Engelmann, the other with Wittgenstein's sister in Gmunden, and for this reason they are often referred to as the Engelmann and Gmunden typescripts respectively.

Wittgenstein then began the process of attempting to have the *Tractatus* accepted for publication, first sending the Gmunden typescript to a publisher called Jahoda. Meanwhile he took the Engelmann typescript with him when he returned to duty on the Italian front in September, and it was therefore this typescript that he had in his possession when he was taken prisoner at the end of the war in November 1918. During the nine months of his captivity in a prisoner of war camp in Italy Wittgenstein made surprisingly few further revisions to it, considering how much spare time he must now have had. (This makes a notable contrast with the fate of his typescripts in the 1930s, which were subjected to innumerable revisions.) Wittgenstein asked Engelmann to send him Frege's *Grundgesetze* in the camp (Engelmann 1967: 17), but we do not know why he wanted to consult it or even whether it arrived. Perhaps he only wanted to convince one of the other prisoners of the glories of its preface, which he had had to recite from memory.

When Jahoda returned the Gmunden typescript to Vienna, Wittgenstein's sister arranged for another copy to be made from it. One reason for this was that Wittgenstein wanted his sister to send a copy to Frege, which she arranged to be done on 24 December 1918. Whether this was, as is generally supposed, what is now known as the Vienna typescript has been questioned by Grasshoff and Lampert in their Introduction to a photographic reproduction of all three surviving typescripts (2004). It would take us too far out of the way to address this issue here.

15 Russell's introduction

Getting a copy of the *Tractatus* to Russell while Britain remained formally at war with Austria was trickier than sending it to Frege. Eventually Russell received the Engelmann typescript at the beginning of July 1919, Keynes having acted as an intermediary in arranging for it to be conveyed by diplomatic bag from Wittgenstein's PoW camp to England. Russell met Wittgenstein in the Hague that December to discuss the book, and promised to write an Introduction, in the hope that this might help get a German or Austrian publisher to agree to publish it. Once it had been confirmed that Wittgenstein had found a publisher (Reclam), Russell wrote his Introduction in February and March 1920, during a period when he was lecturing in London on analysis of mind: for the details see the editorial note on the reprinting of the Introduction in Russell (1988). Russell had just broken his collarbone and could not write, so he composed most of it by dictation to Dorothy Wrinch, sending the completed introduction to Wittgenstein on 19 March.

Wittgenstein's refusal to use the introduction is famous, but should not be overplayed. Of course, the ostensible reason he gave Russell (1995: 153)—that the German translation obscured the elegance of Russell's style—was half-hearted and implausible. (Why, one wonders, did he not translate the piece into German himself?) Wittgenstein clearly did think Russell had misrepresented him in certain ways.

There's so much of it that I'm not quite in agreement with—both where you're critical of me and also where you're simply trying to elucidate my point of view. (Wittgenstein 1995: 152)

On the other hand, Wittgenstein's concerns were probably not just philosophical. More than once Wittgenstein pulled back just at the point when he seemed close to arranging for the book's publication. He rejected Braumüller's publishing offer because it required Wittgenstein to pay the printing costs himself and therefore seemed to him too much like vanity publishing. (Did he know that Whitehead and Russell had had to pay Cambridge University Press part of the cost of typesetting *Principia*?) He was not willing to let Ficker publish the book as a personal favour (Wittgenstein 1969: no. 26). Yet there was no point in asking the opinion of a philosophy professor before deciding whether to publish it since he would not understand it (Wittgenstein 1969: no. 24). Perhaps he was now coming to resent the fact that Reclam was willing to publish it not on its own terms but only because of Russell's Introduction. Yet the ways he chose to describe what 'its own terms' amounted to were scarcely such as to reassure a cash-strapped publisher.

The incident bears an obvious similarity with the earlier difficulties over Wittgenstein's Cambridge dissertation. ('If I'm not worth your making an exception for me *even in some* STUPID *details* then I may as well go to Hell directly'; Wittgenstein 1995: 85–6.) It is almost as if Wittgenstein wanted, perhaps even expected, the circumstances of publication to be wholly exceptional. Russell's response—'I don't care twopence about the introduction but I shall be really sorry if your book isn't printed' (Wittgenstein 1995: 156)—was an astonishing display of patience and restraint in the circumstances.

Nonetheless, even if Wittgenstein did disagree with some of the explanations of his views in the Introduction, the plain fact is that Russell had discussed the book with Wittgenstein in detail for a week just over a year earlier, and hence his explanations of its meaning have a privileged position compared to the writings of other commentators. (Even Ramsey's critical notice cannot claim such privilege, since he wrote it just *before* going to Austria to discuss the book with its author.)

16 The English translation

Eventually Wittgenstein relented and let Russell try to arrange publication of the book along with Russell's Introduction: 'If you feel like getting it printed, it is entirely at your disposal and *you can do what you like with it*' (Wittgenstein 1995: 157). Russell delegated this task to Dorothy Wrinch while he went on an extended trip to China. After the typescript had been rejected by Cambridge University Press, she managed to have it accepted for publication in the series *Annalen der Naturphilosophie* edited by Wilhelm Ostwald, a Nobel prize-winning chemist. Perhaps it is as well that Wittgenstein did not see Ostwald's letter to Wrinch, which made it clear that he was accepting the work not on its own merits—he had at this point not even seen the typescript—but only because of the 'extremely high regard' in which he held Russell. When he eventually saw a copy, Wittgenstein described this first publication of the book as a pirate edition, by which he presumably meant not that it was unauthorized—he had undoubtedly given Russell unambiguous authority to act in the matter—but rather that it had not been properly proof-read before publication and was therefore full of typographical errors.

A little later Wrinch arranged the English publication of the book with C. K. Ogden, who accepted it for The International Library of Psychology, Philosophy, and Scientific Method, a series which he had recently started editing for Kegan Paul, Trench, Trubner & Co. Ogden was a freelance writer in Cambridge, founding editor of the *Cambridge Magazine*, a friend both of Wrinch and of Russell, as well as (like Ostwald) a slightly eccentric figure. It was to be a dual language edition. In the Hague two years earlier Russell had offered to do the translation himself, but he told Wittgenstein in December 1921 that it would be done by 'two young men at Cambridge who know mathematical logic' (Wittgenstein 1995: 174). One of these young men was Frank Ramsey, then in the second year of a mathematics degree. The identity of the other is not known: perhaps it was Max Newman. In the end, though, it was Ramsey on his own who did the first draft, early in 1922, using an offprint of the Ostwald printing as his text: his father later recalled that Ramsey had dictated it at the Cambridge university typing office. Only later was the Engelmann typescript retrieved from Ostwald and sent to Wittgenstein along with a copy of the Ostwald printing for Wittgenstein to correct. The German text in the dual language edition was typeset from this corrected copy of the Ostwald printing.

Russell promised Wittgenstein that he would inform the translators of what he and Wittgenstein had agreed (presumably during their meeting at the Hague) as regards translations of terms (Wittgenstein 1995: 174). Unfortunately, the evidence on this point is somewhat uncertain: Ogden told Wittgenstein that the translation incorporated 'all Russell's M.S. notes and some of his suggestions' (quoted in Grasshoff and Lampert 2004: 48), but it is not clear what these amounted to. So we cannot be sure, for instance, that the decision to translate *Sachverhalt* as 'atomic fact' was originally Wittgenstein's. The most we can say is that Wittgenstein did not actively object to it. The correspondence with Ogden shows that one translation he did not care for was 'state of affairs' for *Sachlage*, but Ogden chose not to adopt his alternative suggestion, *status rerum*.

For many years Ramsey's translation was regarded with disdain by many Wittgensteinians, encouraged perhaps by Anscombe.

> English readers of the *Tractatus* may need to be warned that Ogden's translation is notoriously very bad. Wittgenstein told me that he had not checked the whole of this translation, but only answered a few questions that were put to him about some passages. E.g. I think we can see Wittgenstein's hand in the free but excellent rendering of 4.023. (Anscombe 1959: 17)

Anscombe is right to see Wittgenstein's hand in the English of 4.023.

> The proposition determines reality to this extent, that one only needs to say 'Yes' or 'No' to it to make it agree with reality.

However, her description of the Ogden–Ramsey translation as a whole as 'notoriously very bad' is at best a gratuitous exaggeration, and what she reports Wittgenstein saying about the extent of his involvement in the rest of the translation is certainly wrong. In fact, he revised it in considerable detail, and the surviving typescript bears numerous changes in his hand. That is not to say that the translation is perfect, of course, and in a few places it is certainly mistaken. At 2.11, for instance, Ramsey faithfully translated what is in fact a misprint in the Ostwald printing (*Tatsachen* instead of *Sachlage*): Wittgenstein corrected the misprint in the German but failed to correct Ramsey's translation from 'facts' to 'state of affairs' correspondingly. Wittgenstein gave Ogden instructions to prefer Russell's translations where possible, but for some reason Ogden did not do this. As a result, one significant mistake got through to the final printing: Russell had correctly translated 'der Sprache, die allein ich verstehe' as 'the only language I understand'; perhaps Wittgenstein's grip of English idiom after almost nine years away from Britain was too shaky for him to realize that Ramsey's translation, 'the language which only I understand' (Wittgenstein 1922: 5.62), does not mean the same thing. (When the book was reprinted in 1933, Wittgenstein changed it to '*the* language which I understand', which is not as misleading as Ramsey's translation, but still needlessly obscure.)

In the first week of May 1922 Russell revised the Introduction for the English edition to take account of comments from Ogden (but not from Wittgenstein). Russell returned corrected proofs of the Introduction to Ogden on 29 June. By the time Russell and Wittgenstein met again, in Innsbruck in August 1922, Wittgenstein had just sent back the proofs of the main text to Ogden. The book finally appeared in November 1922.

17 Conclusion

I have tried to piece together what evidence there is of how and when the *Tractatus* was composed. Much of the reconstruction I have offered is undoubtedly conjectural, and it is unlikely that it is correct in every respect. If the chronology and proposals concerning missing documents that I have offered here (summarized in the table below) are to be preferred to those presented by McGuinness, for instance, the reason

Table 1 MSS and TSS referred to in this chapter.

Description	*Label*	*Date*	*Von Wright number*	*Location*
Birmingham TS		October 1913		Lost
Russell's manuscript	R	February 1914		Russell Archive
Summary TS	S	October 1913		Russell Archive
Notes on Logic TS	T	February 1914	TS201a	Russell Archive
Costello TS		March 1914	TS201b	Unknown
Notes dictated to Moore		April 1914	D301	Cambridge UL
Norwegian notebook	NB0	October 1913–August 1914		Lost
1st wartime notebook	NB1	August 1914–October 1914	MS101	Wren Library
2nd wartime notebook	NB2	October 1914–June 1915	MS102	Wren Library
Missing notebook	NB2a	June 1915–April 1916		Lost
Olmütz TS of NB0–2		August 1915		Lost
Final summary		October 1915		Lost
3rd wartime notebook	NB3	April 1916–January 1917		Wren Library
1917 notebook	NB4	January 1917 onwards	MS103	Lost
Bodleianus		Early 1916 onwards		Bodleian Library
Prototractatus TS		February 1918	MS104	Lost
Engelmann TS		August 1918	TS202	Bodleian Library
Gmunden TS		August 1918	TS204	Austrian Nat. Lib.
Vienna TS		December 1918	TS203	Austrian Nat. Lib.
Russell's Introduction		February–March		Russell Archive
English translation TS		January 1922		Austrian Nat. Lib.

is never likely to be more than an accumulation of plausibility: in this territory decisive arguments are rarely to be found.

The reader who has followed me this far is entitled to wonder what point there is in undertaking such detailed reconstruction work on a book's genesis. One response, no doubt the easiest, is narrowly biographical. The account I have offered is a corrective to some of the myths that have grown up around the composition of the *Tractatus*. It was not, as some would like to believe, composed in a short burst of creative genius, but was the product of a highly methodical process that started when Wittgenstein came to Cambridge in 1911. Moreover, although the composition of the book itself certainly took place while Wittgenstein was a serving soldier, there is no need to suppose that any of it happened at the front. The account I have offered suggests instead that although he may have had a working notebook with him at the front, his normal procedure was to leave *Bodleianus* behind in Vienna or Olmütz for safe keeping. From the autumn of 1915 onwards, if not before, he regarded the book he had begun to assemble as something he wanted to outlast him, and he made careful arrangements to ensure that in the event of his death it would be conveyed to Russell, whom he evidently expected to arrange for its publication.

It would be wrong, however, to see an appreciation of Wittgenstein's attitude to his work as the only pay-off to be gleaned from a reconstruction of its compositional history. If the chronology I have offered is correct, it offers us a route to a better understanding of the development of Wittgenstein's conception of the book. It began as a work in what we would now call philosophical logic narrowly conceived, and its concluding sentence specified a recursive procedure by which all propositions are obtainable from elementary propositions. What I have suggested here is that it is consistent with the available evidence to suppose that the widening of the book's ambitions to encompass an appreciation of mathematics, ethics, and religion did not take place until after the summer of 1916. The influence of Schopenhauer on Wittgenstein's thinking during that summer is visible in the surviving notebook and has often been remarked on by commentators. The influence of Kant on the additions made to the *Tractatus* thereafter is a matter of current controversy, although documentary evidence of Wittgenstein's exposure to his works at this stage has proved frustratingly elusive and a discussion of these Kantian themes will have to be left for another time.[1]

References

Anscombe, G. E. M. (1959) *An Introduction to Wittgenstein's Tractatus*. London: Hutchinson.

Bazzocchi, L. (2005) 'The Strange Case of the *Prototractatus* note', in *Time and History: Papers of the 28th International Wittgenstein Symposium*, 24–6.

——(2010) 'The "Prototractatus" manuscript and its corrections', in N. Venturinha, ed., *Wittgenstein after his Nachlass*. London: Palgrave Macmillan.

Engelmann, P. (1967) *Letters from Ludwig Wittgenstein, with a Memoir*. Oxford: Blackwell.

Geschkowski, A. (2001) *Die Enstehung von Wittgensteins Tractatus*. Bern Studies in the History and Philosophy of Science. Books on demand.

Grasshoff, G. and Lampert, T., eds (2004), *Ludwig Wittgensteins Logisch-Philosophische Abhandlung: Entstehungsgeschichte und Herausgabe der Typoskripte und Korrekturexemplare*. Vienna: Springer.

Kang, J. (2005) 'On the Composition of the *Prototractatus*', *Philosophical Quarterly* 55, 1–20.

McGuinness, B. (1989) 'Wittgenstein's pre-Tractatus manuscripts', in *Wittgenstein in Focus—Im Brennpunkt: Wittgenstein*. Amsterdam: Rodopi, 35–47.

[1] Although the conclusions I reach in this chapter differ in a number of respects from those of Brian McGuinness, I hope the extent of my debt to his humane and scholarly study of Wittgenstein's life and work is clear.

——(1990) *Wittgenstein, A Life: Young Ludwig (1889–1921).* Harmondsworth: Penguin.

——(2002) *Approaches to Wittgenstein: Collected Papers.* London: Routledge.

Potter, M. (2009) *Wittgenstein's Notes on Logic.* Oxford: Oxford University Press.

Russell, B. (1988) *Collected Papers*, vol. 9: *Essays on Language, Mind and Matter 1919–26.* London: Routledge.

Venturinha, N. (2010) *Lógica, Ética, Gramática: Wittgenstein e o Método da Filosofia.* Lisbon: Imprensa Nacional-Casa da Moeda.

Von Wright, G. H. (1990) *A Portrait of Wittgenstein as a Young Man: From the Diary of David Hume Pinsent 1912–1914.* Oxford: Blackwell.

Wittgenstein, L. (1922) *Tractatus Logico-Philosophicus*; revised edition, 1933. London: Kegan Paul and Trubner.

——(1969) *Briefe an Ludwig von Ficker*, ed. G. H. von Wright. Salzburg: Müller.

——(1973) *Letters to C. K. Ogden with Comments on the English Translation of the Tractatus Logico-Philosophicus*, ed. G. H. von Wright. Oxford, London: Blackwell, Routledge and Kegan Paul.

——(1979) *Notebooks 1914–1916*, 2nd edition. Oxford: Blackwell.

——(1991) *Geheime Tagebücher 1914–1916.* Vienna: Turia and Kant.

——(1995) *Cambridge Letters: Correspondence with Russell, Keynes, Moore, Ramsey, and Sraffa*, ed. B. F. McGuinness and G. H. von Wright. Oxford: Blackwell.

——(2004) *Gesamtbriefwechsel.* CD-ROM. Intelex.

3

Why does Wittgenstein say that ethics and aesthetics are one and the same?

Hanne Appelqvist

1 Introduction

According to the *Tractatus*, there is no value in the world. If there were a complete description of the world—this being the hypothetical book that Wittgenstein calls *The world as I found it*—this description would not include any value statements (5.631, 6.4). This is because the facts that constitute the world are contingent and therefore of equal value (6.41). Further, Wittgenstein claims that 'states of affairs are independent of each other' (2.061), and that 'superstition is nothing but belief in the causal nexus' (5.1361). Because of this, 'the world is independent of my will' (6.373). I may wish that a certain state of affairs results from my action, but 'even if all we wish for were to happen, still this would only be a favour granted by fate, so to speak: for there is no logical connexion between will and the world which would guarantee it' (6.374). Hence, the contingent facts of the world, facts that are neither good nor bad, include my actions as well, insofar as these actions are taken to be physical or psychological events. These, too, fall into the category of 'indifference for what is higher' (6.432).

This view gives rise to the following dilemma: if the facts of the world are ultimately beyond my control and lack absolute value, then what is the point of ethics, which presumably deals with what I should do or what I should will? I believe that Wittgenstein's claim about the sameness of ethics and aesthetics is his solution to this dilemma.

2 A Kantian framework

Before turning to Wittgenstein's early remarks on ethics and aesthetics, I would like briefly to outline the interpretative approach that has influenced my reading of Wittgenstein's early thought about ethics. This is an approach that emphasizes the continuity of a philosophical tradition from Kant to Wittgenstein. It is difficult, based on the available texts, to map in detail actual historical influences between Kant and Wittgenstein's early thought. It is known that Schopenhauer's transcendental idealism was a philosophical position that attracted the young Wittgenstein and, as has been observed by many scholars, one can easily find echoes of Schopenhauerian views in Wittgenstein's early writing, particularly in his remarks on ethics and action (see e.g. Hacker 1986; Stokhof 2002). It is also known that Wittgenstein had read Kantians such as Herz and Bolzmann and mentions these writers among those who influenced his intellectual development (CV 19). Moreover, there are reasons to believe that Wittgenstein was familiar with Kant's own writing. For example, McGuinness reports that he discussed Kant with friends in Olmütz in 1916 (McGuinness 1988: 252), and we find him reading Kant's *First Critique* with Hänsel at the prisoner of war camp in Italy (McGuinness 1988: 270; Monk 1990: 158). It is also noteworthy that Wittgenstein describes Kant as a 'deep thinker' and does so in explicit contrast to Schopenhauer (Rhees 1981: 95). This, as well as Wittgenstein's comment on Kant's method as 'the right sort of approach in philosophy' (LWL 73), would be surprising comments had he not been familiar with Kant's work.

Whatever the historical explanation may be, many commentators have detected significant similarities between Kant's philosophical system and Wittgenstein's early thought. One of the first to do so was Erik Stenius, who suggested that the *Tractatus* should be read as a Kantian work. According to him, Wittgenstein's early project is a modification of Kant's project of distinguishing between the provinces of theoretical and practical reason. In Stenius' view, the Tractarian logical analysis of language is 'a kind of "transcendental deduction" in Kant's sense, the aim of which is to indicate the *a priori* form of experience which is "shown" by all meaningful language and therefore cannot be "said"' (Stenius 1960: 220). This is to say that, in his notion of meaningful language as limited by logic, the early Wittgenstein offers a linguistic variant of the Kantian idea of the bounds of theoretical reason. However, given that logical form is a presupposition of every meaningful proposition, it cannot be expressed in language. While the resulting lack of a synthetic a priori doctrine may be seen as a significant difference between Kant and Wittgenstein, it is important to notice that the very notions which for Kant belong to the domain of practical reason—namely, the will as the subject of ethical attributes (6.423), God (6.432), and the immortality of the human soul (6.4312)—still occupy a role in Wittgenstein's early thought (Stenius 1960: 214–26).

The Kantianism of the *Tractatus* is not restricted to the attempt to draw a limit between sense and nonsense. The connection, on the view we are considering, lies deeper, in the conception of philosophy that these two thinkers share. Commentators such as Hans-Johann Glock and Heikki Kannisto have argued that, for the early Wittgenstein, philosophy is first and foremost reflective activity in Kant's sense. Like Kant, Wittgenstein takes philosophy to be an enterprise which is qualitatively different from that of the natural sciences. What sets philosophy apart from the natural sciences is its aim of reflecting those preconditions of meaningful discourse that are already presupposed in everyday language and in the language of natural sciences (4.111, 4.112). While Kant aimed at uncovering the necessary formal or structural features of our experiences, the early Wittgenstein is concerned with the logical preconditions of any meaningful representation. Moreover, like Kant, Wittgenstein seems to assume that this reflection of the nature of our ways of experiencing or representing reality begins by studying the structure of our own thinking (Kannisto 1986; Glock 1992, 1997).

I mentioned above that, for the early Wittgenstein, the contingency of facts renders them indifferent for value. However, there is an aspect of necessity in the world of contingent facts, in the way in which the facts are put together. This is because the structure of any particular state of affairs is determined by the logical forms of its constituent objects (2.032, 2.033). These objects are the world's substance which is unalterable and 'subsists independently of what is the case' (2.024). Hence, though there is no a priori order of things, i.e. no particular experience or configuration of objects is a priori, logic itself is a priori (5.4731, 5.552, 5.634).

Both Kannisto and Glock take the Tractarian conception of logic to be central evidence of the Kantianism of Wittgenstein's early thought. They point out that the logical form discussed in the *Tractatus* is not meant to be a contingent feature of the empirical world. Logic is not a descriptive theory of the factual features of the empirical world. Nor is logic prescriptive, i.e. logic does not outline the principles according to which people ought to think. Instead, logic in the *Tractatus* is *transcendental*, in the sense of being concerned with the necessary preconditions of representation. Like Kant's transcendental logic, it treats the formal features of our thinking that are also the formal features of reality. In this respect Tractarian logic is 'not a body of doctrine, but a mirror-image of the world' (6.13) (Kannisto 1986: 114–130; Glock 1992: 14–21).

But what guarantees that the form of thinking is isomorphic with the form of reality? What is the ultimate source of this necessity presupposed by meaningful language? Glock follows commentators such as Pears and Hacker in assuming that the answer to this question lies in the metaphysical essence of reality.[1] Kannisto, however, is willing to take the Kantian interpretation of the *Tractatus* one step further and to suggest that the logical form of both language and reality is of subjective origin. Wittgenstein's writes: 'what makes logic *a priori* is the impossibility of illogical thoughts' (5.4731). According to Kannisto it is difficult to understand this remark if we assume that the isomorphism between thinking and reality is somehow produced by the forms of reality: the processes that would have the required result seem too contingent and

[1] Hacker (1986: 23n.); Pears (1987: 8); Glock (1997: 296). Glock notes that the Tractarian logical form is introduced primarily as a precondition of representation. But in his view, the ontological interpretation advanced by Hacker and Pears is closer to the truth than the one emphasizing the conventions of language (e.g. Ishiguro 1969; McGuinness 1981). While Kannisto argues the source of the logical form to be in some sense subjective, this does not leave room for linguistic conventions in the Tractarian language. This is because, for the metaphysical subject, there is only one language, 'that language which alone I understand' (5.62); linguistic conventions, whatever they are, remain at the level of everyday language (4.002, 5.5563).

unreliable. Hence, insofar as we are not willing to appeal to some sort of *deus ex machina*, the only available explanation is the Kantian one. It is impossible to think illogical thoughts because thinking does not adjust to a reality independent of itself. Instead, the source of the logical form of both thinking and reality is the metaphysical subject (Kannisto 1986: 119–130, 149–155).

In the *Tractatus*, Wittgenstein compares the relation between the metaphysical subject and the world to that of the eye and the visual field (5.633). As pointed out by Peter Hacker, the metaphor of the eye reflects Schopenhauer's description of the subject of transcendental idealism; and of course a similar notion of a transcendental subject is central in Kant's system (Hacker 1986: 86–90). Kant, Schopenhauer, and Wittgenstein all claim that the metaphysical subject does not belong to the world and thus cannot be an object of experience. Rather, the metaphysical subject is that to which all experiences are given. Wittgenstein writes: 'I objectively confront every object. But not the I. So there really is a way in which there can and must be mention of the I in a *non-psychological sense* in philosophy' (NB 80). According to the Kantian interpretation, the kind of philosophy mentioned here does not follow the Tractarian description of the correct philosophical method, namely 'to say nothing except what can be said, i.e., propositions of natural science' (6.53). Rather, here Wittgenstein uses the term 'philosophy' to refer to transcendental reflection on the conditions of the world and its representation, whose expression is ultimately doomed to be nonsensical (6.54).

In the *Tractatus*, the notion of the metaphysical subject is introduced in close connection with the notion of logic. According to Wittgenstein, 'the limits of my language mean the limits of the world' (5.6). These limits are also the limits of logic (5.61). Further, it is precisely the fact that 'the world is my world' which brings the metaphysical subject into philosophy (5.641). And this subject, Wittgenstein claims, *is* the limit of the world (5.632). This connection between logic and the metaphysical subject may be seen as a further link between Kant and Wittgenstein. As argued by Glock, Wittgenstein's view of philosophy is Kantian in the precise sense that, like Kant but unlike Schopenhauer, Wittgenstein is primarily concerned with the formal features of experiencing or depicting reality (Glock 1999: 432). Given that both logic and the metaphysical subject are identified with the limits of the world, it does not seem too far-fetched to assume that the metaphysical subject and logic have a relation similar to the relation between the transcendental subject and the preconditions of experience in Kant's philosophy.

The above considerations are relevant to our theme because of the close analogy between the Tractarian treatments of logic and ethics. According to the *Tractatus*, both logic and ethics are transcendental and neither can be expressed in language (6.13, 6.421). In the *Notebooks*, Wittgenstein links the two by writing: 'Ethics must be a condition of the world, like logic' (NB 77). He also writes: 'As the subject is not part of the world but a presupposition of its existence, so good and evil which are predicates of the subject, are not properties in the world' (NB 79). If logic is transcendental in the

Kantian sense outlined above, then it seems plausible to assume that both logic and ethics originate with the metaphysical subject.

In Kant's system, the realm of experience and the realm of morality fall under different sets of laws that seem to be in disharmony with each other. Morality presupposes that we are free in the sense of being capable of obeying laws prescribed by our own reason, but in the phenomenal world everything happens in accordance with the laws of nature. For Kant, a judgement of beauty gives us a mediating link between these two, seemingly incompatible realms. According to Kant, beauty is a 'symbol of morality' (Kant 2000: §§58–9).

One can see an analogous split between the Tractarian world of facts that are expressible in meaningful language and that which must be passed over in silence. Moreover, there is again an apparent tension between these two realms. In the *Notebooks*, Wittgenstein writes: '[I]t is a fact of logic that wanting does not stand in any logical connexion with its own fulfilment. And it is also clear that the world of a happy man is a *different* world from the world of the unhappy' (NB 77). Could not this be read as Wittgenstein's rather cryptic formulation of the Kantian problem about the discrepancy between concepts of nature and concepts of morality? Wittgenstein discusses ethics in terms of good and bad willing, and in doing so he comes closer to Kant than to Schopenhauer (see Glock 1999: 442–3). Given that the exercise of the will cannot alter what can be expressed by means of language (6.43), the subject seems to be deprived of an ethically relevant connection to the world of facts. But in contrast to Kant, Wittgenstein does not treat beauty as a symbol of the freedom presupposed by morality. Instead, he identifies ethics and aesthetics, and by doing so implies that the ethical outlook on life is nothing but seeing the world of contingent facts from a certain perspective. As argued by Adrian Moore, this move may, in fact, offer a more plausible account of the relation between ethical freedom and the contingencies of the world, and be more in tenor with the tradition of transcendental idealism (Moore 1987: 132–3).

How, then, should we understand the claim about the sameness of ethics and aesthetics? The reading I am proposing draws on Kannisto's and Glock's claim that Wittgenstein's early thought manifests Kantian reflective method. This is to say that, in my view, the beginning of an answer to the problem of the discrepancy between the world of facts and the possibility of ethics is Wittgenstein's remark: 'Here everything seems to turn, so to speak, on *how* one wants' (NB 78). If Wittgenstein's notion of transcendentality is indeed Kantian in the sense that it concerns the preconditions of certain types of experiences or representation, then the transcendentality of ethics means that the 'how' in this quotation may be read as indicating a shift of focus to the formal features or preconditions of the ethical experience or attitude. There is, Wittgenstein holds, a *way of seeing* an object that brings value in, and this way of seeing is one and the same in both ethics and aesthetics. Wittgenstein describes the connection between art and ethics as follows:

> The work of art is the object seen *sub specie aeternitatis;* and the good life is the world seen *sub specie aeternitatis*. This is the connexion between art and ethics.
>
> The usual way of looking at things sees objects as it were from the midst of them, the view *sub specie aeternitatis* from outside.
>
> In such a way they have the whole world as background.
>
> Is this it perhaps—in this view the object is seen *together with* space and time instead of *in* space and time? (NB 83)

As the general form of a proposition, 'This is how things stand' (4.5), gives the essence of description (5.4711), so we might say that the *sub specie aeterni* perspective gives the essence of valuation. This perspective is necessarily present in any experience of value.

As I will argue in the following, the *sub specie aeterni* attitude is primarily aesthetic. However, I do not mean to say that Wittgenstein's early ethics should be understood as a form of aestheticism which reduces ethics to aesthetics and takes 'aesthetic enjoyments' to be a sufficient answer to the lack of objective value in the world of facts (see e.g. Hintikka and Hintikka 1986: 67–8). While this view seems to come close to some of Wittgenstein's remarks,[2] it wrongly suggests that the nature of aesthetic valuation can be taken as an unproblematic given in explaining the ethical. Instead, a proper understanding of Wittgenstein's remarks on ethics and aesthetics demands a closer analysis of the aesthetic attitude itself.

A more serious misunderstanding of the claim about the sameness of ethics and aesthetics would be to take aesthetics or art to be subservient to ethics. This seems to be the view taken by Janik and Toulmin, who claim that, for the early Wittgenstein, the task of art is to provide examples of moral behaviour. Consequently, they argue, the *Tractatus* should be seen as a condemnation of 'art for art's sake' and of formalism, views that treat the arts primarily in terms of their specifically artistic forms (Janik and Toulmin 1973: 197–8). But it is unclear how any form of art could provide examples of moral behaviour if specific behaviours are describable facts in the world. Besides, this reading begs the question of what makes a particular behaviour moral. For one of the striking features of Wittgenstein's remarks on ethics is his reluctance to give examples of specific actions that would count as good or bad. More typical is his remark, written in 1946: 'an action to which one *might* be prompted by good, generous motives is something one may also be doing out of cowardice or indifference' (CV 48). In 1931 he writes about Schubert's brother who, after Schubert's death, cut some of the remaining scores into pieces and then gave these pieces to Schubert's pupils. According to Wittgenstein, we can understand this act as a sign of piety, but a different act, such as keeping the scores untouched or burning them, could be seen as a sign of piety as well

[2] Wittgenstein writes, for example: 'Is it the essence of the artistic way of looking at things, that it looks at the world with a happy eye? Life is grave, art is gay' (NB 86). The mistake underlying the aestheticist reading is to treat art as means of producing happiness. However, in the early 1930s Wittgenstein rejects such a view as repellent (BB 178). As I will argue below, the happiness in question is more like the pleasure that, according to Kant, one experiences in making a judgement of taste, namely, a feature of the judgement itself.

(PO 127). The point here seems to be that only by adopting a certain perspective on the action can we see it as a sign of piety.

Now, if my proposal of the Kantian, reflective nature of the Tractarian ethics is plausible, then the remaining question is whether Wittgenstein's remarks on value include traces of the formal features of the perspective of valuation. Can we, in other words, say something more specific about the preconditions of the perspective shared by ethics and aesthetics? In the following I will try to do just this, using Kant's analysis of judgements of taste as a framework for the discussion. Granted, given the nature of the textual evidence available, it may be impossible to reconstruct Wittgenstein's position. I am also willing to admit that what I am about to say will capture only one aspect of Wittgenstein's view of ethics. Nonetheless, the similarities one can find between Kant's aesthetics and Wittgenstein's remarks on ethics and aesthetics are pervasive and interesting enough to deserve attention.

3 Judgements of taste

3.1 Disinterestedness

According to Kant a judgement of taste must fulfil four conditions. The first of these is disinterestedness. Unlike cognitive judgements, whether theoretical or practical, judgements of taste must be devoid of all interest. This means that, in order for me to appreciate the beauty of an object, I must be indifferent regarding its existence. A judgement of taste should be free in that the pleasure one gets from the contemplation of beauty resides solely in the judgement itself (Kant 2000: §§1–5).

This is not a very controversial claim about the aesthetic attitude. However, while Kant held that the consequences of an action are irrelevant to its moral value—a claim made by Wittgenstein as well (6.422)—he thought that, in contrast to the beautiful, the good always contains a relation between reason and volition, and hence an interest.[3] Wittgenstein, however, denies this. According to him, the renunciation of any interest is essential for the *sub specie aeterni* attitude.

At the outset it may seem as if the ethical perspective described by Wittgenstein contains an interest. This is because Wittgenstein connects happiness and the good exercise of the will by claiming that the good life is happy and the bad life unhappy (NB 78). Taken together with the Tractarian remark that there must, in some sense, be ethical rewards and punishments, it may seem as if Wittgenstein assumes that ethical life will produce happiness which, in turn, would serve as an interest one has in aiming at

[3] Kant writes: 'Both [good for something and good in itself] always involve the concept of an end, hence the relation of reason to (at least possible) willing, and consequently a satisfaction in the **existence** of an object or of an action, i.e., some sort of interest' (2000: 93). 'But despite all this difference between the agreeable and the good, the two still agree in this: that they are always combined with an interest in their object, not only the agreeable . . . , but also that which is good absolutely and in all respects, namely the moral good, which carries the highest interest with it' (2000: 94).

ethical conduct (6.422). However, a closer look at Wittgenstein's notions of action and happiness shows that this reading cannot be quite right. The passage on rewards and punishments reads:

> It is clear, however, that ethics has nothing to do with punishments and rewards in the usual sense of the term. So our question about the *consequences* of an action must be unimportant.—At least those consequences should not be events. For there must be something right about the question we posed. There must indeed be some kind of ethical reward and ethical punishment, but they must reside in the action itself.
>
> (And it is also clear that the reward must be something pleasant and the punishment something unpleasant.) (6.422)

Now, by 'action' Wittgenstein cannot mean actions as ordinarily understood. In his early account of ethics events such as opening the door for an elderly lady, or punching a stranger in the stomach, must be secondary. This is because 'what happens, whether it comes from a stone or from my body, is neither good nor bad' (NB 84). Besides, it may be imagined that the act of will for raising one's arm occurs, but that the arm does not move; perhaps the limb is paralysed or a sinew is torn (NB 86). This possibility leads Wittgenstein to state: 'the act of will is not the cause of action but the action itself' (NB 87). Thus, when Wittgenstein asks, 'Is seeing an activity?', the question at hand must concern the nature of ethical action (NB 77). I think it is clear that Wittgenstein's answer to this question must be affirmative. For him, the ethically relevant action is nothing but seeing the world from the right perspective.

Moreover, Wittgenstein's notion of happiness already presupposes disinterestedness. This is because the good or bad exercise of the will alters the world without altering the facts of the world (6.43). In other words, 'the will is an attitude of the subject to the world' (NB 87). Adopting the *sub specie aeterni* attitude to the world means that one simply accepts the facts of the world, whatever they may be. According to Wittgenstein, 'I can only make myself independent of the world—and so in a certain sense master it—by renouncing any influence on happenings' (NB 73). The only possibility left is to contemplate the world as it is given to me. This means accepting that I am powerless regarding the external circumstances of my life, but also that these circumstances are neither good nor bad. Wittgenstein writes: 'I am either happy or unhappy, that is all. It can be said: good and evil do not exist' (NB 74).

Instead of bringing about a change in the factual circumstances of life, the attitude shared by ethics and aesthetics changes the limits of the world (6.43). Given that these limits are identical with the metaphysical subject (5.632), the conclusion that follows is that the world of the happy man differs from the world of the unhappy man (6.43). In other words, it is the subject itself that changes by looking at the world from a perspective different from the usual: 'What is good and evil is essentially the I, not the world' (NB 80). Another formulation of this point may be found in the *Notebooks*. Recall that, in Wittgenstein's view, the metaphysical subject relates to the world as the eye relates to the visual field (5.633). In the *Notebooks*, Wittgenstein suggests that 'the essence of the artistic way of looking at things is that it looks at the world with a happy eye' (NB 86). Thus, when Wittgenstein claims that 'the beautiful *is* what makes happy' (NB 86), it is unlikely that he means that the subject becomes happy as a result of seeing a beautiful object among the facts of the world. Rather, it is the way of seeing an object in the world that makes the object beautiful, the way of looking at the world that makes it the world of a happy man. As with Kant's notion of beauty that does not belong to objects but is a feature of the judgement itself, Wittgenstein's notion of happiness is included in the perspective the metaphysical subject may adopt towards the world.

The consequences of proposing disinterestedness as a condition of the ethical attitude are striking, at least from the viewpoint of traditional normative accounts.[4] For it is not just that, in order to gain the right perspective on the world, I have to give up hope or interest regarding the happenings of my own life. I also have to give up the hope of influencing the lives of others. In the *Notebooks*, Wittgenstein expresses the resulting tension as follows:

> 'To love one's neighbour' would mean to will!
>
> But can one want and yet not be unhappy if the want does not attain fulfilment? (And this possibility always exists.)
>
> Is it, according to common conceptions, good to want *nothing* for one's neighbour, neither good nor evil?
>
> And yet in a certain sense it seems that not wanting is the only good. (NB 77)

The point is further accentuated by Wittgenstein's claim that the possibility of ethics cannot depend on the existence of living beings besides myself. He asks:

> Can there be any ethics if there is no living being but myself?

And answers:

> If ethics is supposed to be something fundamental, there can. (NB 79)

Thus, while Kant claims that the ontological status of the object is irrelevant for a judgement of taste, Wittgenstein makes an analogous claim about the existence of other living beings for ethics. Seeing the world from the right perspective requires resignation to the possibility that my efforts in trying to do good to others are futile in the ultimate sense of term. Any feature of the world as a foundation or motivation for ethics, even the existence of others, would undermine what Wittgenstein calls the fundamentality of ethics.

Finally, the condition of disinterestedness appears in Wittgenstein's treatment of the immortality of the human soul. In Kant's account, eternal life is a postulate of practical

[4] On the other hand, as noted by Glock (1999: 438), to link ethics and aesthetics by stressing the centrality of a contemplative attitude is not uncommon in Wittgenstein's cultural background. The notion can be found in Schopenhauer, but Kant's view of beauty as a symbol of morality may also be read as exemplifying a similar notion.

reason that is supposed to resolve the disharmony between concepts of nature and concepts of morality. According to Wittgenstein, however, the assumption of eternal life 'fails to accomplish the purpose for which it has always been intended'. If I fail to see my life as significant now, assuming that it will continue forever does not give it any more significance. Hence, Wittgenstein claims, 'the solution of the riddle of life in space and time lies *outside* space and time' (6.4312).

It is unlikely that Wittgenstein's phrase 'outside space and time' refers to a transcendent realm postulated, for example, by Christianity. Again, Wittgenstein seems to point to a change occurring in the metaphysical subject.[5] While he writes that 'God does not reveal himself *in* the world' (6.432), this does not entail that God could be found in some other place or time. Instead, Wittgenstein identifies God with the world and notes that, in this sense, God would simply be fate. Along the same lines, he claims that to believe in God is to understand the question about the meaning of life and that to pray is just to think about the meaning of life (NB 73–4). Religious notions are thus interpreted as attempts to capture the very same experience or attitude that is the essence of ethics and aesthetics. In fact, in the 'Lecture on Ethics' Wittgenstein makes this point explicitly by claiming that all ethical and religious sentences are allegorical attempts to express the mystical experience (PO 42).

In the *Tractatus*, Wittgenstein writes: 'If we take eternity to mean not infinite temporal duration but timelessness, then eternal life belongs to those who live in the present' (6.4311). According to the *Notebooks*, this means living without fear, but also without hope (NB 76). As we have seen, this requires giving up all hope of influencing the circumstances of the world. But what is more, it also requires giving up the hope of receiving an ethical reward in a future life (6.431–6.4312). Like the pleasure one experiences in aesthetic contemplation, that always happens in the present moment and in the subject's reaction to something particular, the ethical reward mentioned in the *Tractatus* must reside in the *sub specie aeterni* attitude itself (6.422).[6]

[5] According to Stenius, Wittgenstein would have used the term 'transcendent' for the inexpressible had he adopted Kant's distinction between the transcendent and the transcendental (Stenius 1960: 222). I disagree. True, in the *Notebooks*, Wittgenstein does use the term 'transcendent' (NB 78–9). But in the *Tractatus* the term 'transcendent' has been replaced by 'transcendental'—and not incidentally, I believe. The point is that the Tractarian notions to which the term applies are limiting conditions, dependent on the metaphysical subject.

[6] Martin Stokhof has read Wittgenstein's early ethics in the light of the Buddhist tradition. A historical connection supporting this move is Schopenhauer's interest in Buddhism. Yet, as Stokhof notes, both Schopenhauer and Wittgenstein draw mostly on the views of early Christianity and specifically its attitude of renunciation of the world. Stokhof notes that these 'Gnostic' views of the first to fourth centuries have 'survived as an undercurrent in Western Christianity and culture' (Stokhof 2002: 33–4, see also 227–34). The early Christian views to which Stokhof refers have survived, and not just as an undercurrent, also in the Eastern Christian tradition. This is because, historically, Eastern Christianity has been more open to the texts, sometimes called 'Gnostic', that include the idea that the Kingdom of Heaven is not a spatio-temporal state of affairs in the future, but rather a contemplative experience in the present moment. Hence, for example, a characteristic feature of Byzantine eschatology is the notion that eschatology relates, not just to future events, but also to present experience. This experience, described as 'gnosis', is not meant to be intellectual, emotional, or physical; it is instead 'mystical' (see e.g. Meyendorff 1974: 212–22). While I do not want to

3.2 *Subjective purposiveness*

Another feature of a Kantian judgement of taste is that, in the judgement, one abstracts away from the purpose of its object. The object's utility or inner perfection as a thing of its kind must be set aside when appreciating its beauty; the focus should be on the form of the object. Yet, in the judgement, there is subjective, formal purposiveness (Kant 2000: §§10–17). According to Kant, this means that 'the agreement of the manifold with a unity (leaving undetermined what it is supposed to be) does not by itself allow any cognition of objective purposiveness at all'. Thus, 'nothing remains but the subjective purposiveness of representations in the mind of the beholder' (2000: 112).

In the *Notebooks*, there are at least two remarks that come close to expressing a similar notion. The first of these is Wittgenstein's analogy between language and music, between a proposition and a musical tune. Wittgenstein writes: 'Musical themes are in a certain sense propositions', and adds: 'A tune is a kind of tautology' (NB 40). Unlike meaningful propositions that are pictures of states of affairs, tautologies do not say anything about reality. They lack sense by being unconditionally true. Yet, tautologies do not fall into the category of nonsensical sentences, as they show their own logical form. They show that they say nothing (4.46–4.463). *Contra* Janik and Toulmin, the implications of these remarks are formalistic. That Wittgenstein takes tautologies to be the linguistic analogy for musical tunes shows that, in his view, music neither pictures nor refers to reality. As tautologies, musical themes do not say anything extra-musical; they merely show their own form. Yet, while a tune does not say anything about the world, it still reveals something: '[A tune] is complete in itself; it satisfies itself' (NB 40). The more Kantian formulation of this point is to say that the musical form of a tune has purposiveness of a formal kind, purposiveness without any specific purpose, without any extra-musical content.[7]

claim that Wittgenstein has been directly influenced by Byzantine theology, one should not forget that Wittgenstein read Dostoevsky and Tolstoy, both of whom come from a culture heavily marked by Eastern Orthodox Christianity. In fact, in the *Notebooks*, the discussion of immortality includes a reference to Dostoevsky: 'Dostoievsky is right when he says that the man who is happy is fulfilling the purpose of existence' (NB 73).

[7] Here I disagree with B. R. Tilghman, who claims that Wittgenstein's early view on aesthetics should be read as a variant of expressivism, a view that treats works of art as expressions of spirit, character, or intention (Tilghman 1991: 46–7). True, Wittgenstein writes: 'Art is a kind of expression. Good art is complete expression' (NB 83). However, given what he writes on the topic elsewhere—for example, his statement that Tolstoy's expressivist theory of art is 'false' and 'absurd'—it seems odd to place him in the expressivist tradition (CV 58–9). As I have suggested elsewhere, Wittgenstein's remarks on art, music in particular, seem to reflect a position that could be characterized as formalist (Ahonen 2005). In Wittgenstein's cultural background this view was defended by writers such as Eduard Hanslick and, to some extent, Franz Grillparzer. Hanslick, for example, expresses a notion strikingly similar to Wittgenstein's analogy between a tune and a tautology: 'The independent, aesthetically not further reducible unit of musical thought in every composition is the theme. . . . So what shall we say is the content? The tones themselves? Of course. But they are already formed. What shall we say is the form? Again, the tones themselves, but they are forms already fulfilled' (Hanslick 1986: 80).

According to Kant, to see formal purposiveness in an object involves seeing it as a unified whole. I believe Wittgenstein aims at a similar point:

> As a thing among things, each thing is equally insignificant; as a world each one equally significant.
>
> If I have been contemplating the stove, and then am told: but now all you know is the stove, my result does indeed seem trivial. For this represents the matter as if I had studied the stove as one among the many things in the world. But if I was contemplating the stove *it* was my world, and everything else colourless by contrast with it. (NB 83)

The attitude described here sets aside what can be known about the stove as a fact. As a fact the stove has no value. But even an object as mundane as a stove can be seen as an aesthetically significant phenomenon when approached from the right perspective, i.e. when seen as a unified whole. Moreover, as Wittgenstein stresses, in the contemplation of the stove, the stove becomes the world for the subject. If this is taken literally, then it seems that, though *art* aims at presenting objects *sub specie aeterni* (NB 83), in the attitude itself there is no difference between ethics and aesthetics in terms of their respective objects.

Again, the feature that Kant attributes to aesthetic judgements finds an expression in Wittgenstein's characterization of the ethical. In the *Tractatus*, he writes: 'To view the world sub specie aeterni is to view it as a whole—a limited whole' (6.45). But this just means that one sees the world—or, what is the same, one's life (5.621)—not as a totality of facts, but rather as a work of art, from the same perspective that transforms the stove into an aesthetic object. Later, in 1930, Wittgenstein makes this claim more explicitly. He discusses the possibility of seeing one's life from the point of view that we adopt when we are engaged in looking at works of art. This would mean, for example, that we look at our simple everyday activities as if we were watching a play at a theatre. 'Nothing could be more remarkable', Wittgenstein writes, for this would amount to seeing one's life 'as a work of art created by God and, as such, . . . worth contemplating' (CV 4).

As noted by Wittgenstein, seeing one's life as a work of art is not an easy task. We see life every day, but usually it does not make the slightest impression on us (CV 4). Yet, this is exactly what the ethical attitude requires. Ethics is concerned with 'the problem of life', and this is the problem of the sense or purpose of life. Wittgenstein writes:

> What do I know about God and the purpose of life?
>
> I know that this world exists.
>
> That I am placed in it like my eye in its visual field.
>
> That something about it is problematic, which we call its sense. (NB 72–3)

Studying the facts of the world does not help us discover what the purpose of life might be, because even if 'all possible scientific questions have been answered, the problems of life remain completely untouched' (6.52). All we can expect to gain from a scientific investigation is an inventory of contingent facts (4.11). Once these have been put to words, there are no questions left. And this, Wittgenstein claims, *is* the answer to the problem of life (6.52).

In my view, this response by Wittgenstein is Kantian, though it is one given along the lines of the *Third*, not the *Second Critique*. We cannot find out what the purpose of life is, for there is none. Yet, the *sub specie aeterni* perspective shows life as purposive. Wittgenstein writes:

> We could say that the man is fulfilling the purpose of existence who no longer needs to have any purpose except to live. This is to say, who is content. The solution of the problem of life is to be seen in the disappearance of this problem. (NB 73; cf. 6.521)

To see the world *sub specie aeterni* alters its limits: the world waxes or wanes as a whole (6.43). In the *Notebooks* this remark continues, 'as if by accession or loss of meaning' (NB 73). The word Wittgenstein is using in the German text is *Sinn*, sense. But it is clear that in this context the term cannot refer to the bipolarity of a proposition that is the essence of the notion of sense in the context of the picture theory of meaning. Rather, this seems to be Wittgenstein's attempt to show something about the mystical experience, namely, about the experience of seeing sense where technically speaking there cannot be any, purposiveness in life while recognizing that objectively it has none.

3.3 Subjective universality

According to Kant, judgements of taste are not based on concepts. From this it follows that it is impossible to compel another person to acknowledge that an object is beautiful by giving arguments for its beauty. There can be no rules of taste, nor can I acquire taste by imitating someone else's taste. Instead I always have to make the judgement for myself. In contrast to moral judgements, judgements of taste are based on nothing but my feeling of pleasure or displeasure. Nevertheless, in contrast to judgements about the agreeable (for example, the agreeableness of canary wine or the taste of coffee[8]), which rest on empirical generalizations and are comparative in nature, judgements of beauty are universal in the sense that we demand that others agree with us (Kant 2000: §§6–9).

In the *Tractatus*, the transcendentality of ethics is explained by the statement that ethics cannot be expressed in language (6.42, 6.421). Language depicts states of affairs, but given that the states of affairs are contingent, the propositions depicting them are of equal value (6.4). Wittgenstein writes:

[8] Kant (2000: §7) illustrates the distinction between judgements about beauty and judgements about the agreeable by using art as an example of the former, canary wine as an example of the latter. In the *Lectures on Aesthetics*, where Wittgenstein argues that aesthetic explanations cannot be causal, he draws a similar distinction between questions about art and questions about pleasant tastes and smells; his example of the latter is the taste of coffee (LA 11).

> If there is any value that does have value, it must lie outside the whole sphere of what happens and is the case. For all that happens and is the case is accidental.
>
> What makes it non-accidental cannot lie *within* the world, since if it did it would itself be accidental.
>
> It must lie outside the world. (6.41)

There can be no ethical concepts, propositions, principles, nor a theory of ethics. The fundamentality of ethics means that, if it were possible to express an ethical principle or a value judgement in language, this would immediately render it accidental. Rather, the attitude shared by ethics and aesthetics is a feeling: 'Feeling the world as a limited whole—it is this that is mystical' (6.45).

In the 'Lecture on Ethics', Wittgenstein repeats the claim about the impossibility of a theory of ethics. He also repeats the claim about the sameness of ethics and aesthetics by pointing out that he will use the term ethics in a sense that includes what he takes to be the essential in aesthetics. Wittgenstein's lecture begins by drawing a distinction between two different ways of using the terms 'good', 'right' 'valuable', and 'important'. These are the relative use and the absolute use. An example of the relative use of the word 'good' is the sentence 'he is a good pianist'. In spite of the appearance, this sentence is not an example of an aesthetic judgement in any absolute sense, as the sentence can be analysed into facts about the relative skill and dexterity of the pianist. This is to say that the grounds of the relative use are, as the grounds of Kantian judgements about the agreeable, reducible to empirical facts that are comparative in nature. The second, absolute use of words such as 'good' and 'right' is the one characteristic of ethical expressions. In his attempt to illustrate what might be meant by the absolute use, Wittgenstein brings up the expression 'the absolutely right road'. According to him, if it were possible to meaningfully use such a notion, it would mean: 'the road which *everybody* on seeing it would, *with logical necessity*, have to go, or be ashamed for not going' (PO 40). However, Wittgenstein claims, there are no possible semantic grounds for statements of the absolutely valuable, for 'no state of affairs has, in itself, what I would like to call the coercive power of an absolute judge' (PO 40). Hence, the absolute use of the word 'good' is nonsensical.

While the immediate argumentative goal of the 'Lecture on Ethics' seems to be an explication of the nonsensicality of ethical propositions, Wittgenstein ends up pointing to what he calls the absolutely valuable. He does this by mentioning three personal experiences: the experience of wondering at the existence of the world, the feeling of being absolutely safe, and the feeling of guilt. In the *Notebooks*, the first of these experiences is called a 'miracle' (NB 86); in the *Tractatus* it is called 'mystical' (6.44). These experiences have two important characteristics. First, as soon as a mystical experience is understood as a fact about the world including the mental states of some individual, the experience becomes a contingent, insignificant event. As soon as it is put to words, it loses its character as a miracle. This is due to the conditions of meaningful representation. Regarding absolute value, Wittgenstein states, 'I would reject every significant description that anybody could possibly suggest, *ab initio*, on the ground of its significance' (PO 44). Second, as Wittgenstein stresses, an experience of absolute value is 'an entirely personal matter and others would find other examples more striking' (PO 41). In this respect, experiences of absolute value are like judgements of beauty that rest on nothing but the feeling of pleasure the subject experiences. In Wittgenstein's words: 'Things acquire significance only through their relation to my will' (NB 84).

By arguing that any proposition about absolute value necessarily lacks semantic grounds and must thus be nonsensical, Wittgenstein is giving value an autonomous status. He is protecting value from 'babbling'.[9] But here, too, the autonomy in question comes closer to the freedom that Kant attributes to judgements of beauty. Instead of relying on general ethical principles or rules, one has no other option but to face one's life as a unique phenomenon that requires a response (see NB 79). In Wittgenstein's words: 'At the end of my lecture on ethics, I spoke in the first person. I believe that is quite essential. Here nothing more can be established, I can only appear as a person speaking for myself' (Wittgenstein 1965: 16).

3.4 The ethical 'ought'

The last, peculiar feature of a Kantian judgement of taste concerns the nature of the 'ought' embedded in the judgement. When I find something beautiful, I take it to have a necessary reference to liking. I take my judgement of beauty to be an example of a universal rule, albeit a rule that I am unable to state. While I cannot argue that something is beautiful, I still take my judgement to be exemplary and demand that others ought to agree with me. However, given that my judgement is based on my own feeling, not on shared concepts, this common sense is just an ideal standard (Kant 2000: §§18–22).

In the *Tractatus*, Wittgenstein writes: 'When an ethical law of the form, "Thou shalt . . . ", is laid down, one's first thought is, "And what if I do not do it?"' (6.422) What is the point of such a remark, reminiscent of the response of a 3-year-old who is told to brush her teeth? In my view, a natural way to read the remark is to take it as a comment on a tradition that connects ethics to necessity, exemplified, for example, in Kant's assumption that the moral law is apodeictic, i.e. bound up with the awareness of its necessity. However, as stated in the *Tractatus* only a few remarks before the one just quoted, 'the only necessity that exists is logical necessity, . . . the only impossibility that exists is logical impossibility' (6.375). Logical necessity certainly cannot apply to ethics,

[9] 'My work consists of two parts: one of which is here, and of everything which I have *not* written. And precisely this second part is the important one. For the Ethical is delimited from within, as it were, by my book; and I'm convinced that, *strictly* speaking, it can ONLY be limited in this way. In brief, I think: All of that which *many* are *babbling* today, I have defined in my book by remaining silent about it. Therefore the book will, unless I'm quite wrong, have much to say which you want to say yourself, but perhaps you won't notice that it is said in it' (Luckhardt 1979: 94–5).

as ethics cannot be expressed in meaningful language and does not treat the facts of the world.

If the transcendentality of Wittgenstein's early ethics means that ethics concerns the formal preconditions of the *sub specie aeterni* attitude, and if these preconditions are closer to those we would normally expect of an aesthetic attitude, then can this ethics include an 'ought'? The problem is, in Cyril Barrett's words, that 'happiness in aesthetics... has little or nothing to do with conscience, the fear of death, the will or voice of God' (Barrett 1984: 19).

Given Wittgenstein's claim that the existence of others is irrelevant for ethics, then the 'ought' cannot arise from a Kantian presupposition of common sense. Instead, Wittgenstein's early thought is egocentric, even solipsistic: 'What has history to do with me? Mine is the first and only world!', he exclaims (NB 82; cf. 5.62). In this respect, Wittgenstein fits Weininger's characterization of a genius for whom ethics is primarily a duty to himself, a duty to guard his own integrity in spite of what others might think.[10]

Nevertheless, Wittgenstein uses expressions such as 'absolute', 'fundamental', and 'condition of the world' to describe ethics (PO 40, NB 77, 79), all of which point to an element of necessity. Moreover, in the *Notebooks*, there are remarks that seem to imply an ethical 'ought'. Wittgenstein talks about conscience as the voice of God, about fear of death as a sign of a bad life, and about evil as a predicate of the will (NB 75, 79). However, insofar as these notions are given specific substance, this is done by reference to the notion of happiness as the essence of the *sub specie aeterni* attitude. For example, Wittgenstein claims that to do the will of God means that one is in agreement with the world (NB 75). But given that God is the world, or the meaning of the world (NB 74, 75, 79), doing the will of God turns out to be nothing but the acceptance of the world as it is. Moreover, given that world and life are identical (5.621, NB 77), to be in agreement with the world just means that one's life is a happy life. The pressing question here is the one asked by Wittgenstein himself: 'Can one say: "Act according to your conscience whatever it may be"?'—His answer is: 'Live happy!' (NB 75)

The example that comes closest to a moral judgement in any ordinary sense of the term is Wittgenstein's discussion of suicide. He suggests that suicide could be understood as a kind of elementary sin:

> If suicide is allowed then everything is allowed.
> If anything is not allowed then suicide is not allowed. (NB 91)

[10] Weininger (1906: 160) writes: 'It is certainly true that most men need some kind of a God. A few, and they are the men of genius, do not bow to an alien law. The rest try to justify their doings and misdoings, their thinking and existence (at least the mental side of it), to some one else, whether it be the personal God of the Jews, or a beloved, respected, and revered human being. It is only in this way that they can bring their lives under the social law'. 'It is certainly true that the greater an artist or philosopher may be, the more ruthless he will be in keeping faith with himself, in this way often disappointing the expectation of those with whom he comes in contact in every-day life' (1906: 173).

It is significant, I think, that these claims are hypothetical. For it is unclear what, in Wittgenstein's early system, would entail the sinfulness of suicide. At best we can assume that, if the motive for the suicide were fear, i.e. the inability to accept whatever the facts of the world might be, then the sin in question would be nothing but the failure to see one's life *sub specie aeterni*. And indeed, in line with his general reluctance to offer any moral principles, Wittgenstein ends this line of thought by asking: 'Or is even suicide in itself neither good nor evil?' (NB 91)

Martin Stokhof has suggested that Wittgenstein's discussion of suicide should be read in the light of the Tractarian remark on the problem of life: 'The facts all contribute only to setting the problem, not to its solution' (6.4321). Stokhof claims that the point Wittgenstein wants to make is that everyday life presents us with an ethical task or a challenge, and that refusing to accept this challenge is an act of taking a negative ethical stance (Stokhof 2002: 189–99). Insofar as this is intended as a general moral judgement, then in my view it says a bit too much. At the same time, however, I believe that Stokhof is absolutely right to read Wittgenstein's discussion of suicide in the light of this Tractarian remark. For the problem of life is one of seeing life as meaningful, and the will to commit suicide may result from a failure to see one's life as such. However, this failure is more like the failure to see inner perfection in a work of art.

Most of those who discuss Wittgenstein's ethics correctly emphasize the gravity with which Wittgenstein approached ethical matters in his own life. His attitude seems to be well captured by his response to Fania Pascal, who inquired whether Wittgenstein wanted to be perfect: '*Of course* I want to be perfect', Wittgenstein declared (Rhees 1981: 50). What is less often stressed is that the same seriousness is characteristic of Wittgenstein's aesthetic responses. One need only recall the story of the despair of the locksmith with whom Wittgenstein worked while building the Stonborough house in Vienna, and to whom Wittgenstein declared that a millimetre does indeed make a big difference, to realize that nothing less than perfection was good enough for Wittgenstein in matters of an aesthetic nature (Rhees 1981: 7). The same quest for perfection can be seen in his attempt to find the right proportions for a door, the proper accentuation in the reading of a poem, and the right loudness for the bass so that a musical passage would harmonize (LA 4, 7, 13). This is to say that, in spite of underrating his own artistic contributions –'in my artistic activities I really have nothing but *good manners*' (CV 25)—Wittgenstein's approach to aesthetic questions shows the kind of conscientiousness and commitment one would normally expect only from an artist. And his ideal of artistic expression is nothing less than perfection. He writes: 'Bach said that all his achievements were simply the fruit of industry. But industry like that requires humility and enormous capacity for suffering, hence strength. And someone who, with all this, can also express himself perfectly, simply speaks to us in the language of a great man' (CV 71).

The ethical 'ought' that may be found in Wittgenstein's early remarks does not have the normative force of a Kantian moral judgement. Nor can it be explained by reference to consequences, utility, psychological dispositions, or other mental states.

Rather, the difference between the good and bad exercise of the will is nothing but a difference in the perspective of the metaphysical subject towards the world: 'The world of the happy man is a different one from that of the unhappy man' (6.43). In terms of facts, the world seen by these men is the same but the men differ in the perspective they have towards the world. Their difference is like the difference between two men who are listening to a piece of music. While one recognizes the harmony and the musical completeness of the melodies, the other hears nothing but unintelligible noise. And there is nothing the first man, the happy man, can say to the second in order to convey how he sees the world. Or at best he can say, like Wittgenstein: 'Aesthetically, the miracle is that the world exists. That there is what there is' (NB 86).

Still, this attitude carries the kind of necessity which, according to Kant, connects beauty and liking. There is a necessary connection between the *sub specie aeterni* attitude and happiness, and this shows itself in the nonsensicality of asking for any further explanation or foundation for the attitude. Wittgenstein writes:

> I keep coming back to this! simply the happy life is good, the unhappy bad. And if I *now* ask myself: But why should I live *happily,* then this of itself seems to me to be a tautological question; the happy life seems to be justified, of itself, it seems that it *is* the only right life. . . . We could say: 'The happy life seems to be in some sense more *harmonious* than the unhappy'. But in what sense? (NB 78)

I have tried to show that this question is exactly like the question, 'What makes a melody complete?' If this is indeed the case, then I should not be able to offer any answer.

References

Ahonen, Hanne (2005) 'Wittgenstein and the Conditions of Musical Communication', *Philosophy* 80 (314), 513–29.

Barrett, Cyril (1984) '"(Ethics and Aesthetics Are One)"?', in *Aesthetics: Proceedings of the 8th International Wittgenstein Symposium*, part I. Vienna: Holde-Pichler-Tempsky.

Glock, Hans-Johann (1992) 'Cambridge, Jena or Vienna? The Roots of the *Tractatus*', *Ratio* 5 (1), 1–23.

——(1997) 'Kant and Wittgenstein: Philosophy, Necessity and Representation', *International Journal of Philosophical Studies* 5 (2), 285–305.

——(1999) 'Shopenhauer and Wittgenstein: Language as Representation and Will', in C. Janaway, ed., *The Cambridge Companion to Schopenhauer*. Cambridge: Cambridge University Press, 422–58.

Hacker, P. M. S. (1986) *Insight and Illusion: Themes in the Philosophy of Wittgenstein*. Oxford: Clarendon Press.

Hanslick, Eduard (1986) *On the Musically Beautiful*, trans. G. Payzant. Indianapolis: Hackett.

Hintikka, Merryll and Hintikka, Jaakko (1986) *Investigating Wittgenstein*. Oxford: Blackwell.

Ishiguro, Hidé (1969) 'Use and Reference of Names', in P. Winch, ed., *Studies in the Philosophy of Wittgenstein*. London: Routledge, 20–50.

Janik, Allan and Toulmin, Stephen (1973) *Wittgenstein's Vienna*. New York: Simon & Schuster.

Kannisto, Heikki (1986) *Thoughts and Their Subject.*: *Acta Philosophica Fennica* 40, Helsinki.

Kant, Immanuel (2000) *The Critique of the Power of Judgment*, ed. P. Guyer, trans. P. Guyer and E. Matthews. Cambridge: Cambridge University Press.

Luckhardt, C. G. (1979) *Wittgenstein: Sources and Perspectives*. Ithaca, NY: Cornell University Press.

McGuinness, Brian (1981) 'The So-called Realism of the *Tractatus*', in I. Block, ed., *Perspectives on the Philosophy of Wittgenstein*. Oxford: Blackwell, 60–73.

——(1988) *Wittgenstein: A Life—Young Ludwig (1889–1921)*. London: Duckworth.

Meyendorff, John (1974) *Byzantine Theology: Historical Trends & Doctrinal Themes*. New York: Fordham University Press.

Monk, Ray (1990) *Ludwig Wittgenstein: The Duty of Genius*. London: Vintage.

Moore, A. W. (1987) 'Beauty in the Transcendental Idealism of Kant and Wittgenstein', *British Journal of Aesthetics* 27 (2), 129–37.

Pears, David (1987) *The False Prison*, vol. 1. Oxford: Clarendon Press.

Rhees, Rush, ed. (1981) *Ludwig Wittgenstein: Personal Recollections*. Oxford: Blackwell.

Stenius, Erik (1960) *Wittgenstein's Tractatus: A Critical Exposition of its Main Lines of Thought*. Oxford: Blackwell.

Stokhof, Martin (2002) *World and Life as One: Ethics and Ontology in Wittgenstein's Early Thought*. Stanford, Calif.: Stanford University Press.

Tilghman, B. R. (1991) *Wittgenstein, Ethics and Aesthetics: The View from Eternity*. London: Macmillan.

Weininger, Otto (1906) *Sex and Character*. New York: Howard Fertig.

Wittgenstein, Ludwig (1958) *The Blue and Brown Books* (BB). Oxford: Blackwell.

——(1961) *Notebooks 1914–1916* (NB), ed. G. H. von Wright and G. E. M. Anscombe, trans. G. E. M. Anscombe. Oxford: Blackwell.

——(1961) *Tractatus Logico-Philosophicus* (TLP), trans. D. F. Pears and B. F. McGuinness. London: Routledge.

——(1965) 'Lecture on Ethics' and Notes on Talks with Wittgenstein (LE), *Philosophical Review* 74 (1), 3–16.

——(1966) *Lecture and Conversations on Aesthetics, Psychology and Religious Belief* (LA), ed. C. Barrett. Berkeley: University of California Press.

——(1980a) *Culture and Value* (CV), ed. G. H. von Wright, trans. P. Winch. Oxford: Blackwell.

——(1980b) *Wittgenstein's Lectures, Cambridge 1930–1932* (LWL), from the notes of J. King and D. Lee, ed. D. Lee. Oxford: Blackwell.

——(1993) *Philosophical Occasions* (PO), ed. J. Klagge and A Nordmann. Indianapolis: Hackett.

4

Kierkegaard and the *Tractatus*

Genia Schönbaumsfeld

1 Introduction

Wittgenstein once said to Drury, 'Kierkegaard was by far the most profound thinker of the last century. Kierkegaard was a saint' (Drury 1981: 87). This chimes with G. H. von Wright's remark that 'Wittgenstein received deeper impressions from some writers in the borderland between philosophy, religion, and poetry than from the philosophers, in the restricted sense of the word. Among the former are St. Augustine, Kierkegaard, Dostoevsky, and Tolstoy' (Malcolm 2001: 19). This seems to me an accurate assessment. Indeed, it has almost become a commonplace among Wittgenstein and Kierkegaard scholars to pay some form of lip-service to the thought that Wittgenstein held Kierkegaard in incredibly high regard. But although commentators like to mention this fact, most are at a loss about what else to do with it.[1] An excellent example of this kind of response is Bertrand Russell's surprise, when he found that after the First World War Wittgenstein had, in Russell's words, 'become a complete mystic'. Russell wrote to Lady Ottoline Morrell on 20 December 1919, 'I had found in his book a flavour of mysticism, but was astonished when I found that he has become a complete mystic. He reads people like Kierkegaard and Angelus Silesius, he seriously contemplates becoming a monk' (Russell 2001: 198). Even more recent and more penetrating critics of Wittgenstein's work, however, often either fail to see the extent to which Kierkegaard influenced Wittgenstein's views on philosophy and religion or else misdescribe the points of contact between their respective authorships.

The problem is particularly acute as regards discerning any sort of parallel between the austere monolithic character of Wittgenstein's early work and the astonishing prolixity of Kierkegaard's pseudonymous production. For this reason, philosophers keen to make a connection between the *Tractatus* and Kierkegaard's extensive oeuvre have traditionally focused on Wittgenstein's conception of the ineffable and have sought to read this idea back into Kierkegaard's central notion of 'paradox'[2]—after

[1] Notable exceptions are Charles Creegan (1989), James Conant (1993), D. Z. Phillips (1993, 1999), Jamie Ferreira (1997), Stephen Mulhall (2001), and Mariele Nientied (2003).

[2] See e.g. Evans (1983).

all, as Wittgenstein himself once said to Waismann: 'Man has the urge to run up against the limits of language . . . Kierkegaard too recognized this running up against something and even described it in much the same way (as running up against paradox)' (Wittgenstein 1984: 68).[3] James Conant has recently turned the tables on this kind of interpretative strategy by claiming that the 'ineffability doctrine' that commentators attribute to both authors alike, far from being endorsed by either of them, actually features only as a target in their work. In other words, Conant construes the central parallel between Kierkegaard and Wittgenstein as consisting not in 'their sharing some mystical doctrine of ineffable truth but rather in their sharing a common twofold project of exposing the incoherence of any such doctrine and diagnosing the sources of its attraction' (Conant 1993: 218).

Although I cannot argue it here,[4] both accounts seem to me to be mistaken. Conant's interpretation does have the virtue of not needing to ascribe incoherent doctrines to either Wittgenstein or Kierkegaard, but it does no justice to the intentions, ethical or otherwise, of either author and in fact comprises a severe distortion of their texts. There are, to be sure, interesting parallels between Kierkegaard and the early Wittgenstein, but the exposure (or, indeed, the endorsement) of a doctrine of 'substantial nonsense' (or of ineffable truth) is not one of them. For whilst, as it seems to me, Wittgenstein clearly *is* committed to the notion of ineffabilia in the *Tractatus*, Kierkegaard's concerns are not with the limits of language and what might lie beyond them at all. Wittgenstein himself therefore appears guilty of imputing his own vision to Kierkegaard when he says that 'Kierkegaard too recognized [this urge] and even described it in much the same way'. Kierkegaard neither recognized this urge nor described it in a similar way and hence the temptation to read him in Tractarian vein as either repudiating—in the manner of Conant—or as espousing—in the manner of Wittgenstein himself—an ineffability conception needs very much to be resisted.

The points of contact that can therefore be perceived between the early Wittgenstein and the Danish philosopher cannot be located in the actual endorsement (or rejection) of similar views about 'ineffable truth', but are rather to be found in a certain commonality of vision as regards ethics and religion. My main aim in this chapter is to offer a fuller account of this commonality. First, though, I would like to say something about the biographical evidence available for early Wittgenstein's engagement with Kierkegaard.[5]

2 Historical background

There is every reason to suppose that Wittgenstein was introduced to the writings of Kierkegaard from a very early age. During his childhood and adolescence his elder

[3] Except where an English translation is cited, translations are my own.

[4] For the missing argument, see my 2004, as well as chapter 3 of my (2007). See also: Lippitt and Hutto (1998); McGinn (1999); Conant (2000 and 2002); Hacker (2000); Proops (2001); Sullivan (2002).

[5] For a fuller account, see my (2007: ch. 1).

sister Margarete ('Gretl') served as his 'philosophical' mentor. In the words of Ray Monk, 'Gretl was acknowledged as the intellectual of the family, the one who kept abreast of contemporary developments in the arts and sciences, and the one most prepared to embrace new ideas and to challenge the views of her elders' (Monk 1991: 16). Given that Kierkegaard was Gretl's favourite author (Wuchterl and Hübner 1979: 30) and was generally very much en vogue in turn-of-the-century Vienna, it would be very surprising indeed if Gretl had not drawn her younger brother's attention to Kierkegaard's works.

Be that as it may, direct evidence is certainly available that Wittgenstein was exposed to some Kierkegaard subsequently. In 1914, while spending some time in Norway, Wittgenstein first came across Ludwig von Ficker, the editor of the literary journal *Der Brenner*, which published the work of Theodor Haecker, whose German translations of Kierkegaard first introduced the Danish philosopher to an Austrian audience. Among the Kierkegaard texts that featured in *Der Brenner* between 1913 and 1921 are the preface to *Prefaces*, the introduction to *Johannes Climacus*, the discourse 'At a Graveside' from *Three Discourses on Imagined Occasions*, the discourse 'The Thorn in the Flesh' from *Four Upbuilding Discourses*, *A Critique of the Present Age*, some journal entries from 1835 and 1836, as well as the discourse 'God's Strength in Human Weakness' ('*Die Kraft Gottes in der Schwachheit des Menschen*'). Wittgenstein clearly read this journal and even decided, through Ficker, to donate some of his family money to Austrian artists in need. Theodor Haecker was one of the beneficiaries (Monk 1991: 106–9).

Further evidence of Wittgenstein's engagement with Kierkegaard can be gleaned from the extensive correspondence of Wittgenstein's during the war period. Wittgenstein's sister, Hermine, for example, writes in her letter to Ludwig of 20 November 1917: 'Thank you very much for your lovely card from 13th November. You were perfectly correct in supposing that I did not receive the earlier one with your request for books, but I've just been out for them and a number of Kierkegaard volumes are already on the way. I hope they are the ones you want, because, given that I don't know anything about him and his writings, I simply chose a few at random. The Diary of a Seducer, which I bought in a different bookshop, will follow' (Wittgenstein 1996: 48). It is unfortunate that Hermine does not say which volumes she sent, but the fact that Wittgenstein had her send them to him at the Front suggests the importance they had for him.

Kierkegaard also comes up in the correspondence between Wittgenstein and his close wartime friend, Paul Engelmann,[6] who quotes from the *Stages on Life's Way* in one of his letters to Wittgenstein. Engelmann says, '"If I had had faith, I would have

[6] Wittgenstein first met Engelmann when he was in Olmütz. They became very close friends and Engelmann later collaborated with Wittgenstein on the house he built for his sister Gretl in the Kundmanngasse in Vienna; see Monk (1991: 150–1, 235).

stayed with her"[7] ... It seems to me that you are lacking in faith' (Engelmann 1970: 8.1.1918). To which Wittgenstein replies, 'When you say that I do not have faith, you are quite right, except that I did not have it previously either. It is obvious that someone who wants to invent a machine in order to become a better person, that such a one has no faith. But what should I do? One thing is clear to me: I am much too bad to ponder about myself; I will either remain a swine or I will improve and that's that! No transcendental twaddle when everything is as clear as a slap' (Engelmann 1970: 16.1.1918).

From the notebooks and the coded diaries dating from this period it becomes clear that Wittgenstein was constantly preoccupied with spiritual matters. Although there are no direct references to Kierkegaard in these materials, it would, I think, not be an exaggeration to say that Wittgenstein was constantly suffering from a form of 'Kierkegaardian despair'. The diaries reveal that Wittgenstein was continuously tormented by his moral worthlessness and his sense of being at odds with the world. Wittgenstein believed that this kind of unhappiness—what Kierkegaard would call despair—is the sign of a bad life, the mark of someone who is incapable of doing God's will and who, as the previously cited letter suggests, lacks faith (Wittgenstein 1993a: 168–9).

In the *Notebooks* for 8.7.16 Wittgenstein equates doing God's will with coming to terms with the facts, and at *Tractatus* 6.4321 he writes, 'The facts all belong only to the task, not to the solution.' On this conception of things, where life is seen as a task to be mastered (a conception that Kierkegaard shared), a lack of faith is therefore regarded as a *moral* failing. This is also the reason why Wittgenstein accepted the Dostoevskyan thought that if suicide is allowed, then everything is allowed. On 10.1.1917 Wittgenstein writes, '... suicide is, so to speak, the elementary sin'. This is so, because suicide is an evasion of the task that is life, a sign that one's life is not in harmony with the facts, or, to put it religiously, is in rebellion against the will of God. An exposition of this view can be found in the writings of Anti-Climacus, the Christian author of *The Sickness unto Death*: 'That is why the pagan ... judged suicide with such singular irresponsibility, yes, praised suicide, which for spirit is the most crucial sin, escaping from existence in this way, mutinying against God ... The point that suicide is basically a crime against God completely escapes the pagan.'[8] So it seems that Wittgenstein would also have accepted the central Kierkegaardian notion that the opposite of sin is not virtue, but *faith*, and that only Christianity provides a real solution to the 'problem of life'.[9] In one of his coded diary entries from 1914 Wittgenstein says: 'Bought Nietzsche volume 8 [containing *The Antichrist*] and read around in it. Was deeply

[7] This is an allusion to Kierkegaard's tortured relationship with his ex-fiancée, Regine Olsen.

[8] Kierkegaard (1980: 46); subsequent references to this work will be abbreviated 'SUD'.

[9] Of course this is also an idea to be found in Tolstoy and we know that Wittgenstein was constantly reading Tolstoy's *The Gospel in Brief* at this time. However, it is no part of my argument, here or elsewhere, to claim that Wittgenstein was only influenced by Kierkegaard in religious matters.

impressed by his antagonism towards Christianity. For also in his writings there is a grain of truth. To be sure, Christianity is the only *certain* way to happiness. But what if someone rejected this happiness?! Might it not be better to be ground to the dust [*zu Grunde gehen*] in the hopeless struggle against the external world? But such a life is meaningless. But why not lead a meaningless life? Is it unworthy?' (Wittgenstein 1991: 49)

It seems, therefore, that at this point in his life, although he could not come to have faith himself, Wittgenstein accepted the idea that a life without faith is meaningless, that without it we are mere playthings of contingency, doomed to a life of despair and, as Kierkegaard would doubtlessly add, spiritlessness [*Geistlosigkeit*]. In *Fear and Trembling*, for example, Kierkegaard's pseudonym, Johannes de Silentio, expresses the same idea as the quotation above, though more poetically: 'If a human being did not have an eternal consciousness, if underlying everything there were only a wild, fermenting power that writhing in dark passions produced everything, be it significant or insignificant, if a vast, never appeased emptiness hid beneath everything,[10] what would life be then but despair?'[11]

That Wittgenstein was no stranger to the religious phenomenology of 'fear and trembling' and generally couched his religious thoughts and feelings in Kierkegaardian terms is corroborated by a recently found journal entry from 1922 where Wittgenstein recounts the following experience:

> I suddenly felt my complete nothingness and saw that God could demand of me what He wills on the condition that my life would immediately become meaningless if I didn't comply . . . I felt totally annihilated and in the hands of God who could at every moment do with me as He wills. I felt that God could at any time force me immediately to confess my crimes [*Gemeinheiten*]. That he could at any moment force me to take the worst upon myself and that I am not prepared to take the worst upon myself. That I am not now prepared to renounce friendship and all earthly happiness . . . As I said, tonight I saw my complete nothingness. God has deigned to show it to me. During the whole time I kept thinking about Kierkegaard and that my condition is 'fear and trembling'. (Wittgenstein 2004, entry for 13.1.1922)

This is quite remarkable—not only does the entry contain an obvious reference to the Danish philosopher, it is almost as if Kierkegaard himself had written it. Wittgenstein here identifies doing God's will with what de Silentio, in FT, calls the 'last stage before faith': 'infinite resignation'—renouncing all finite (relative) ends. But although Wittgenstein believes that disobedience will make his life meaningless, he cannot force himself to comply with God's commands. Nevertheless, Climacus, the pseudonymous author of *Concluding Unscientific Postscript*, would probably say that Wittgenstein is poised on the brink of faith, as the feeling of complete self-annihilation before God that Wittgenstein mentions is, according to Climacus, one of the most decisive features

[10] This is also reminiscent of Pascal's oft-quoted aphorism: 'Le silence éternel de ces espaces infinis m'effraie'.

[11] Kierkegaard (1983: 15); subsequent references to this work will be abbreviated 'FT'.

of religiousness: 'Religiously, the task is to comprehend that a person is nothing at all before God or to be nothing at all and thereby to be before God, and he continually insists upon having his incapability before him, and its disappearance is the disappearance of religiousness.'[12]

It seems, therefore, that Wittgenstein read much Kierkegaard both during and immediately after the First World War—during which time he wrote the *Tractatus*—and it should therefore come as no surprise to find a certain Kierkegaardian flavour permeating that work. It is to an explication of this that I shall now turn.

3 Kierkegaard and the mysticism of the *Tractatus*

Paul Engelmann describes Wittgenstein's attitude to life in the following way: 'an ethical totalitarianism in all questions, a single-minded and painful preservation of the purity of the uncompromising demands of ethics, in agonizing awareness of one's own permanent failure to measure up to them' (Engelmann 1970: 109). Not only would this description be equally true of Kierkegaard, it also provides us with a key for understanding what James Conant finds so mystifying about early Wittgenstein—how he could have found the notion of 'ineffabilia' intelligible. But if we bear in mind that Wittgenstein was, as it were, an 'ethical totalitarian', then his motivations will no longer seem quite so bizarre. For, as Kant had already seen, if the 'absolute value' of value is not to be compromised, value can have no place in a merely contingent empirical world, as it would otherwise only be one more part of that world and *eo ipso* on the same plane as 'the facts'. Consequently, there can be no propositions of ethics, since these would necessarily be true, if, *per impossibile*, they could be expressed. But, as Wittgenstein says, 'The only necessity that exists is *logical* necessity' (6.37).

The upshot of this is that the early Wittgenstein's view of ethics, like Climacus' in CUP, is radically anti-consequentialist: the ethical must be willed purely for its own sake and whatever effects my actions might have in the external world are completely irrelevant, since everything that happens in that world is utterly contingent and therefore, no matter how good my intentions are, they can always bring about a bad result. Consequently, what actually comes to pass is out of an individual's control and no blame or praise can legitimately attach to it. This is how Climacus puts it:

> True ethical enthusiasm consists in willing to the utmost of one's capability, but also, uplifted in divine jest, in never thinking whether or not one thereby achieves something. As soon as the will begins to cast a covetous eye on the outcome, the individual begins to become immoral ... —the individual demands something other than the ethical itself. A truly great ethical individuality would consummate his life as follows: he would develop himself to the utmost of his capability; in the process he perhaps would produce a great effect in the external world, but this would not

[12] Kierkegaard (1992: 461); subsequent references to this work will be abbreviated 'CUP'.

occupy him at all, because he would know that the external is not in his power and therefore means nothing either *pro* or *contra*. (CUP 135–6)

That the external means nothing either *pro* or *contra* is also a consequence of Wittgenstein's Tractarian view of ethics. Given that Wittgenstein believes, in the *Tractatus*, that value is located outside the world and is, in this respect, transcendental like logic (6.421), it follows that ethical actions cannot be events in the world, for if they were, this would *eo ipso* strip them of all *ethical* content. In the 'Lecture on Ethics'—given as late as 1929—Wittgenstein explains:

Suppose one of you were an omniscient person and therefore knew all the movements of all the bodies in the world dead or alive and that he also knew all the states of mind of all human beings that ever lived, and suppose this man wrote all he knew in a big book, then this book would contain the whole description of the world; and what I want to say is, that this book would contain nothing that we would call an *ethical* judgement or anything that would logically imply such a judgement . . . all the facts described would, as it were, stand on the same level and in the same way all propositions stand on the same level. There are no propositions which, in any absolute sense, are sublime, important or trivial . . . Ethics, if it is anything, is supernatural and our words will only express facts; as a teacup will only hold a teacup full of water [even] if I were to pour out a gallon over it. (Wittgenstein 1993b: 39–40; cf. 6.41)

In other words, given Wittgenstein's dogmatic distinctions between what can and cannot be said and between fact and value, it cannot but follow that the ethical must either be a chimera or an inexpressible, mystical something. For Wittgenstein is trying to square the circle here by attempting to wed a positivistic view of language—sentences can only picture facts—to a romanticized absolutist conception of ethics. But, as Wittgenstein himself says, given that 'no state of affairs has, in itself, what I would like to call the coercive power of an absolute judge' (1993b: 40), such an attempt is *ex hypothesi* doomed to failure. And this is why Wittgenstein ends up drawing the same conclusion as Climacus when he says that 'the question about the *consequences* of an action must be unimportant.—At least those consequences should not be events. For there must be something right about the question we posed. There must indeed be some kind of ethical reward and punishment, but they must reside in the action itself' (6.422). Notice how much this chimes with CUP:

Victory in the outer demonstrates nothing at all ethically, because ethically the question is only about the inner. Punishment in the outer is negligible, and far from insisting with aesthetic busyness on visible punishment, the ethical proudly says: I shall punish, all right, namely, in the inner, and it is plainly immoral to class punishment in the outer as something comparable to the inner. (CUP 297)

Hence, although the two philosophers have got there in slightly different ways—Climacus is here criticizing the Hegelian notion of the identity of the inner and the outer while Wittgenstein is following through on his logical commitments—they nevertheless end up arriving at the same destination. That is to say, in order to defend the absolute demands of ethics against the encroachment of the claims of contingency, the ethical is relegated, in both authors, to the realm of inwardness. The 'coercive power of an absolute judge' that both Kierkegaard and Wittgenstein require of the ethical is therefore to be found, on their conception, not in some external sanction, but rather within the individual himself, in the sense that his good or evil willing becomes constitutive of the kind of world that individual inhabits. This is why Wittgenstein says, 'if the good or bad exercise of the will does alter the world, it can alter only the limits of the world, not the facts—not what can be expressed by means of language. In short the effect must be that it becomes an altogether different world. It must, so to speak, wax and wane as a whole. The world of the happy man is a different one from that of the unhappy man' (6.43).

Kierkegaard seems to have something similar in mind when he distinguishes, for example, between the different existence spheres. For these, too, are in a relevant sense entire 'worlds' that require the use of different concepts which only become available once an individual has undergone an inner transformation. As Climacus says, 'In the world of spirit, the different stages [spheres] are not like cities on a journey, about which it is quite all right for the traveller to say directly, for example: We left Peking and came to Canton and were in Canton on the fourteenth. A traveller like that changes place, not himself; and thus it is all right for him to mention and to *recount* change in a direct, unchanged form. But in the world of spirit to change place is to be changed oneself, and therefore all direct assurance of having arrived here and there is an attempt *à la* Münchhausen' (CUP 281). Although the target here is Hegel and his 'direct assurance', in the system, of having reflected himself out of immediacy, the underlying idea is clearly the same as Wittgenstein's: ethical (or religious) transformation takes place in the inner and can't be cashed out directly, as it were, in philosophical propositions.

In other words, Kierkegaard's and Wittgenstein's emphasis on the inner is the result of a common moral absolutism that is to guard the sanctity of the ethical, as well as of a shared sense of powerlessness in the face of the contingency of the world which can, at any moment, reduce my ethical striving to nothing. Wittgenstein speaks, in this context, of being completely dependent on an 'alien will' and in this way the ethical very quickly acquires a religious countenance:

> To believe in God is to understand the question of the meaning of life. To believe in God is to see that the facts of the world are not the end of the matter. To believe in God is to see that life has meaning. The world is *given* me, i.e. my will approaches the world completely from the outside as something finished... That is why we have the feeling that we depend on an alien will. *However this may be*, we *are*, in any case, in a certain sense dependent and what we are dependent on, we can call God. God would, in this sense, simply be fate or, what is the same: the world independent of our will. I can make myself independent of fate . . . In order to live happily, I have to be in agreement [*Übereinstimmung*] with the world... I am then, as it were, in agreement with that alien will on which I seem dependent. This means: 'I am doing the will of God'... When my conscience throws me off balance, I am not in agreement with something.

But what is this? Is it *the world*? It must be correct to say: conscience is the voice of God. (Wittgenstein 1993a; 8.7.1916)

Sections 6.43–6.4321 of the *Tractatus* are clearly modelled on these remarks from the *Notebooks*, without, however, identifying being in agreement with the world with doing the will of God. In fact, the only reference to God (in a religious context) occurs at *Tractatus* 6.432 where Wittgenstein says: '*How* things are in the world is a matter of complete indifference for what is higher. God does not reveal himself *in* the world.' But given that Wittgenstein goes on to say, at 6.4321, that the facts all belong to the task not to the solution, this makes it clear, I think, that he must have had in mind something similar to the *Notebooks* passage. Indeed, as 6.43 cited above tells us, Wittgenstein, in the *Tractatus*, seems explicitly to identify the good will with being in agreement with the world (the facts), with being happy, and with doing the will of God. In other words, what he seems to be saying is that when my conscience does not, as it were, throw me off balance, i.e. when my will is good, I am in agreement with the world; I am consequently doing the will of God, and the world—to use his Tractarian metaphor—waxes as a whole. Hence, the important thing for Wittgenstein is not *how* the world is or *what* I do, but rather the *spirit* in which I view it and in which I perform my duties.

Climacus has a similar conception in mind when he writes, echoing Kant:

But freedom, that is the wonderful lamp. When a person rubs it with ethical passion, God comes into existence for him. And look, the spirit of the lamp is a servant . . . , but the person who rubs the wonderful lamp of freedom becomes a servant—the spirit is the Lord . . . So the resolving person says: I will—but I also want to have world-historical importance—*aber* [but]. So there is an *aber*—and the spirit vanishes again, because the rubbing has not been done properly, and the beginning does not occur. But if it has occurred or has been done properly, every subsequent *aber* must again be renounced, even if existence in the most flattering and inveigling way did everything to force it upon one. (CUP 138–9)

Climacus here voices many of the same ideas as Wittgenstein above: the notion that to serve the ethical is to do the will of God as well as the thought that this only becomes a possibility once I renounce my stake in worldly affairs—once, as Wittgenstein would put it, I make myself independent of the facts. For, as Wittgenstein says in another *Notebook* entry, 'Even if everything we desired happened, this would only be the luck of the draw, as there is no logical connection between will and world . . . ' (Wittgenstein 1993a: 5.7.1916). And a little later he adds, in Stoic vein, 'only that life is a happy one which is able to renounce the comforts of the world. For such a life these comforts are just so many mercies of fate' (Wittgenstein 1993a: 13.8.1916).

For both Wittgenstein and Climacus, then, the ethical life is a life of renunciation and in this sense it is very close to the religious. One would consequently be very much misled if one were to think that because the two authors emphasize, like Kant, the 'inner', or, as Climacus would say, 'subjectivity', as of paramount importance in ethics, the force of the ethical 'ought' is in any way diminished thereby. Quite to the contrary:

if one's conscience is, indeed, conceived to be the voice of God, then its wrath is more terrible than any external punishment could be, since, as Climacus notes so perceptively, 'If in my relationship with God I regard what I am doing as good and do not keep watch over myself with the infinite's mistrust of me, then it is just as if God, too, were content with me, because God is not something external that quarrels with me when I do wrong but the infinite itself that does not need scolding words, but whose vengeance is terrible—the vengeance that God does not exist for me at all, even though I pray' (CUP 162–3). Hence, on the conception that Wittgenstein and Kierkegaard share, there is an internal connection between the good will (moral virtue) and faith and a lack of the latter is therefore tantamount to a moral failing. This is why Wittgenstein says in the letter to Engelmann mentioned earlier that 'it is obvious that someone who wants to invent a machine in order to become a better person, that such a one has no faith' (Engelmann 1970: 16.1.1918). For someone who shirks the painful process of inner transformation, and would rather create a *deus ex machina* in order to do the work for him, is clearly still enslaved, in Climacus' words, to the '*abers*' of this world and consequently the spirit of the Lord vanishes again. The result is the hopelessness and despair that Wittgenstein felt—as his diary entries confirm—during much of the time of writing the *Tractatus* and which Anti-Climacus depicted so ably in *The Sickness unto Death*.

In other words, Kierkegaard's and Wittgenstein's quasi-religious conception of ethics implies what Anti-Climacus calls the 'upbuilding thought' of being a self 'directly before God': a self, that is, that has God and not the merely human as a criterion. As Anti-Climacus puts it, the 'self takes on a new quality and qualification by being a self directly before God. And what infinite reality the self gains by being conscious of existing before God, by becoming a human self whose criterion is God!...Just as only entities of the same kind can be added, so everything is qualitatively that by which it is measured, and that which is its qualitative criterion [*Maalestok*] is ethically its goal [*Maal*]; the criterion and goal are what define something, what it is, with the exception of the condition in the world of freedom, where by not qualitatively being that which is his goal and his criterion a person must himself have merited this disqualification. Thus the goal and the criterion still remain discriminatingly the same, making it clear just what a person is not—namely, that which is his goal and criterion' (SUD 79–80). Naturally, on such a conception, human guilt cannot but be infinitely magnified into sin, which perhaps explains the intense feelings of moral worthlessness that plagued Wittgenstein throughout his life. For with God or God's will as the measure of moral conduct, every failing becomes a sin against God. As Anti-Climacus confirms, 'because sin is against God it is infinitely magnified. The error [of older Christian dogmatics] consisted in considering God as some externality and in seeming to assume that only occasionally did one sin against God. But God is not some externality in the sense that a policeman is. The point that must be observed is that the self has a conception of God and yet does not will as he wills, and thus is disobedient. Nor does one only occasionally sin before God, for every sin is before God, or, more correctly, what really makes human guilt into sin is that the guilty one has the consciousness of existing before God' (SUD 80). Here Anti-Climacus repeats, and religiously accentuates, the point made by Climacus above that, because God is not something external, or, as Wittgenstein says in the *Tractatus*, does not reveal himself in the world, it is impossible ever to escape his scorching gaze, for this is located within. So, to continue with Climacus' metaphor, the wonderful lamp of freedom must be handled with care, as the spirit of the Lord, once conjured up, is here to stay. That is to say, once a self recognizes that it is before God, Wittgenstein and Kierkegaard seem to agree, it can never divest itself of this knowledge on pain of losing itself or, what amounts to the same, plunging into despair.

Anti-Climacus' definition of faith as a state in which 'the self in being itself and in willing to be itself rests transparently in God' (SUD 82) can therefore be seen to parallel Wittgenstein's identification, in the *Notebooks* and in the *Tractatus*, of being in agreement with the world (the facts), with doing the will of God and with being happy. For the 'faithful' self that wills to be itself is, in Wittgenstein's parlance, in agreement with itself and the world—i.e. it 'happily' accepts the limitations and constraints imposed upon it from outside without viewing this as cause for despair. So the self, in faith, despite constantly struggling to better itself, bears a positive relation towards itself and the world. In this respect, the self's particular constitution and the way the world happens to be (the facts) becomes irrelevant in the sense that both these things, as Wittgenstein says, only belong to the task, not the solution. The self must therefore continue to work on itself—must, as Climacus puts it, continue to strive—but without regarding the way it is made or the tasks that the world sets it as reasons, as it were, to pick a quarrel with God. The attainment of such a perspective is consequently marked by not feeling at odds with the world (or alienated from oneself) and by ceasing to perceive life as a problem. This is why Wittgenstein says, 'the solution of the problem of life is seen in the vanishing of the problem' (6.521).

In order for such a perspective of faith to be a genuine possibility, Anti-Climacus would argue, real trust in God is necessary and this involves the conviction, on the side of the believer, that for God everything is possible: 'The *believer* sees and understands his downfall, humanly speaking . . ., but he believes. For this reason he does not collapse. He leaves it entirely to God how he is to be helped, but he believes that for God everything is possible . . . The believer has the ever infallible antidote for despair—possibility—because for God everything is possible at every moment. This is the good health of faith that resolves contradictions' (SUD 39–40). Hence, with this armour, so to speak, the self, in faith, avoids both possibility's despair, what Anti-Climacus characterizes as the failure to obey, 'to submit to the necessity in one's life', and necessity's despair, a kind of fatalism that can see nothing but unalterable facts (determinism). Wittgenstein's account in the *Tractatus*—although containing all other aspects of Anti-Climacus' formulation—is devoid of this dimension of possibility, for what it proposes is merely what Johannes de Silentio calls infinite resignation (the renunciation of all relative ends). And although de Silentio would agree that the dying to immediacy that early Wittgenstein proposes—renouncing the comforts of the world—is a necessary condition for faith, it is only the last step before faith and not equivalent to faith itself, as the knight of faith, in the end, regains the world after having renounced it. Wittgenstein's identification of the will of God with the world seems to preclude this, however, as, on his conception, happiness can only be achieved by making oneself completely independent of the facts. But if God and the world (the facts) are equated, as in Wittgenstein's *Notebooks*, then this would seem to imply also having to make oneself independent of God. So, it seems that Engelmann's diagnosis of a lack of faith has turned out to be correct: Wittgenstein, given his ethical-religious conception, has contracted Anti-Climacus' disease—the sickness unto death—but without, in the end, being able to reach for the 'radical cure' that Christianity would provide.

So far we have seen, then, that Kierkegaard and Wittgenstein share a Kant-inspired absolutist conception of ethics, which, however, ultimately goes beyond Kant's, for it is inextricably bound to the notion of God's inexorable presence and constant admonishment. Furthermore, Wittgenstein and Kierkegaard both reject the German philosopher's rationalistic view that even God is subject to the decrees of human reason.[13] In a conversation with the Vienna Circle from 1930 Wittgenstein makes this clear:

> Schlick says that there are two conceptions of the essence of the good in theological ethics: according to the more superficial view, the good is good because God wills it; according to the more profound view, God wills the good because it is good. I believe that the first conception is the more profound: good is what God commands. For it cuts off the way to any kind of explanation of why the good is good, whereas the second conception is precisely the superficial, rationalistic one which pretends that reasons can be given for this. The first conception clearly states that the essence of the good has nothing to do with the facts and can consequently not be explained by a proposition. If there is a sentence that expresses precisely what I mean, it is this: good is what God commands. (Wittgenstein 1984: 115)

That God is the *terminus ad quem* or final arbiter on what is good or absolute, and that no further grounds can be given for this, is also Johannes de Silentio's view in FT, where the issue is presented in a particularly radical way using the biblical story of Abraham and Isaac as a paradigmatic example. In the section entitled 'Is There an Absolute Duty to God?' de Silentio says:

> The ethical is the universal, and as such it is also the divine. Thus it is proper to say that every duty is essentially duty to God, but if no more can be said than this, then it is also said that I actually have no duty to God. The duty becomes duty by being traced back to God, but in the duty itself I do not enter into relation to God . . . God comes to be an invisible vanishing point, an impotent thought, his power is only in the ethical, which fills all of existence.[14] Insofar, then, as someone

[13] See e.g. Kant (1974: 201–6, 245–53).

[14] Notice how this echoes the Kantian (1974: 202) view: 'There are no special duties to God in a common [*allgemeinen*] religion; for God cannot receive anything from us; we cannot act upon or for him . . . '.

might wish to love God in any other sense than this, he is a visionary, is in love with a phantom, which, if it only had enough power to speak, would say to him: I do not ask for your love—just stay where you belong. (FT 68)

If Kant and Hegel are right, in other words, then, as de Silentio goes on to point out, Abraham cannot be the father of faith he is commonly hailed to be, but ought rather to be regarded as an especially abhorrent type of potential murderer. For Abraham conceived his duty to God as absolute, i.e. as something over and above all his other duties, even the duty to his son. And on such a conception, *pace* Kant and Hegel, it must at least in principle be possible that a collision can occur between one's duty to God and all one's other ethical commitments. If this is deemed impossible *ab initio*, then, as de Silentio says, it implies that there actually is no duty to God, but then of course there is also no such thing as faith, as the 'coercive power of the absolute judge' then becomes nothing more than the impotent power of an underlabourer of the ethical (and hence, on the Kantian view, of reason).

Consequently, what both Kierkegaard and Wittgenstein are rejecting here—and this is a feature that remains constant throughout the latter's authorship—is the craving for explanation and the idea that everything can be justified by appeal to the high court of reason. And this temptation, both philosophers believe, is all-pervasive: it is a deep-seated *malaise* that is not only to be found in ethics and religion, where, both would argue, it is especially pernicious, but in philosophy and science as well. It is surely not coincidental that in the sections directly preceding Wittgenstein's remarks on ethics and the mystical in the *Tractatus*, he says:

There is no compulsion making one thing happen because another has happened. The only necessity that exists is *logical* necessity. The whole modern conception of the world is founded on the illusion that the so-called laws of nature are the explanations of natural phenomena. Thus people today stop at the laws of nature, treating them as something inviolable, just as God and Fate were treated in past ages. And in fact both are right and both wrong: though the view of the ancients is clearer in so far as they have a clear and acknowledged terminus, while the modern system tries to make it look as if *everything* were explained. (6.37–6.372)

Compare this with what Climacus says in CUP:

An objective religious person in the objective human mass does not fear God; he does not hear him in the thunder, because that is a law of nature, and perhaps he is right. He does not see him in events, because they are the immanental necessity of cause and effect, and perhaps he is right. But what about the inwardness of being alone before God? Well, that is too little for him; he is not familiar with it, he who is on the way to accomplish the objective. (CUP 544)

In other words, what both Kierkegaard and Wittgenstein are saying in these passages is that it is possible to see the world in two different ways: in terms of the 'so-called laws of nature' or as manifestations of the will of God or of Fate. Neither conception, both authors believe, can be finally justified or absolutely grounded; and this, at least, the religious perspective makes clear by acknowledging, as Wittgenstein says, 'a clear terminus'. In this respect the religious conception is ethically superior to the scientific one, for it curbs philosophical (or scientific) *hubris* by recognizing a clear limit. What is more, both authors seem to urge, although there need not necessarily be a conflict between the two points of view, the scientific conception can stand in the way of seeing things from a religious perspective. For in order for a religious vantage point to be possible, one needs, as Wittgenstein says, to awaken to wonder, and in its desire to explain everything, 'science is a way of sending him [man] to sleep again' (1980: 5). And this is also the reason why, according to Climacus, the 'objective man' knows nothing about the inwardness of being alone before God, as the *spirit* in which science is carried on these days is not, in the end, compatible with such a conception.

So, it is precisely in order not to be the kind of philosopher who, as Climacus says in his other book, makes 'supernatural things ordinary and banal' (Kierkegaard 1985: 53) that early Wittgenstein banishes all that is great and important—ethics, aesthetics, and religion—into the realm of the ineffable and eschews all theorizing about it. For in the sphere of value, Wittgenstein thinks, theories will necessarily bypass the problems which trouble us: 'Whatever one were to tell me, I would reject it, and precisely not because the explanation is wrong, but because it is an *explanation* . . . One cannot teach the ethical. If I could only explain the nature of the ethical to someone else by means of a theory, then the ethical would have no value . . . *For me* a theory has no value. A theory does nothing for me' (Wittgenstein 1984: 116–17). Or, as Climacus would say, the solution to the problem of life is not to be found in a *doctrine*—in a sum of propositions that gives me a recipe for how to live—but, rather, 'in the vanishing of the problem' (6.521), which, Kierkegaard and Wittgenstein agree, can only be brought about by the right way of life.

4 Conclusion

We can now understand the spirit in which the final two 'propositions' of the *Tractatus* are intended: 'My propositions serve as elucidations in the following way: anyone who understands me eventually recognizes them as nonsensical, when he has used them—as steps—to climb up beyond them. (He must, so to speak, throw away the ladder after he has climbed up it.) He must transcend these propositions, and then he will see the world aright. What we cannot speak about we must pass over in silence' (6.54–7). I would like to suggest that these sentences constitute an *ethical* revocation of the work which shows, as Wittgenstein writes in the letter to von Ficker, that the *Tractatus* consists of two parts: of the one which is here, and of everything that Wittgenstein has *not* written. 'And precisely this second part,' Wittgenstein insists, 'is the important one. For the Ethical is delimited from within, as it were, by my book; and I'm convinced that, *strictly* speaking, it can ONLY be delimited in this way. In brief, I think: All of that which *many* are *babbling* today, I have defined in my book by remaining silent about it' (quoted in Monk 1991: 178).

In other words, the *Tractatus* highlights the importance of the ethical precisely by having next to nothing to say about it. In this sense, silence is not just the upshot of Wittgenstein's restrictive conception of the sayable, it is also a *moral* injunction: where only babble or transcendental twaddle is possible, it is a sin not to hold one's tongue.[15] And here there is an interesting parallel to be discerned not only with the name of the author of *Fear and Trembling* (Johannes *de Silentio*), but also with the revocation that John, the Ladder (Kierkegaard's Johannes Climacus), offers us in an 'Understanding with the Reader' at the end of CUP:

> Just as in Catholic books, especially from former times, one finds a note at the back of the book that notifies the reader that everything is to be understood in accordance with the teaching of the holy universal mother Church, so also what I write contains the notice that everything is to be understood in such a way that it is revoked, that the book has not only an end but has a revocation to boot. (CUP 619)

That is to say, Climacus is intimating that a personal appropriation of the book's message is more important than its content: once the perspective that the book is trying to convey has been grasped—once the book has, that is, been 'understood'—author and text 'annul' themselves and are, in this sense, 'revoked'. Consequently, the reader is to see that 'the understanding is a revocation—the understanding with him as the sole reader is indeed the revocation of the book. He [the reader] can understand that to write a book and to revoke it is not the same as refraining from writing it, that to write a book that does not demand to be important for anyone is still not the same as letting it be unwritten' (CUP 621). This also seems to be the reason why Wittgenstein says in the Preface to the *Tractatus*: 'Perhaps this book will be understood only by someone who has himself already had the thoughts that are expressed in it—or at least similar thoughts.—So it is not a textbook.—Its purpose would be achieved if it gave pleasure to one person who read it and understood it.' (Notice how Climacus speaks of 'the sole reader' and Wittgenstein of 'one person'.)

What this means is that it is the purpose of neither the *Tractatus* nor of CUP to communicate new knowledge, but rather to achieve the change in perspective that, in the end, makes all philosophical talk—for example (though not exclusively) about the ethical or the religious—superfluous. In this respect, as Wittgenstein says in the letter to von Ficker, 'the point of the book is ethical.' Or, indeed, as Kierkegaard puts it in 'A First and Last Explanation':

> What I in one way or another know about the pseudonymous authors of course does not entitle me to any opinion, but not to any doubt, either, of their assent, since their importance... unconditionally does not consist in making any new proposal, some unheard-of discovery, or in founding a new party and wanting to go further, but precisely in the opposite, in wanting to have

[15] Which implies, of course, that, strictly speaking, Wittgenstein should not have written the *Tractatus* at all or, at any rate, should not have written more than proposition 7: 'what we cannot speak about we must pass over in silence.'

no importance, in wanting, at a remove that is the distance of double-reflection, once again to read through solo, if possible in a more inward way, the original text of individual human existence-relationships, the old familiar text handed down from the fathers. (CUP 629–30)

And this, I would like to think, is also the *Tractatus*' aim, however implausible this might, at first blush, appear.

References

Conant, James (1993) 'Kierkegaard, Wittgenstein and Nonsense', in Ted Cohen, Paul Guyer, and Hilary Putnam, eds, *Pursuits of Reason*. Lubbock: Texas Tech University Press.

—— (2000) 'Elucidation and Nonsense in Frege and Early Wittgenstein', in Crary and Read (2000), 174–217.

—— (2002) 'The Method of the *Tractatus*', in E. Reck, ed., *From Frege to Wittgenstein*. Oxford: Oxford University Press, 374–462.

Crary, Alice and Read, Rupert, eds (2000) *The New Wittgenstein*. London: Routledge.

Creegan, Charles (1989) *Wittgenstein and Kierkegaard: Religion, Individuality and Philosophical Method*. London: Routledge.

Drury, M. O'C. (1981) 'Some Notes of Conversations with Wittgenstein', in Rhees (1981), 90–170.

Engelmann, Paul (1970) *Ludwig Wittgenstein: Briefe und Begegnungen*. Vienna-Munich: Oldenbourg.

Evans, C. Stephen (1983) *Kierkegaard's Fragments and Postscript: The Religious Philosophy of Johannes Climacus*. New York: Humanities Press International.

Ferreira, Jamie (1997) 'The Point outside the World: Kierkegaard and Wittgenstein on Nonsense, Paradox and Religion', *Wittgenstein Studies* 2/97 (on-line publication).

Hacker, P. M. S. (2000) 'Was He Trying to Whistle It?', in Crary and Read (2000), 353–88.

Kant, Immanuel (1974) *Die Religion innerhalb der Grenzen der bloßen Vernunft*. Stuttgart: Reclam.

Kierkegaard, Søren (1980) *The Sickness unto Death*, ed. and trans. Howard and Edna Hong. Princeton: Princeton University Press.

—— (1983) *Fear and Trembling*, ed. and trans. Howard and Edna Hong. Princeton: Princeton University Press.

—— (1985) *Philosophical Fragments*, ed. and trans. Howard and Edna Hong. Princeton: Princeton University Press.

—— (1992) *Concluding Unscientific Postscript to* Philosophical Fragments, ed. and trans. Howard and Edna Hong. Princeton: Princeton University Press.

Lippitt, John and Hutto, Daniel (1998) 'Making Sense of Nonsense', *Proceedings of the Aristotelian Society* 98, 263–86.

Malcolm, Norman (2001) *Ludwig Wittgenstein, A Memoir*. Oxford: Oxford University Press.

McGinn, Marie (1999) 'Between Metaphysics and Nonsense: Elucidation in Wittgenstein's *Tractatus*', *Philosophical Quarterly* 49, 491–513.

Monk, Ray (1991) *The Duty of Genius*. London: Vintage.

Mulhall, Stephen (2001) *Inheritance and Originality*. Oxford: Clarendon Press.

Nientied, Mariele (2003) *Kierkegaard und Wittgenstein: 'Hineintäuschen in das Wahre'*. Berlin: de Gruyter.

Phillips, D. Z. (1993) *Wittgenstein and Religion*. London: Macmillan.

—— (1999) *Philosophy's Cool Place*. Ithaca, NY: Cornell University Press.

Proops, Ian (2001) 'The New Wittgenstein: A Critique', *European Journal of Philosophy* 9, 375–404.

Rhees, Rush, ed. (1981) *Recollections of Wittgenstein*. Oxford: Blackwell.

Russell, Bertrand (2001) *The Selected Letters of Bertrand Russell: The Public Years, 1914–1970*, ed. N. Griffin. London: Routledge.

Schönbaumsfeld, Genia (2004) 'No New Kierkegaard', *International Philosophical Quarterly* 44 (4), 519–34.

—— (2007) *A Confusion of the Spheres: Kierkegaard and Wittgenstein on Philosophy and Religion*. Oxford: Oxford University Press.

Sullivan, Peter (2002) 'On Trying to be Resolute: A Response to Kremer on the *Tractatus*', *European Journal of Philosophy* 10, 43–78.

Wittgenstein, Ludwig (1971) *Tractatus Logico-Philosophicus*, trans. D. F. Pears and B. F. McGuinness, 2nd edition. London: Routledge.

—— (1980) *Culture and Value*, ed. G. H. von Wright, trans. P. Winch. Oxford: Blackwell.

—— (1984) *Ludwig Wittgenstein und der Wiener Kreis*, ed. B. F. McGuinness, in Werkausgabe, vol. 3. Frankfurt: Suhrkamp.

—— (1991) *Geheime Tagebücher*, ed. W. Baum. Vienna: Turia and Kant.

—— (1993a) *Tagebücher 1914–16*, in Werkausgabe, vol. 1. Frankfurt: Suhrkamp.

—— (1993b) *Philosophical Occasions*, ed. by J. Klagge and A. Nordmann. Indianapolis: Hackett.

—— (1996) *Wittgenstein Familienbriefe*, ed. B. F. McGuinness, M. Ascher, and O. Pfersmann. Vienna: Verlag Hölder-Pichler-Tempsky.

—— (2004) *Licht und Schatten*, ed. I. Somavilla. Innsbruck: Haymon Verlag.

Wuchterl, Kurt and Hübner, Adolf (1979) *Ludwig Wittgenstein in Selbstzeugnissen und Bilddokumenten*. Hamburg: Rowohlt.

5

What is Frege's 'concept *horse* problem'?

Ian Proops

Elizabeth Anscombe once complained of the 'wild irrelevance' of almost all that had previously been published on the *Tractatus*. This unfortunate state of affairs, she supposed, arose from a neglect of Frege (1959: 12). Sensitive to these concerns, Peter Geach went on to urge Frege's importance as the source of much of what the *Tractatus* has to say about the distinction between what can be said in language and what can only be shown, and about the inexpressibility of logical category distinctions in particular. The *Tractatus*, he says, simply 'takes over' certain Fregean doctrines in this area, adding its own modifications and extensions where appropriate (1976: 55–6).

These claims seem worth evaluating. To do so it will be necessary to look carefully at Frege's thought to see whether we can find there the seeds of the morals Wittgenstein drew, and, if so, to gauge how far Frege himself developed them. The ultimate purpose of such a detailed investigation is to distinguish the possible claim that Wittgenstein *derived* his views on the inexpressibility of logical category distinctions and the say/show distinction from Frege from the view that he merely *arrived at* these views by reflecting on issues to which Frege's work gives rise. I will argue that while the second claim is plausibly correct, the first is not supported by the texts.

Although this is our ultimate goal, our immediate purpose is to get clearer about the nature of the concept *horse* problem, and to gauge how far Frege thought it could be solved. The concept *horse* problem is a natural starting point for any discussion of Frege's views on logical category distinctions, for here our classificatory practices seem to run into trouble. Something we should like to categorize as a concept turns out—in Frege's view—not to be so categorizable. The puzzle, however, has numerous aspects or sub-problems that must be carefully distinguished if its philosophical and historical significance is to be properly gauged. In what follows I shall attempt to distinguish these sub-problems more sharply than has been the custom in the scholarly literature. By so doing I hope to work towards a better understanding of which sub-problems Frege took seriously and which he regarded as superficial—or even just ignored. With those matters clarified, it will be easier to judge how far these problems can be addressed, and—most importantly for the purposes of this volume—how far, and in what ways, Frege's reflections in this area are likely to have influenced Wittgenstein.

1 First problem: a breach of custom

Frege introduces the concept *horse* problem in his article 'On Concept and Object' (1892b). This article was written as a response to certain criticisms that had been leveled against Frege's position in *The Foundations of Arithmetic* by a prominent student of Brentano's, the Strasburg philosopher, Benno Kerry. Kerry wrote a series of eight articles appearing during the years 1885 to 1891 in the journal *Vierteljahrsschrift für Wissenschaftliche Philosophie* under the title 'Über Anschauung und ihre psychische Verarbeitung' ('On intuition and its psychical elaboration'). The concept *horse* problem is posed in the fourth of these articles (Kerry 1887) in the context of a discussion of Frege's fundamental tenet that the distinction between concepts and objects is exclusive. The article 'On Concept and Object', which contains Frege's defence of that thesis, then appeared in the same journal.

Kerry maintains against Frege's exclusivity thesis that the properties of being an object and of being a concept are relative rather than absolute: just as someone can be a father relative to one person and a son relative to another, something can be a concept relative to one way of regarding it and an object relative to another (1887: 272). He offers two arguments to back up this contention. The first seeks to demonstrate the incompatibility of the exclusivity thesis with Frege's criterion for being an object. That criterion takes the form of a sufficient condition: if an expression results from pre-fixing the definite article or a demonstrative pronoun to a singular occurrence of a concept word, it is a proper name (1884: §§51, 66), hence the name of an 'object' in Frege's sense of that term (1884: §68, n. 1). To pose his problem Kerry draws attention to the judgment, 'The concept I am now talking about is an individual concept.' He regards it as obvious that in this judgment, which he treats as true, the subject—i.e. what we would call the referent of the subject term—is a concept, hence, by the exclusiveness of the distinction, no object. But the subject term is of the form 'the *F*', and so by Frege's criterion its referent must be an object. We thus arrive at a contradiction, and so one of Frege's two assumptions must be given up.

By Kerry's own lights this first argument is not decisive against the exclusivity thesis, since he in any case rejects Frege's criterion on the ground that logical distinctions cannot be founded upon linguistic distinctions (1887: 273). Accordingly, he offers a second direct or 'positive' argument for the thesis that some concepts are objects. Kerry adduces the judgment, 'The concept "horse" is a concept easily attained,'[1] and assumes (without argument) that whatever falls under the concept expressed by the words 'is a

[1] Kerry uses quotation marks where Frege uses italics.

concept easily attained' must be an object. He concludes that the concept 'horse' must be both a concept and an object (1887: 274).

In response Frege allows that the subject term in each of Kerry's two examples may indeed refer to a 'concept', but not in the logical sense with which he is concerned (1892b: 182, 185). Kerry's argument, he thus suggests, falls at the first fence.

As directed against the two examples just discussed, Frege's response is convincing. The sort of concept that is 'attainable' is (something like) a psychological capacity, and the 'individual concepts' of which Kerry speaks would seem to be a species of psychological representation. Both of Kerry's examples therefore involve 'concepts' only in a psychological sense, and these would clearly count as 'objects' in the logical sense Frege is concerned with. Nonetheless, this response may not be altogether fair. For Frege neglects to mention that in his second argument Kerry offers a better example for his critical purposes. This involves one of Frege's own examples from the *Foundations*, viz., the second-level concept '[is] equinumerous with the concept *F*' (Kerry 1887: 274). Obviously, if Frege were to accept Kerry's assertion that the entities that fall under this concept are concepts, he would have to accept that they are concepts in the logical sense.

Although Frege fails to credit Kerry with this more promising criticism, he clearly recognizes its force. For he proceeds to acknowledge as a problem for his view an example that is naturally taken to involve concepts in the logical sense, viz., 'The concept *horse* is a concept'. His response is to deny the truth of this sentence, and beyond that, to assert its negation. The cost of this response, as he acknowledges, is an unavoidable 'awkwardness of language' manifested in a 'breach of custom'. We customarily say, 'The city Berlin is a city' and 'The volcano Vesuvius is a volcano', but if Frege is right, we must also say, 'The concept *horse* is not a concept'. More fully, the difficulty we face, if we follow Frege, is that, as a rule, we take instances of the following schema to be true: 'The *K T* is a *K*', where '*K*' is replaceable by a kind term and '*T*' by a term for something belonging to the kind *K*. But when '*K*' is replaced by the word 'concept' and '*T*' by the word '*horse*' the result is a falsehood. Let us call this the 'the breach of custom problem'.

Perhaps surprisingly, Frege does not try to solve this problem by arguing that the breach of custom is *only* apparent. Instead, he accepts it at face value, commenting, somewhat meekly, that 'Language is here in a predicament that justifies the departure from custom' (1892b: 186). The nature of the 'predicament' is clear enough: when I try to make a concept the subject of discourse the sentence I end up with turns out, against my intentions, to have an object as its subject. But while it might be reasonable to think that the predicament *necessitates* the departure from custom, it is hard to see how it could be supposed to justify it. After all, as Frege would later himself observe (1979: 177), the expression 'the concept *prime*' is (plausibly) 'constructed in a way which precisely parallels [the expression] "the poet Schiller"'. If that is so, then, since the sentence 'the poet Schiller is a poet' seems to be true in virtue of its form, the same would have to apply to the sentence 'the concept *prime* is a concept', in which case the departure from custom would be wholly unjustified.

The unfortunate fact is that in the essay in which he introduces the breach of custom problem Frege fails to engage with it in any sustained or convincing way.[2] Nonetheless, we may ask on Frege's behalf how the problem might be addressed within a Fregean setting. To this end it will be useful to begin by considering an interesting proposal once made by Terence Parsons.

1.1 Parsons on the proxy object solution

According to Parsons (1986), a consistent application of Frege's ideas actually requires him to have said that the concept horse *is* a concept. Parsons is prompted to say this by Frege's remark in 'On Concept and Object' that in order to say something about a concept 'an object must go proxy [*vertreten*] for it' (1892b: 186). What is meant to justify the idea that the object referred to by 'the concept *horse*' goes proxy for the concept is Frege's view that in asserting something of this object one thereby expresses the very same thought that would be expressed by asserting something else of the corresponding concept. So, to use Frege's (1892b: 188) own example, in asserting something of a certain *object* in:

[1] The concept *square root of 4* is realized,

one thereby expresses the very same thought that one would express by asserting something else of a *concept* in:

[2] There is at least one square root of four.

And for precisely the same reasons, according to Frege's way of thinking, the first-level predicate expressed by the last two words of [1] can be thought to represent or go proxy for the second-level predicate expressed by the first five words of [2]. That being so, the question naturally arises: Why not treat the sentence 'the concept *horse* is a concept' as making an assertion about a proxy object in order to express a truth about a concept? That is to say, why not treat the first-level predicate expressed by 'is a concept' as being used to make an assertion about an object, which assertion expresses precisely the same thought that would be expressed by asserting the corresponding

[2] Joan Weiner disagrees. She judges Frege's reply an 'appropriate and adequate' response to Kerry (1990: 246). I am not entirely sure what she takes that reply to be, but her main idea would seem to be that Frege is entitled to be untroubled by the paradoxical nature of sentences such as 'the concept *horse* is not a concept', since this would only be a problem, from Frege's point of view, if it prevented the sentence from playing what she (1990: 256) takes to be its elucidatory role of '[getting] his readers to understand [Begriffsschrift] regimentation'. And Weiner seems to suppose that this sentence is *not* hampered in its elucidatory role. One has to wonder, however, how this sentence could help one to appreciate (for example) that the Begriffsschrift's second-order variables range over the referents of such expressions as 'is a horse'. It bears mentioning, moreover, that when Frege claims that 'the concept *horse* is not a concept' his immediate aim is not to get his readers to understand his Begriffsschrift regimentation, but rather to rebut Kerry's charge that the distinction between concept and object is not exclusive.

second-level predicate directly of a concept. Let's call this proposal 'the proxy object solution to the custom breach problem'.

In order to evaluate it[3] we will need to consider, first, whether we can really make sense of this second way of expressing the thought in question, and, second, whether such a solution, if strictly thought through, would be consistent with Frege's other commitments. There is reason for pessimism on both scores, but these misgivings will take some time to develop.

As a preliminary, let's return to the simple illustrative case of claim [1]. If [1] is genuinely to express a true or false thought, it must, according to Frege, be expressible in his formalized language or Begriffsschrift. For the Begriffsschrift is supposed to be a tool for the expression of scientific thought: it is supposed to be capable of expressing any thought that has a claim to truth or falsehood. But how would [1] be expressed in a Begriffsschrift? It has seemed to some commentators—notably among them Jean van Heijenoort (1977: 104)[4]—that it would go over into a Begriffsschrift as: '$\exists x$ (x is a square root of 4)'. Such a formulation, after all, would seem to bring out perspicuously Frege's view that 'existence is a property of concepts' (1884: §53). But this is not, in fact, Frege's view. As he sees it, [1] is a significant sentence containing a proper name in the argument place of a first-level concept word. Accordingly, since Frege holds that every genuine concept must be true or false of any object whatsoever,[5] [1]'s significance implies that the result of replacing its component phrase 'the concept *square root of 4*' with any other Fregean proper name must also make sense. And, indeed, consistently with this demand, Frege treats the sentence

[3] Julius Caesar is realized

as *false* rather than nonsensical (1892b: 189). If van Heijenoort were right, this sentence could not be rendered in Begriffsschrift, for in that case a predication of non-emptiness would be paraphrasable by a Begriffsschrift sentence only when the proper name of which the predication was being made was of the form 'the concept *F*'. And so, on the plausible assumption that all significant, truth-valued sentences should be capable of being rendered in a Begriffsschrift, we would have to deem [3] nonsensical. The fact that Frege treats it as making sense therefore suggests that he would have taken the first-level predicate 'ξ is realized' to be significantly predicable even of objects that (for all we know) do not proxy any concept. Frege would have seen this predicate as holding

[3] That is to say, in order to evaluate it *as a historical proposal*. On its own terms it runs very quickly into the paradoxes. Or rather it does so if it is taken to involve the claim that *every* concept has a proxy object associated with it. For Russell's antinomy, in its Fregean setting, shows that some concepts do not have objects associated with them. Consider, for example, the concept of being the proxy object of a concept that fails to hold of its own proxy object. That concept cannot have a proxy object associated with it, on pain of contradiction.

[4] Less notably, before thinking hard about 'On Concept and Object', I held this view myself.

[5] This is just the requirement that concepts should be sharply defined. In the review of Schoenflies Frege (1979: 179) glosses the requirement as saying: 'for every object it holds that it either falls under the concept or does not do so'.

true of just those objects that are the proxy objects associated with non-empty concepts. A natural suggestion for a Begriffsschrift expression that would express its sense perspicuously is therefore:

[4] $\exists f [\xi = \grave{\epsilon} f(\epsilon) \mathbin{\&} \exists x\, fx]$.

Here '$\grave{\epsilon} f(\epsilon)$' is the name of the extension of the concept *f*, and the predicate as a whole, informally speaking, expresses the property of being the extension of some realized concept.

It must be conceded that Frege nowhere suggests that the proxy object associated with a concept is its extension, and indeed he seems to have remained deliberately agnostic on that point.[6] All the same, there can be no doubt that in a Fregean setting it is at least extremely natural to equate the object proxying a concept with its extension. Such a decision seems to do no obvious violence to Frege's views. At any rate, in order to develop the proxy object solution, I shall suppose that the expression 'the concept *horse*' may regarded as adequately expressed in a Begriffsschrift by the singular term: '$\grave{\epsilon}$ (*horse* ϵ).' The success of the proxy object solution will depend on whether we can suppose that the first-level predicate 'ξ is a concept' is itself expressible in a Begriffsschrift. If the answer should turn out to be 'no', it will be unclear with what right we take ourselves to have expressed a genuine truth in uttering the words 'the concept *horse* is a concept'.

Using as our model the Begriffsschrift rendering of the first-level predicate that goes proxy for the second-level predicate of 'being realized', we arrive at the following idea. Why not take the second-level predicate that we unsuccessfully[7] try to express in English with the words 'Φ is a first-level concept of one argument' to be proxied by the first-level predicate expressing the concept of being the value range of a first-level concept of one argument. In symbols:

[5] $\exists f [\xi = \grave{\epsilon} f(\epsilon) \mathbin{\&} C_x(f_x)]$

Here '$C_x(\varphi_x)$' is some Begriffsschrift expression for the proxied second-level concept, and contains an argument place open to first-level functional expressions of one argument. It does not matter that the concept to be proxied forms a part of the complex concept that proxies it. Our goal is not to eliminate the predicate that expresses the proxied concept, but only to earn the right to suppose that in natural language an assertion about an object could serve to effect talk about the proxied concept. The idea would be that the sentence '$\exists f [\grave{\epsilon}\ (\text{horse}\ (\epsilon)) = \grave{\epsilon} f(\epsilon) \mathbin{\&} C_x(f_x)]$' expresses the very same thought as '$C_x(\text{horse}_x)$' despite its radically different logical articulation.

This solution would be acceptable only if '$C_x(\varphi_x)$' could itself could be framed in the vocabulary of a Begriffsschrift. But *can* such a predicate be framed? It is often said that

[6] A detailed discussion of this point is contained in chapter 3 of Sullivan (1989).

[7] The reasons why this predicate is unsuccessful are discussed below.

Frege's Begriffsschrift contains no symbols corresponding to the English idioms 'is a function' or 'is a concept'.[8] That observation, for what it's worth, is correct, since the Begriffsschrift obviously does not provide for the expression of anything corresponding to such would-be *type-neutral* expressions. But the more interesting question is whether Frege would have taken the type-specific concept of 'being a first-level concept of one argument' to be expressible in a Begriffsschrift. And the answer to this question is that, at one stage, he would.

In his letter to Russell of the 29 June 1902 Frege introduces a Begriffsschrift symbol for a primitive second-level function that holds only of first-level functions of one argument, namely, the so-called rough-breathing function '$\grave{\epsilon}\varphi(\epsilon)$'.[9] After acknowledging the unavoidable inadequacy—or as he says 'imprecision'—of the *natural-language* expression '... is a function',[10] Frege continues:

> In a conceptual notation we can introduce a precise expression for what we mean when we call something a function (of the first level with one argument), e.g.:
> '$\grave{\epsilon}\ \varphi\ (\epsilon)$'. Accordingly, '$\grave{\epsilon}\ (\epsilon\ .\ 3 + 4)$' would express precisely what is expressed imprecisely in the proposition '$\zeta\ .\ 3 + 4$ is a function'. (1980: 136)

It is straightforward along these lines to frame a Begriffsschrift symbol for the concept of being a first-level concept of one argument. One simply frames a symbol for being a first-level *function* of one argument whose value for any object as argument is a truth-value, thus:

[6] $\grave{\epsilon}\varphi(\epsilon)\ \&\ \forall x\ (\varphi x = \text{——}\varphi x)$

The predicate [6] succeeds in expressing the concept we unsuccessfully try to express in natural language with the words 'φ is a first-level function of one argument whose value for any argument is a truth-value'. The latter is a failed formulation of a second-level predicate because there is no satisfactory way of generating an instance. Suppose, for example, that we tried to use such a predicate to say that 'plus two' is a first-level function of one argument. We will find that replacing 'φ' with either of the obvious

[8] See Weiner (1990: 259) and Conant (2000: 181).

[9] Notice that the diacritic here is a Greek rough breathing, which faces in the opposite direction to the smooth breathing familiar from its occurrence in Frege's terms for value ranges. This use of the rough breathing sign is not to be confused with another that occurs in Frege's correspondence with Russell. In his letter to Russell of 13 November 1904, for the purposes of discussing Russell's proposal that names of functions should be regarded as interchangeable with names of objects, Frege uses '$\grave{\epsilon}\varphi(\epsilon)$' for a definite description purporting to refer to a function. Here the rough-breathing sign is treated as combining with a predicate to form a definite description. Thus, for example, '$\grave{\epsilon}(\epsilon + 1)1$' would be read as 'the result of applying to the number one that function whose value for any argument ϵ is $\epsilon + 1$' (see Frege 1980: 160–2).

[10] The context makes clear that when Frege says an imprecise expression is 'unavoidable' he just means 'unavoidable in language', not 'unavoidable in a Begriffsschrift', and it is equally clear that by 'language' here he means 'natural language'. A typical passage in which Frege plainly uses 'language' to mean 'natural language' occurs in his article 'Negation': 'The criterion for distinguishing between a negative and an affirmative judgment... cannot be derived from language; *for languages are unreliable on logical questions*. It is indeed not the least of the logician's tasks to indicate the pitfalls laid by language in the way of the thinker' (1918: 381, emphasis added).

candidate substituends—viz., '+ 2' and 'ζ + 2'—will yield only nonsense.[11] The Begriffsschrift predicate [6], on the other hand, does not face this problem because it carries with it variables to fill the argument place of any first-level functional expression that occupies its argument place. It turns out, then, that by Frege's lights, at least, the predicate [5] *can* be expressed in Begriffsschrift and Parsons's proposal must therefore be taken seriously. Nonetheless, it faces several difficulties, presented here in order of increasing acuteness.

First, although the exercise we have just gone through shows that Begriffsschrift is not a filter that will consign the natural-language sentence 'the concept *F* is a concept' to the bin of nonsense, nonetheless what we have come up with as its rendering in Begriffsschrift involves a wholly redundant functional expression. Even to grasp the predicate '$\forall x$ (φx =—φx)' we must already understand that 'φ' has the feature we express using the predicate '$\acute{\epsilon}\varphi(\epsilon)$'; so [6] could be replaced simply by '$\forall x$ (φx =—φx)' on its own. The expression '$\acute{\epsilon}\varphi(\epsilon)$', in short, is a mere redundancy: it could never do anything for you in a proof.

The second difficulty is that the proxy predicate, once rendered in Begriffsschrift, turns out to be true of a lot of things we didn't want it to be true of. The expression '...is a concept' is treated as meaning '...is the value range of a first-level concept of one argument', and—given Frege's definition of number in *The Basic Laws of Arithmetic*—this holds, among other things, of the number 1. So numbers turn out to be concepts (or, if we refuse to acquiesce in the home language, we might say that the proxy predicate turns out to hold of numbers), which is precisely not a result Frege desires.

Third, and finally, if the proxy solution solves the narrow custom-breach problem, it does so only at the cost of sharpening the general threat to the exclusivity thesis. For now it comes out true both that the concept horse is an object and that it is a concept.

In light of these problems it seems worth trying another tack.

1.2 *A second attempt at a solution*

A natural line of response to the breach of custom problem would involve pressing the question whether 'the concept *horse* is a concept' is really an instance of the schema 'The *K T* is a *K*'. If not, then in affirming that the concept *horse* is not a concept Frege will not after all be committing a breach of custom. What, then, is Frege's view of the structure of this sentence? Since he takes the parallel with 'The volcano Vesuvius is a volcano' seriously, one must suppose that he sees the phrase 'The concept *horse*' as involving two noun phrases in apposition, namely: 'The concept' and '*horse*'. And one must suppose that Frege takes the italicization of '*horse*' to signal that it functions as a proper name, presumably a name of the same object that is named by the expression 'The concept *horse*'. But this reflection is already enough to show that we do not really

[11] Cf. '[A] function-name can never occupy the place of a proper name, because it carries with it empty places that answer to the unsaturatedness of the function' (1893: §21).

have something of the form: 'The *K T* is a *K*', *where T is a member of the kind K*. For objects do not belong to the kind 'concepts'. This seems like a promising thing to say about the breach of custom problem. But, even if it should prove successful, matters cannot be left there, for Frege's criterion creates deeper problems of which the one currently under consideration is just a symptom.

2 Second problem: self-stultification

Consider the claim: 'The concept *horse* is unsaturated'. This is just the sort of utterance to which Frege might resort in attempting to convey his idea that concepts differ fundamentally from objects in having a kind of gappiness. But, by his own lights, it must be viewed as false, since the entity referred to by 'The concept *horse*' is an object and so *not* unsaturated. So Frege is not out of the woods yet. Although he doesn't raise it explicitly himself, he faces a wider problem, illustrated by this claim, of which the custom-breach problem is just an instance: the form in which we attempt to refer to concepts sometimes makes what we want to say about them false. Let's call this more general difficulty 'the problem of self-stultification'.

2.1 The 'exception' solution

The next solution to be considered attempts to address this broader problem. It begins with the observation that when, in 'On Concept and Object', Frege discusses his criterion for being a proper name he acknowledges at least one class of exceptions to it. He recognizes that when they occur in generic constructions definite descriptions cannot be taken to refer to objects. So, for example, in the sentence 'The horse is a four-legged animal'—used with its ordinary sense—the phrase 'the horse' cannot with any plausibility be regarded as a singular term referring to an individual horse. Recognizing this problem, Frege in effect proposes that generic occurrences of definite descriptions be treated as what Russell would later call 'incomplete symbols': that is to say, expressions that have no meaning in isolation but which are paraphrastically eliminable from every sentential context in which they occur. In this spirit 'The horse is a four-legged animal' is to be taken to mean: 'All properly constituted horses are four-legged animals' (1892b: 185).

Frege claims that counter-examples to his criterion for proper name-hood are so easily recognized as special cases that the value of that criterion is not impaired. He is right, I think, to suppose that we are good at recognizing generic uses of definite descriptions, even if, at the same time, he is wrong in thinking that his analysis of generics will work across the board. (His treatment will not work, for example, in familiar problem cases such as 'The duck-billed platypus lays eggs'.) But because Frege fails to consider any possible exceptions beyond generics,[12] we cannot suppose him to

[12] I read Frege's (1892b: 185) example of 'The Turk besieged Vienna' as one to which he *does* take the criterion to apply, since 'The Turk' in Frege's view is the proper name of a people.

have successfully stated even a sufficient condition for being a proper name. And of course, there are several other apparent exceptions that do need to be considered, for example, cases in which a definite description gets its meaning from an adverb of quantification, e.g. 'The owner of a Porsche is often smug'.[13]

That said, it may be that the best way to view Frege's criterion is as making a point about the onus of proof: we might take him to be saying that, in the absence of powerful reasons for thinking that a phrase of the form 'the *F*' is not functioning as a singular term, it ought to be treated as one.

If that is how Frege is thinking, then the question immediately arises why he should not have sought to address the breach of custom problem by treating the appearance of a breach as itself evidence that 'the concept *horse*' is functioning in the relevant context not as a proper name but rather as an incomplete symbol. The answer, I take it, is that Frege must have realized that we cannot in fact treat expressions of the form 'the concept *F*' as everywhere eliminable by paraphrase. For while there are many contexts from which this idiom can be eliminated, the sentences 'The concept *horse* is unsaturated' and 'The concept *horse* is not an object' are not among them.

Geach once suggested that we ought to write off such recalcitrant contexts as so much 'philosopher's nonsense,' on a par with 'The concept *man* is timeless' (1981: 221, 229–30). The suggestion is appealing, if rather draconian: it would counsel throwing overboard, along with the obvious junk, certain sentences that seem to have a serious use in semantics, among them: 'What "is a horse" stands for is the *concept* horse'. Against this, Geach might reply that what is threatened by his proposal is not so much semantics per se, as one way of doing it. For one might reject these formulations as nonsensical while claiming that the semantic truths they are the misfired attempts to convey are properly expressed, in Davidsonian fashion, by such claims as: 'For all x, x satisfies the predicate "y is a horse" iff x is a horse.' The solution, however, comes at a cost. At the end of the day, it seems just too high-handed to consign whole realms of discourse to the litter-bin of nonsense just to avoid the concept *horse* problem. So we must conclude that a wholly satisfying, broadly Fregean solution to the self-stultification problem remains elusive.

3 Third problem: the frustration of referential intentions

The third member of the family of difficulties associated with the concept *horse* problem is seldom properly distinguished from the problem of self-stultification. This problem, although it is illustrated by statements about whose truth-value we may be unsure, such as 'The concept *F* is not a concept,' and 'The concept *F* is unsaturated', is not confined to them. It is, instead, a general problem for any statement containing an

[13] For an interesting discussion of this case, see Fara (2001).

expression of the form 'the concept *F*'. Frege presents this more general problem in a celebrated passage occurring toward the close of 'On Concept and Object'. He says:

> ... with the use of my terms ['concept'] and ['object'] I have got hold of a distinction of the highest importance. I admit that there is a quite peculiar obstacle in the way of an understanding with my reader. By a kind of necessity of language, my expressions, taken literally, sometimes miss my thought, in that [*indem*] I mention an object, when what I intend is a concept. I fully realize that in such cases I was relying upon a reader who would be ready to meet me half-way—who does not begrudge a pinch of salt. (1892b: 193)

There is, of course, room for doubt about whether Frege is really entitled to claim that he has successfully latched on to the distinction between concepts and objects. But there can be no doubt that he *believes* he has done so. The difficulty he describes here does not involve grasping that distinction so much as communicating an understanding of certain things—*which* things being not immediately clear. That point is plain enough, but it is a rather more subtle matter to identify the kind of obstacle to understanding that Frege has in mind. It has seemed to some commentators that he may be suggesting that there are inexpressible thoughts which would give expression to the distinction between concept and object if, *per impossibile*, they could be expressed.[14] But such an interpretation is undermined by a careful reading of the passage—a fact that becomes apparent when one attends to Frege's account of what precisely the phenomenon of his words' 'missing' his thought consists in. My words, he says, miss my thought *in that* I mention an object when what I intend (to mention) is a concept. Frege's all-important explanation is obscured by the standard translation of this passage, which suppresses the German word *indem*. But when that nuance is restored it becomes clear that the obstacle Frege sees as standing in the way of an understanding with his reader *consists in* the fact that he fails to refer to what he intends to refer to.

Two points about this 'obstacle to understanding' require emphasis. First, and most obviously, the obstacle Frege describes is specific to sentences containing singular terms. So Frege is *not* discussing difficulties that attend general sentences, not even a sentence that tends nonetheless to be treated by commentators as a possible focus of this passage, viz: 'no concept is an object'. Second, the obstacle is present even in those cases in which a thought *is* expressed. It is present, for example, in the sentence 'The concept *horse* is realized' even though this sentence, expressing as it does the same thought as 'There are horses', is, by Frege's lights, uncontroversially significant. Oddly, then, my words can 'miss my thought' in the precise sense explained in the pinch of salt

[14] One commentator who has flirted with such a reading of the 'pinch of salt' remark—insofar as he describes it as a 'possible reading'—is James Conant. After quoting that remark, Conant (2000: 188) continues: '[Frege's] words miss his thought (and end up being nonsense); so there is a thought they are aiming at: an understanding of what his words intend to say depends upon his reader latching onto the thought his words fail properly to express. This failure is due, according to Frege, to "a kind of necessity of language". If he is to convey the thought he here seeks to convey he has no alternative but to have recourse to (elucidatory) nonsense'. I shall argue, to the contrary, first, that for Frege elucidations are *not* nonsense, and, second, that the expressions he is referring to here are not, by his lights, nonsense.

passage in spite of the fact that in the ordinary sense they *succeed* in expressing the thought intended. It appears that Frege must be using the phrase 'miss my thought' in a rather idiosyncratic way. He cannot mean *in general* that when the obstacle he mentions is present his words fail to express his intended thought. Still less can he mean that they are nonsense. What he does mean, surely, is just that the thought expressed by a sentence containing the phrase 'the concept *F*' is expressed misleadingly because the sentence expressing it creates the misleading impression that the subject of discourse is a concept when it is really an object. In requesting his pinch of salt in the above passage Frege is just asking leave to say things about concepts by mentioning objects. He is not asking us somehow to latch on to absolutely inexpressible thoughts.

In 'On Concept and Object' this problem, which, in order to do justice to its generality, I shall refer to as 'the frustration of referential intentions problem', is presented as inevitable. But in his unpublished paper 'Comments on Sense and Meaning', which is either contemporary with or—more likely—post-dates, 'On Concept and Object',[15] Frege no longer presents it this way. He writes:

> [A concept being essentially unsaturated] there is now a great obstacle in the way of expressing ourselves correctly and making ourselves understood. If we want to speak of a concept, language, with an *almost* irresistible force, compels me to use an inappropriate expression which obscures—I might *almost* say falsifies—the thought. (1979: 119, emphases added)

The obscuring tendency of language (i.e. natural language)[16] is here presented as powerful but not *wholly* irresistible. Frege neither acknowledges nor explains this softening in his stance, but it may well be connected with his willingness in these comments to take seriously the possibility that natural language might after all contain expressions that are capable of referring to a concept while occupying subject position. For having remarked that we ought to outlaw the expression 'the meaning of the concept-word *A*,' he continues:

> It would be better to confine ourselves to saying 'what the concept word *A* means,' for this at any rate is to be used predicatively: [as in] 'Jesus is what the concept word "man" means', [which has] the sense of 'Jesus is a man'. (1979: 122)

Geach once proposed (1973: 156) that this suggestion—which constitutes a candidate solution to the failure of referential intentions problem—should be used to provide a solution to the self-stultification problem (though he didn't call it that). The idea

[15] The exact date of the comments is unknown, but we know they must have been written after 1892, the date of publication of 'On Sense and Meaning'. On 'Concept and Object' was also published in 1892.

[16] Notice that here, in speaking of an obstacle in the way of making oneself understood, Frege's considered view is that language is *obscuring* rather than *falsifying* (if one might *almost* say that *p*, then one can't quite say that *p*). This suggests that those cases in which the obstacle arises are principally those in which a true thought is expressed, albeit in a potentially misleading way. And these, I take it are such cases as 'The concept *horse* is realized' and 'The concept *mammal* is wider than the concept *whale*'. I take this as some evidence supporting a similar reading of the obstacle mentioned in 'On Concept and Object'.

would be to treat the thought we attempt to express when we utter the sentence 'The concept *horse* is a concept' as properly expressed by the sentence:

[7] What the concept word 'horse' means is a concept,

where here the predicate '. . . is a concept' is understood to refer to a concept of *second* level.

Were this proposal to prove successful, it could double as a solution to the custom-breach problem, for then the sentence 'The concept *horse* is a concept' could be regarded merely an abbreviation of its form-revealing paraphrase [7], and so could be treated as straightforwardly true.

But the proposal, unfortunately, is not persuasive. It relies on our being able to hear the phrase 'what *X* means' as meaning something other than 'the meaning of *X*'. And that is not an easy thing to do. Frege, by alleging the meaningfulness of

[8] Jesus is what the concept word 'man' means,[17]

tries to make it seem as if the phrase following the word 'is' in [8] occurs both in the role of a predicate and, at the same time, as the subject of an assertion. For in [8] the phrase 'Jesus is' might be taken to express a second-level predicate true of just those concepts that hold of Jesus, which is here predicated of the first-level concept expressed by 'what the concept word "man" means'. But in practice it is hard to hear [8] as making sense, unless it is construed—against Frege's intentions—as meaning 'Jesus falls under the concept *man*.'

It is unclear whether Geach's idea ever occurred to Frege, but even if it did, it cannot have been his considered view. For, when discussing the inappropriateness of the expressions 'the concept' and 'the function' in an unpublished article of 1914, Frege reverts to the unqualified claim that 'Language forces an inappropriate expression on us' (1979: 239). And before that, in 1906, he had said that the *only* way language can fit the designation for a concept into its grammatical structure is as a proper name (1979: 177). But although, on reflection, the problem came to seem inevitable to Frege, it is equally clearly something he thought we could learn to live with. The passage from 1914 continues: 'This is a situation which, unfortunately, can hardly be avoided, *but we can render it harmless* by always bearing the inappropriateness of language in mind' (1979: 239, emphasis added). The inappropriateness of language, then, poses no more of a threat to the philosophy of mathematics than does the moon illusion to astronomy. So long as we bear the predicament of language in mind, we are no more likely to be misled in our *judgments* in the former case than we are in the latter—or so Frege seems to suppose.

[17] I have removed the comma the translators leave in place, since it is required in German but not in English. Cf. 'Mann ist, was er isst.'

4 Fourth problem: the inexpressibility of logical category distinctions

Frege's difficulties over the concept *horse* are often associated with a worry about our inability to put logical category distinctions into words. The connection is usually established as follows: One observes that if the expression 'the concept *horse*' is a proper name, then the predicate 'is a concept' must be a first-level predicate true or false of objects. But—if one rejects the proxy object solution—it follows that this predicate must be false of anything of which it can be significantly predicated. And then one realizes that there are claims one might wish to make using this predicate, for example, that no first-level concept is an object, which cannot be made. Or rather, one realizes that, although such claims can be made, they fail to do the kind of work we wanted them to do. The claim that no first-level concept is an object for example, comes out as *vacuously* true if analyzed as meaning, 'for any object, if it is a first-level concept then it is not an object'. But someone who utters the sentence 'no first-level concept is an object' typically intends to express some non-vacuous truth; so this proposition cannot be what is intended; and we can know this even though, obviously, we are unable to cite any particular proposition as the one that is intended.[18]

Frege does not himself explicitly discuss the claim, 'no first-level concept is an object', but his writings abound with remarks that face a similar predicament. In 'On Concept and Object', for example, he says:

> [9] Second-level concepts, which concepts fall under, are essentially different from first-level concepts, which objects fall under. (1892b: 190)

The intended contrast can be spelled out more precisely in the claim:

> [10] Any first-level concept, but no second-level concept, is such that any object either falls under that concept or under its negation.

Frege's remarks in §21 of the *Basic Laws of Arithmetic* suggest another example of this broad kind, one which attempts to distinguish a different pair of logical categories:

> [11] Objects are fundamentally different from first-level functions because any object is saturated but no [first-level] function is.[19]

Each of these claims aims to ground a distinction between logical categories in a difference between their properties. The canonical criticism of such an ambition is found in the *Tractatus*:

[18] The steps just rehearsed are essentially those followed by Weiner (1990: 254–5).

[19] Frege actually says: '*Functions of two arguments* are just as fundamentally different from *functions of one argument* as the latter are from objects. For whereas objects are wholly *saturated*, functions of two arguments are saturated to a lesser degree than functions of one argument, which too are already *unsaturated*' (1893: §21). Cf. '[functions] may be called "unsaturated", and in this way we mark them out as fundamentally different from numbers' (1904: 292).

One cannot distinguish forms from one another by saying that one has this property, the other that: for this assumes that there is sense in asserting either property of either form. (4.1241)

Here the word 'form' operates as a synonym for 'type'. Wittgenstein is drawing attention to a consequence of the thesis he advances in the *Notes Dictated to Moore* when he says that 'a THEORY *of types* is impossible' (NM 109). The consequence is that one cannot theorize about types by using (the contrapositive of) the indiscernibility of identicals.

The argument turns on the idea that any predicative symbol—as the *Tractatus* puts it—'contains the prototype of its argument' (3.333). In other words, predicative symbols cannot be identified and individuated without paying attention to the logical type of their argument places. Taking that idea seriously, we find that each of the sentences [9] to [11] involves an equivocation. In [11], for example, what is denied of first-level functions is not the same as what is asserted of objects. It therefore *fails* to say that objects have a property that first-level functions lack—or rather, strictly speaking, although it does perhaps say *something*, there is nothing it tries but fails to say. The words in [11] do not miss a thought as an arrow might miss its target. Instead, there is only the illusion of a thought at which one who utters [11] imagines himself to aim.

Properly speaking, of course, *Tractatus* 4.1241 is only the misfired attempt to amalgamate all such particular criticisms into one general argument.[20] It attempts to say[21] that there is no establishing the distinctness of any two entities at different levels of Frege's hierarchy by citing a property that one of them has but that the other lacks. But by the lights of the *Tractatus* this very claim is itself nonsensical, since the phrase beginning 'any two entities' attempts to quantify across types. So here is one place at which the reader of the *Tractatus* finds himself standing on a ladder that must, sooner or later, be kicked away.

The reasoning of *Tractatus* 4.1241—which for convenience we shall call 'the forms argument'—grants for the sake of argument the meaningfulness of such phrases as 'is saturated', 'is a first-level function of one argument', and 'is an object'. It does not show—and does not purport to show—that such claims as [9]–[11] are nonsense or ill-formed. The argument's conclusion is rather just that these sentences fail to draw the contrasts their authors intend—or rather, that they *imagine* they intend. One who utters them, if he is reflective, will not recognize them as expressions of any thought he intended, though he may come to regard them as expressing some other thought.

Although I take this (i.e. the inexpressibility of logical category distinctions) to be a deep problem, it can seem otherwise. It can appear to arise merely from an ill-judged adherence to Frege's criterion for proper names. If that criterion is abandoned, the

20 The alert reader will have noticed that even the description just offered of the ambitions of *Tractatus* 4.1241 is itself nonsensical, because there is no sense to be made of the generality involved in the phrase 'all such particular criticisms'.

21 Or, because the 'attempt' is made in the clear knowledge that it must fail, we might say that it 'makes as if to say' this.

problem—or one manifestation of it—can seem to be trivially resolvable. For if we allow that definite descriptions may, after all, refer to functions, then the contrastive claims we seek to express can seem straightforwardly expressible. Suppose we were to say, for example, that what we loosely refer to as 'the function $x + 2$' can be more perspicuously referred to by a definite description '$\lambda x.x + 2$' (read as 'that function whose value for any argument x is $x + 2$'). This singular term would then be able to fit in the argument place of the predicates 'ξ is an object' and 'ξ is a function', so that the claim 'Socrates is an object but $\lambda x.x + 2$ is not', for example, would make full sense. And there would be no difficulty in expressing this logical category distinction in words.

One limitation of this proposal is that it is unclear how it could be extended to deal with those functions that are concepts. The phrase 'the function whose value for any argument x is x is a horse' may make sense in Fregean syntax (as, indeed, '$2 + 2 = 4$ is a horse' does), but it makes no sense in English. A second, deeper, problem with the proposal is that it involves abandoning the Fregean hierarchy. For, if we no longer have syntactic criteria for recognizing an expression as a name of an object, it will be unclear how the types of expressions (and so of their referents) are to be fixed. Certainly, it will be of no use to say that types are fixed by the nature of the referent of an expression, for then the statement 'The concept *horse* is a function but Caesar is not a function', since its two occurrences of the predicate 'is a function' would have different types, would still fail to express the intended contrast. But if the hierarchy is dismantled, the paradox whose resolution it provides for—namely, the 'propositional functions' version of Russell's paradox—will be reinstated.[22]

Frege, as we have seen, makes remarks that are vulnerable to the criticism presented in *Tractatus* 4.1241. This may not amount to a criticism of *him*, for it may be that his remarks are made with a certain degree of self-consciousness. But just how self-conscious are they? If we go on the evidence of the texts, rather than assumptions about what Frege *must* have thought, the answer is: 'not very'. He does note that what is said of a concept can never be said of an object, and he says that the attempt to make such predications is 'impossible' and 'without sense' [*sinnlos*] (1892b: 189). But, surprisingly, he fails to infer from this that statements such as [9] and [11], to the extent that they make sense, fail to draw the contrasts they seem intended to draw. Nor does he explicitly register any difficulty with so starkly problematic a claim as:

> *Objects* stand opposed to functions. Accordingly I count as *objects* everything that is not a function. (1893: §2)

The trouble with this claim is that it aspires to a kind of generality that, according to Frege, does not exist. For, as we have already noted, Frege has no general conception

[22] Wittgenstein's resolution of that paradox, as commentators have observed, makes use of nothing beyond the simple Fregean hierarchy; see Sullivan (2000).

of 'being a function' but only one of 'being an *n*-level function of *m* arguments,' for particular *n* and *m*.

Might Frege's admission that at times he has to rely on 'hints' and 'elucidations' be evidence that he does register the problem of the inexpressibility of type distinctions? Again, a careful examination of the relevant passages suggests not. In 'What is a Function?' Frege says:

> The peculiarity of functional signs which we here called 'unsaturatedness', naturally has something answering to it in the functions themselves. They too may be called 'unsaturated', and in this way we mark them out as fundamentally different from numbers. Of course this is no definition; but likewise none is here possible. I must confine myself to hinting at what I have in mind by means of a metaphorical expression, and here I rely on my reader's agreeing to meet me half-way. (1904: 292)

There is no suggestion here that Frege's appeal to his reader to 'meet [him] half-way' is prompted by the belief that he has failed to draw the intended contrast. It is prompted, rather, by his acknowledgment that the characterization he employs to establish the contrast—a contrast he supposes he has *succeeded* in drawing—fails to amount to a *definition*. This characterization cannot qualify as a definition, Frege makes clear, on account of the *metaphorical* character of the 'hints' deployed. This is exactly what Frege *should* say, since metaphors according to him are expressions that lack sharp boundaries. (That is so, presumably because they have the sense of their corresponding similes, and one thing may be like another without there being any definiteness about *how much* like the other it is.[23]) Frege's frequent recourse to hints, no less than his requests for cooperative understanding, is prompted by the need to convey the meaning of a word for something simple (or primitive) without the use of a definition. Thus toward the beginning of 'On Concept and Object' he remarks: 'On the introduction of a name for something logically simple, a definition is not possible; there is nothing for it but to lead the reader or hearer, by means of hints, to understand the word as is intended' (1892b: 183). And in this essay too, the impossibility of framing a definition is attributed to the need to resort to *metaphorical* hints: '"Complete" and "unsaturated" are of course only figures of speech; but all that I wish or am able to do here is to give hints' (1892b: 194).

There is no evidence here, then, that Frege regards hints as what we come up with when we attempt get across *absolutely* inexpressible logical category distinctions. They are rather presented as something we must resort to when attempting to convey the meaning of primitives, and Frege takes it that these meanings *can* be conveyed—so long as the reader is sufficiently cooperative.

Precisely parallel observations may be made in connection with Frege's discussion of the role played by 'elucidations' in his article 'The Foundations of Geometry II'. There elucidation is recommended as an appropriate way of bringing it about that

[23] I am grateful to Tom Ricketts for this point.

investigators attach a shared meaning to a term which, being logically primitive, is incapable of being explained through a definition:

> Once the investigators have come to an understanding about the primitive elements and their designations, agreement about what is logically composite is easily reached by means of definition. Since definitions are not possible for primitive elements, something else must enter in. I call it elucidation. It is this, therefore, that serves the purpose of mutual understanding among investigators, as well as the communication of the science to others. (1906: 300)

He goes on to say we may need to rely on a little goodwill and cooperative understanding because 'frequently we cannot do without a figurative mode of expression' (1906: 301).

Elucidations are, in general, supposed to contain the primitive terms to be explained. So if Frege is not to be committed—in violation of the context principle—to nonsensical sentences that contain meaningful terms, he cannot have regarded elucidations as nonsensical. He must have regarded them, instead, as significant explanations that lack *scientific* standing because, being metaphorical, they lack the precision of definitions. This conclusion receives some confirmation both from Frege's remark in the 'Foundations of Geometry II', that 'we can demand from the originator of an elucidation that he himself know for certain what he means' (1906: 301), and from his remark in a letter to Hilbert that, 'Unlike definitions, . . . elucidatory propositions cannot be used in proofs *because they lack the necessary precision*' (1980: 37, emphasis added). Although elucidations are non-scientific, there is no suggestion in Frege that they are the misfired attempts to put into words distinctions that cannot be drawn.

All in all, then, it seems that neither Frege's remarks about hints nor what he has to say about elucidations provides any evidence of self-consciousness on his part about the failure of claims such as [9] and [11] to draw the contrasts at which they aim.

I conclude that Geach overstates his case when he claims that Wittgenstein 'accept[ed] from Frege' the view that 'there are logical category-distinctions which will clearly show themselves in a well-constructed formalized language, but which cannot properly be asserted in language' (1976: 55–56). In his published works, Frege had not gone so far as even to raise the problem, let alone provide for its solution by appealing to the say–show distinction. Indeed, a more likely source of the showing idea—or one aspect of it—is Russell's remark in the *Principles of Mathematics* that 'to mention anything is to show that it is' (1903: §427). That idea is, after all, closely related to the idea of *Tractatus* 4.1211, that a proposition '*fa*' shows that the object *a* occurs in its sense.

To be sure, Frege did eventually come to have qualms about the natural language words 'function' and 'concept'. In his letter to Russell of 28 July 1902 he recommends their rejection on the grounds that 'they present themselves linguistically as names of first-level functions' whereas 'logically they should be names of second-level functions' (1980: 141). And in his draft review of Schoenflies, from 1906, he rejects the word 'concept' as 'defective' on the ground that 'the phrase "is a concept" requires a proper name as a grammatical subject; and so, strictly speaking, it requires something contradictory, since no proper name can designate a concept; or perhaps better still, something nonsensical' (1979: 177–8).[24] But it is important to keep in mind—and especially so when considering Frege's possible influence on Wittgenstein—that remarks of this kind occur only in Frege's unpublished writings—and so in writings to which Wittgenstein is unlikely to have had access.[25]

There is, then, scant reason to suppose that Wittgenstein would have *derived* his views on the inexpressible directly from Frege. That being so, it is nonetheless perfectly possible that *reflection* on certain remarks in Frege's writings—statements [9] and [11], among them—would have led him to these ideas. Indeed, in the light of Wittgenstein's comments in the Preface to the *Tractatus*, acknowledging his indebtedness to the 'great works' of Frege for the stimulation of his thoughts, such an hypothesis is plausible. At any rate, in the course of our discussion we have seen one route by which reflecting more thoroughly on the more general problems that receive only narrow solutions in Frege can lead in the direction of Wittgenstein's thought.

5 Conclusion

We have distinguished four sub-problems that might reasonably be taken to fall under the general rubric 'Frege's concept *horse* problem'. It has been argued that the first—'the custom-breach problem'—is plausibly soluble, but that this is a local and shallow solution, since a second, deeper and more general problem—the problem of 'self-stultification'—remains unsolved. The third problem, concerning 'the frustration of referential intentions', is one about which Frege's most considered position is just that it is an inevitable, yet harmless, awkwardness of natural language. The last problem—that of the inexpressibility of logical category distinctions—is a deep and deeply intractable problem. It is not one, however, to which Frege paid much, if any, attention. Although this problem is intractable, the related problem of how strictly inexpressible logical category distinctions may nonetheless somehow be indirectly communicated or got across—and of what that achievement consists in—is one on which some progress has been made in recent work.[26] A discussion of that progress will, however, have to wait for a future occasion.[27]

[24] Frege's meaning is clear enough: in order for the sentence 'the concept *F* is a concept' to be true what is required is that the expression 'the concept *F*' should be the proper name of a concept, but a concept is just the kind of thing that cannot be given a proper name, so the condition the predicate attempts to impose is contradictory. Frege's afterthought is that it is perhaps better to say that what is required is something nonsensical because we can't really make sense of the idea of a symbol designating a concept in the manner of a proper name: after all, a proper name just is something that, if it designates at all, designates an object.

[25] It is possible, of course, that Russell might have mentioned these views to him, or that Frege might have communicated them to him directly in conversation, but conjectures of this kind would be a slim reed on which to hang Geach's interpretation.

[26] Notably by Adrian Moore (2003).

[27] Thanks to Imogen Dickie, Warren Goldfarb, Bernard Katz, Boris Kment, Martin Lin, Ray Monk, Michael Potter, Thomas Ricketts, Göran Sundholm, and Jessica Wilson. I owe a special debt of gratitude to Peter Sullivan for his advice and encouragement. Versions of this chapter were presented at the University of

References

Anscombe, G. E. M. (1959) *An Introduction to Wittgenstein's Tractatus*. London: Hutchinson University Library.

Conant, James (2000) 'Elucidation and Nonsense in Frege and the Early Wittgenstein', in A. Crary and R. Read, eds, *The New Wittgenstein*. London and New York: Routledge, 174–217.

Fara, Delia Graff (2001) 'Descriptions as Predicates', *Philosophical Studies* 102, 1–42. (Originally published under the name 'Delia Graff'.)

Frege, Gottlob (1884) *Die Grundlagen der Arithmetik*. Trans. J. L. Austin as *The Foundations of Arithmetic*. Evanston, Ill.: Northwestern University Press, 2nd revised edition 1978.

——(1892a) 'On Sense and Meaning'. Reprinted in Frege (1984), 157–77.

——(1892b) 'On Concept and Object'. Reprinted in Frege (1984), 182–94.

——(1893) *Grundgesetze der Arithmetik*, vol. 1. Part trans. M. Furth as *The Basic Laws of Arithmetic*. Berkeley and Los Angeles: University of California Press, 1964.

——(1904) 'What is a Function?'. Reprinted in Frege (1984), 285–92.

——(1906) 'On the Foundations of Geometry: Second Series'. Reprinted in Frege (1984), 293–340.

——(1918) 'Negation'. Reprinted in Frege (1984), 373–89.

——(1979) *Posthumous Writings*, ed. Hans Hermes *et al.*, trans. P. Long and R. White. Oxford: Basil Blackwell.

——(1980) *Philosophical and Mathematical Correspondence*, ed. G. Gabriel *et al.*, trans. H. Kaal. Chicago: University of Chicago Press.

——(1984) *Collected Papers on Mathematics, Logic, and Philosophy*, ed. B. F. McGuinness. Oxford: Basil Blackwell.

Geach, Peter (1973) 'Frege', in G. E. M. Anscombe and P. T. Geach, *Three Philosophers*. Oxford: Basil Blackwell.

——(1976) 'Saying and Showing in Frege and Wittgenstein', in J. Hintikka *et al.*, eds, *Essays in Honour of G. H. von Wright, Acta Philosophica Fennica* 28, 54–70.

——(1981) *Logic Matters*. Blackwell: Oxford.

Kerry, Benno (1887) 'Über Anschauung und ihre psychische Verarbeitung', Part 4, *Vierteljahrsschrift für Wissenschaftliche Philosophie* 11, 249–307.

Moore, A. W. (2003) 'Ineffability and Nonsense', *Proceedings of the Aristotelian Society*, suppl. vol. 77, 169–93.

Parsons, Terence (1986) 'Why Frege should not have said "The concept *horse* is not a concept"', *History of Philosophy Quarterly* 2, 449–65.

Russell, Bertrand (1903) *The Principles of Mathematics*. Cambridge: Cambridge University Press.

Sullivan, Peter (1989) *Formal Concepts*, Oxford D.Phil. thesis.

——(2000) 'The Totality of Facts', *Proceedings of the Aristotelian Society* 100, 175–92.

Van Heijenoort, Jean (1977) 'Frege on Sense Identity', *Journal of Philosophical Logic* 6, 103–8.

Weiner, Joan (1990) *Frege in Perspective*. Ithaca, NY: Cornell University Press.

Stirling, the University of Toronto, the University of Southampton, and University of Utrecht. I am grateful to these audiences for many stimulating comments and criticisms.

Wittgenstein, Ludwig (1914) 'Notes Dictated to G. E. Moore in Norway' (NM), in *Notebooks 1914–1916*, 2nd edition, ed. G. H. von Wright and G. E. M. Anscombe, trans. G. E. M. Anscombe. Oxford: Basil Blackwell, 1979.

——(1922) *Logisch–Philosophische Abhandlung/Tractatus Logico–Philosophicus*, trans. C. K. Ogden, London: Kegan Paul Ltd, 2nd edition 1933; trans. D. F. Pears and B. F. McGuinness, London: Routledge & Kegan Paul Ltd, 1961, 2nd edition 1974.

6

Tractatus 5.4611: 'Signs for logical operations are punctuation marks'

Peter Milne

1 Introduction

Wittgenstein's gnomic and little discussed[1] pronouncement at *Tractatus* 5.4611[2] paints starkly—too starkly, as we shall see—a feature of the nature of the logical constants in a truth-conditional account of sentential meaning. Wittgenstein was sensitive, far more so than Frege and Russell, to the subtleties in providing an account of their meaning within a truth-conditional setting. Granted compositionality of meaning and the primacy of sentence meaning, one readily comes to the thought that the meaning of a subsentential expression is the contribution it makes to the meanings of the sentences in which it occurs. And in some sense, granted compositionality and the primacy of sentence meaning, that must be right; but what concerned Wittgenstein was the nature of that contribution in the case of a certain class of subsentential expressions, the logical constants, especially against the backdrop of his picture-theoretic version of a truth-conditional theory of meaning.

Wittgenstein's fundamental thought, as he tells us in the *Tractatus* (4.0312), in a sentence lifted straight from a Christmas Day 1914 notebook entry (NB 37), is that the logical constants are not representatives, are not proxies, do not stand for anything.[3]

[1] 'Punctuation' doesn't make it into the index of the Pears–McGuinness translation. Neither Stenius' fifty-year-old commentary on the *Tractatus* (Stenius 1960) nor Michael Morris's recent one (Morris 2008) so much as mentions 5.4611.

[2] Except where indicated, I follow the Pears–McGuinness translation throughout.

[3] 'Mein Grundgedanke ist, daß die "logischen Konstanten" nicht vertreten' (4.0312). Scare quotes apart, wording in both sources is identical. In the text, I have compromised between Anscombe's and Pears and McGuinness's Englishing. In the *Tractatus* Wittgenstein sometimes uses 'logical constants' for linguistic signs (which, on his view, is what they are), sometimes—especially when discussing Frege and Russell—for what they (putatively) stand for. Russell came to accept this view of logical constants. In the preface to the second edition of *The Principles of Mathematics* from 1937, he says (p. xi), 'Logical constants, therefore, if we are to be able to say anything about them, must be treated as part of the language, not as part of what the language speaks about. In this way, logic becomes much more linguistic than I believed it to be at the time when I wrote the "Principles".'

The groundwork for this key thought was laid at least as early as the 1913 'Notes on Logic'. In *Grundgesetze* Frege had taken declarative sentences to be names of truth-values and his negation- and conditional-strokes to stand for functions with truth-values as values, other connectives being analysed in terms of these; the peculiarities of Frege's system require the functions to allow all objects as arguments (Frege 1967: §§6 and 12). In *The Principles of Mathematics* Russell took (material) implication to be a relation holding between objects of all kinds, 'the relation in virtue of which it is possible for us validly to infer' (Russell 1903: 33).[4] In *Principia Mathematica* the logical constants are introduced as denoting functions taking propositions as arguments (Whitehead and Russell 1910: 6–7). In the 'Notes on Logic', and in contrast to all this, we find Wittgenstein saying:

> Frege said 'propositions are names'; Russell said 'propositions correspond to complexes'. Both are false; and especially false is the statement 'propositions are names of complexes'. (NL 97)

> Propositions are not names. (NL 98)

> The false assumption that propositions are names leads us to believe that there must be logical objects: for the meanings of the logical propositions will have to be such things. (NL 107)

The logical connectives cannot then stand for functions or relations, for they would have no arguments to act on or relata to relate. Hesitantly in 1913 Wittgenstein moves towards what was to be his fundamental thought:

> This is the first thing that indicates that there *may* not be logical constants. (NL 98, emphasis in the original)

He means this in the sense that there may be nothing the logical connectives stand for, as is made clear by the follow-on to the sentence just quoted:

> A reason against them is the generality of logic: logic cannot treat of a special set of things. (Ibid.)

In particular, if there is nothing logical constants (connectives) stand for, the contribution of a logical constant to the sense expressed by a sentence in which it occurs is not a component part of the truth conditions of the sentence, not a component part of the way the world must be in order for the sentence (or proposition it expresses) to be true.

> Whatever corresponds in reality to compound propositions must not be more than what corresponds to their several atomic propositions. . . .
>
> Molecular propositions contain nothing beyond what is contained in their atoms; they add no material information above that contained in their atoms. (NL 98)

There are no items in reality represented by the connectives that occur in a true compound proposition.

[4] On the *Principles of Mathematics* view, see Stevens (2003: §3). As Ian Proops observes, with a reference to 5.5351, *The Principles of Mathematics* is one of the few works of philosophy we know Wittgenstein to have read (Proops 2000: 84).

In the 'Notes on Logic' Wittgenstein has already grasped the essential point that will lead to the fundamental idea:

All that is essential about molecular functions is their T-F schema (i.e. the statement of the cases when they are true and the cases when they are false). (NL 98)

But since he has not yet drawn the crucial distinction between functions and operations—on which more below—he is forced into a dark doctrine:

In regard to notation, it is important to note that not every feature of a symbol symbolizes. In two molecular functions which have the same T-F schema, what symbolizes must be the same. In 'not-not-*p*', 'not-*p*' does not occur; for 'not-not-*p*' is the same as '*p*', and therefore, if 'not-*p*' occurred in 'not-not-*p*', it would occur in '*p*'. (NL 99)

All this is very much at odds with the view exemplified in this passage from §32 of the *Grundgesetze*:

Every such name of a truth-value *expresses* a sense, a *thought*. Namely, by our stipulations it is determined under what conditions the name denotes The True. The sense of this name—the *thought*—is the thought that these conditions are fulfilled. . . .

The names, whether simple or themselves composite, of which the name of a truth-value consists, contribute to the expression of the thought, and this contribution of the individual [component] is its *sense*. If a name is part of the name of a truth-value, then the sense of the former name is part of the thought expressed by the latter name. (Frege 1967: 89–90)

This is no isolated occurrence in Frege's thought. In the posthumously published 'Logic in Mathematics' from Spring 1914, we find this remarkable paragraph:

It is remarkable what language can achieve. With a few sounds and combinations of sounds it is capable of expressing a huge number of thoughts, and, in particular, thoughts which have not hitherto been grasped or expressed by any man. How can it achieve so much? By virtue of the fact that thoughts have parts out of which they are built up. And these parts, these building blocks, correspond to groups of sounds, out of which the sentence expressing the thought is built up, so that the construction of the sentence out of parts of a sentence corresponds to the construction of a thought out of parts of a thought. And as we take a thought to be the sense of a sentence, so we may call a part of a thought the sense of that part of the sentence which corresponds to it. (Frege 1979: 225)[5]

[5] Peter Hylton quotes an extract from this paragraph but (citing Baker and Hacker 1984) urges a little caution in ascribing to Frege the thought that the sense of a sentence is made up of the senses of its parts—'and made up of them almost in the way a wall is made up of bricks' (Hylton 1997: 100–1, n. 17, and 103). In the opening paragraph of 'Compound Thoughts', Frege does say, 'To be sure, we really talk figuratively when we transfer the relation of whole and part to thoughts; yet the analogy is so ready to hand and so generally appropriate that we are hardly even bothered by the hitches that occur from time to time' (Frege 1984: 390). Dummett knows no caution. Of what he labels *Thesis 1*, namely the thesis that 'a thought is isomorphic with the sentence whose sense it is', he says, 'Not only was Thesis 1 held by Frege: he regarded it as a key component of his theory' (Dummett 1989: 1 and 2).

A year earlier, in Frege's lectures for the course *Begriffsschrift II* in the Spring semester of 1913, the first entry in Carnap's notes reads:

> The *meanings* of the parts of a sentence are not parts of the meaning of the sentence. However: The sense of a part of the sentence is part of the sense of the sentence. (Frege 2004: 87)

Wittgenstein denies that what a logical constant contributes to the sense expressed by a sentence it occurs in is a *part* of the sense expressed by that sentence. As Peter Hylton emphasizes (1997: 103), although Wittgenstein uses the Fregean terms *Sinn* and *Bedeutung*, he uses them in a most unFregean way, so that the parts of a simple sentence have *Bedeutung* but not *Sinn*, whilst the sentence itself has *Sinn* but no *Bedeutung*. However we are to understand Wittgenstein's use of *Sinn* and *Bedeutung*, from the perspective of a truth-conditional theory of meaning Wittgenstein is right to deny that what a logical constant contributes to the sense expressed by a sentence it occurs in is a *part* of the sense expressed by that sentence.

In the *Tractatus* Wittgenstein's discussion is carried out within the framework of the picture theory of meaning developed there. If one does not attend closely to some of the details of that theory, it looks very much as though Wittgenstein will end up with a quite untenable account of the logical constants. Attending more closely, we see his route past the threatening incoherence. This route shows us what must be said about the meanings of the logical constants in a truth-conditional theory of meaning (always granted compositionality and the primacy of sentence meaning). This in turn teaches us something about how the notion *picture* is to be understood in the picture theory, and something about the kinds of ways in which things may be said to stand. It will also allow us to see in what way Wittgenstein goes too far when he asserts that signs for logical operations *are* punctuation marks.

2 Truth-conditionality and compositionality

To say that in knowing the meaning of a (declarative) sentence one knows how the world must be in order for the sentence to be true is, of course, in some sense correct. But 'how the world must be' is hardly a perspicuous phrase. Famously Wittgenstein propounded a seductive reading of 'knowing how the world must be' in the *Tractatus*:

> A proposition is a picture of reality, for if I understand a proposition, I know the situation that it represents. (4.021)

> A proposition *shows* how things stand, *if* it is true. (4.022)

> A proposition states something only in so far as it is a picture. (4.03)

> A proposition can be true or false only in virtue of being a picture of reality. (4.06)

In the *Tractatus* this picture theory is held in conjunction with two other theses, both of which are, as I interpret them here, almost commonplaces of analytic philosophy: that the meaning of a sentence is given by its truth-conditions and that meaning is compositional.

To understand a proposition means to know what is the case if it is true. . . .
It is understood by anyone who understands its constituents. (4.024)

A proposition is the expression of its truth-conditions. (4.431)

[I]n order to be able to say '"*p*" is true (or false)', I must have determined in what circumstances I call '*p*' true, and in so doing I determine the sense of the proposition. (4.063)

Compositionality of meaning is the point being made at 4.026, where words and sentences are contrasted:

The meaning of simple signs (words) must be explained to us if we are to understand them.
With propositions, however, we make ourselves understood.

The point is immediately amplified at 4.027:

It belongs to the essence of a proposition that it should be able to communicate a *new* sense to us.[6]

Evidence for *Tractarian* commitment to the Fregean point that the meaning of a subsentential expression is what it contributes to the meanings of the sentences in which it occurs can be drawn from what Wittgenstein says about expressions, symbols, and meaningful sentences, i.e. propositions.

Only propositions have sense; only in the nexus of a proposition does a name have a meaning. (3.3)

I call any part of a proposition that characterizes its sense an expression (or symbol). . . .
Everything essential to their sense that propositions can have in common with one another is an expression. (3.31)

An expression presupposes the forms of all the propositions in which it can occur. (3.311)

An expression has meaning only in a proposition. (3.314)

A sign is what can be perceived of a symbol. (3.32)

In order to recognize a symbol by its sign we must observe how it is used with sense. (3.326)

Expressions are picked out by their contributing to the senses of the sentences in which they occur. Moreover, as only propositions have sense, to observe how a symbol is used 'with sense' must be to observe how it is used in the propositions in which it occurs.

[6] Peter Hacker lists among the requirements a theory of the proposition must meet (and which the picture theory does meet), 'the apprehension of the generative powers of language, now a well-known thesis due to Chomsky, but clearly enunciated by Wittgenstein in 1913 (*Notebooks 1914–1916*, 98) and hammered home repeatedly in the *Tractatus* (4.02, 4.027)' (Hacker 1981: 87; cf. Proops 2000: 62, 67–8).

The focus of our attention here is Wittgenstein's account of the logical connectives. In the course of a very helpful analysis of *Tractatus* 5.4611 to which we shall return, Ian Proops says,

> Despite lacking their own sense or reference, the logical connectives do have a meaning of their own insofar as they make a *uniform* contribution to the meanings of the sentences in which they occur. (Proops 2000: 14)

Nothing so far threatens incoherence. To get to that point we must look more closely at what Wittgenstein says about pictures. We must also defend the reading of 'proposition' as meaningful sentence. We turn to this first.

3 Propositions are meaningful sentences

Problematic for the identification of propositions with meaningful sentences are assertions such as:

> If p follows from q and q from p, then they are one and the same proposition. (5.141)

> If one proposition follows from another, then the latter says more than the former, and the former says less than the latter. (5.14)

> A proposition is a truth-function of elementary propositions. (5)

> The logical product of a tautology and a proposition says the same thing as the proposition. This product, therefore, is identical with the proposition. (4.465)

> An operation can vanish (e.g. negation in '$\sim\sim p$': $\sim\sim p = p$). (5.254)

These all appear to point in the direction of identifying propositions with the same truth-conditions.

Before explaining these away, what is the evidence for thinking of propositions as meaningful sentences? The evidence is plentiful, stemming right from the first reference to propositions in the *Tractatus* (2.0122):

> It is impossible for words to appear in two different rôles: by themselves and in propositions.

This remark would be utterly pointless if propositions by their very nature are not the sort of things words can appear in. And 4.22,

> An elementary proposition consists of names. It is a nexus, a concatenation, of names

would be not pointless but outrageous in its blatant falsity. Another remark that would be pointless if propositions expressing the same sense, i.e. having the same truth-conditions, were identical is

> [W]hat is essential in a proposition is what all propositions that can express the same sense have in common. (3.341)

Other compelling evidence:

A proposition . . . does not actually contain its sense . . . A proposition contains the form, but not the content, of its sense. (3.13)

The simple signs employed in propositions are called names. (3.202)

In the proposition, 'Green is green'—where the first word is the proper name of a person and the last an adjective—these words do not merely have different meanings: they are *different symbols*. (3.323)

When translating one language into another, we do not proceed by translating each *proposition* of the one into a *proposition* of the other, but merely by translating the constituents of propositions. (4.025)

Wittgenstein speaks of propositions as variously having (3.13, 3.144, 3.3, 4.061, 4.062, 4.064), expressing (3.13, 3.34, 3.341) and communicating (4.027, 4.03) senses. Time and again he speaks of the sense *of* a proposition. That sense expressed and proposition expressing it are distinct is made clear by a careful reading of 4.1211:

Thus one proposition '*fa*' shows that the object *a* occurs in its sense, two propositions '*fa*' and '*ga*' show that the same object is mentioned in both of them.[7]

Objects occur in senses but are mentioned in propositions.[8]

How to explain away the embarrassing list that apparently runs counter to the claim that a proposition is a meaningful sentence of a language? Careful attention to what Wittgenstein says about identity turns the trick:

When I use two signs with one and the same meaning, I express this by putting the sign ' = ' between them.

So '$a = b$' means that the sign 'b' can be substituted for the sign 'a'. (4.241)

Now, it's true that in the passages around this one Wittgenstein only speaks of identities involving names, not propositions, but: (1) he states how he uses '=' for signs in general; (2) we have seen that he uses the identity sign between signs for propositions; (3) when, famously, he repudiates use of the identity sign, he is careful to do so only for objects:

Identity of object I express by identity of sign, and not by using a sign for identity. (5.53)

Thus we make best sense of Wittgenstein's identity claims concerning propositions when we take them to be, as 4.241 indicates we should, claims about sameness of

[7] That '*fa*' shows that *a* occurs in its sense is cast in doubt by 4.465: 'The logical product of a tautology and a proposition says the same thing as the proposition. This product is, therefore, identical with the proposition. For it is impossible to alter what is essential to a symbol without altering its sense.'

[8] Proops (2000: 64) unaccountably muddies the waters here, for he says in close proximity, '[Wittgenstein's] account seeks to close the gap between a sentence and the proposition it expresses', and 'For Wittgenstein, the proposition just *is* the propositional sign in its projective relation to the world (3.12)' (loc. cit., emphasis in the original). The projective relation breathes life into the propositional sign, but the proposition is the sign vivified, not the life breathed into it.

meaning. So read, 5.141, 'If *p* follows from *q* and *q* from *p*, then they are one and the same proposition', follows from 5.122, 'If *p* follows from *q*, the sense of "*p*" is contained in the sense of "*q*".'[9]

I have spent some time on this because it is essential to what follows that different propositions may express the same sense, have the same truth-conditions. The pernickety reader may doubt that I have made my case. I have just equated sameness of sense and sameness of truth-conditions. Plausible as that may indeed seem, on what textual grounds? These:

> The sense of a proposition is its agreement and disagreement with possibilities of existence and non-existence of states of affairs. (4.2)

> A proposition is an expression of agreement and disagreement with truth-possibilities of elementary propositions. (4.4)

> The expression of agreement and disagreement with the truth-possibilities of elementary propositions expresses the truth-conditions of a proposition. (4.431)

4 The picture theory

At this point we must take a closer look at the picture theory of meaning. What does it mean to say that a proposition is a *picture* of reality? We get some sense of this from what Wittgenstein says about pictures. These theses are crucial:

> In a picture objects have the elements of the picture corresponding to them. (2.13)

> In a picture the elements of the picture are the representatives of objects. (2.131)

> What constitutes a picture is that its elements are related to one another in a determinate way. (2.14)

> The fact that the elements of a picture are related to one another in a determinate way represents that things are related to one another in the same way. (2.15)

> A picture agrees with reality or fails to agree; it is correct or incorrect, true or false. (2.21)

> What a picture represents is its sense. (2.221)[10]

> The agreement or disagreement of its sense with reality constitutes its truth or falsity. (2.222)

[9] Notice that when restricted to elementary propositions 5.141 holds good read straight but does so vacuously, for, as Wittgenstein says at 5.134, 'One elementary proposition cannot be deduced from another' (and, 'If *p* follows from *q*, I can make an inference from *q* to *p*, deduce *p* from *q*', 5.132). Thomas Ricketts (1996: 82) takes 5.141 to say that if *p* and *q* follow from each other, they are notational variants of the same sentence. But given that notational variants are different propositional signs and that a proposition just is a propositional sign in its projective relation to the world (3.12), I fail to see how this reading can be maintained.

[10] As we have seen, at 3.3 Wittgenstein says, 'Only propositions have sense'. Does Wittgenstein mean us to infer from this and 2.221 that pictures are propositions? No, I think not. At 3.3 he is contrasting propositions with sub-sentential expressions and names in particular.

From these it follows that all that can play a role in determining the sense represented by a picture are elements of the picture and that what is represented is isomorphic to what represents it. Component parts of a picture (and the way they are related within the picture) match component parts of the sense it represents (and the way the latter are related). It's the matching of parts that is bothersome in what follows.

If propositions are pictures, as the picture theory says they are, these claims have direct implications for propositions. They are implications Wittgenstein does not hesitate to draw:

> Instead of, 'This proposition has such and such a sense', we can simply say, 'This proposition represents such and such situation'. (4.031)

> In a proposition there must be exactly as many distinguishable parts as in the situation it represents. (4.04)

5 The problem posed by logical constants

Let us, for a moment, return to Frege. As we have seen, on occasion he declares the senses of parts of a declarative sentence to be parts of the sense (thought) expressed by the sentence. And on occasion he took logical equivalence to be sufficient for sameness of thought (as fits with the sense expressed by a sentence being that its truth-conditions are fulfilled). For examples of the latter, in a letter to Husserl of 9 December 1906, Frege quite explicitly takes logical equivalence as sufficient for sameness of sense (1980: 70–1); and in the late article 'Compound Thoughts' he says, '"*A* and *A*" has the same sense as "*A*"' (1984: 393), and that '*A* and *A*', '*A* or *A*', 'not (not *A*)', and 'not[(not *A*) and (not *A*)]' all 'express the same thought as "*A*"' (1984: 403–5). For a very clear example of the former, in 'Negation', published five years before 'Compound Thoughts', Frege says of negation that it 'needs to be completed by a thought'; that 'the two components, if we choose to employ this expression, are quite different in kind and contribute quite differently towards the formation of the whole', 'one completes, the other is completed'; and that 'the relation of completing, in the realm of thoughts and their parts, has something similar corresponding to it in the realm of sentences and their parts' (1984: 386). He says of 'the negation of the negation of *A*' that it is 'to be regarded as directly composed of the parts "the negation of——" and "the negation of *A*"' and, equally, 'as made up of the parts "the negation of the negation of——" and "*A*"' (1984: 387).

These claims are in tension. The tension that troubles us concerns parts and wholes on the one hand and sameness of thought on the other. As Jeffrey King states the problem,

> Frege claims that 'A' and 'not [not A]' have the *same* sense. But how could this be, given that Frege clearly claims that the sense of 'not' is part of the sense of 'not A' and 'not [not A]', whereas one would think that it is not in general part of the sense of 'A'? (2007: 14, original emphasis)

Likewise Kevin Klement finds that the view expressed in 'Compound Thoughts'

> conflicts with the composition principle. The Gedanke expressed by ⌜A ∨ A⌝ must be composed of the Sinn of the incomplete symbol '$\xi \vee \xi$' and the complete Sinn of A, which would mean that the Sinn of A is a partial component of the Sinn of ⌜A ∨ A⌝. But if these two expressions express the same Sinn, then the Sinn of A is a partial component of itself. This is simply impossible. (Klement 2002: 92)

King regards the tension in what Frege says as irresolvable (2007: 18). Klement holds that it cannot have been Frege's considered view that logically equivalent contingent propositions express the same thought (2002: 92).

Wittgenstein's recognition of the incompatibility is what leads him to the claim in the *Notebooks* that 'not-*p*' does not occur in 'not-not-*p*'. In the *Tractatus* he is not forced to such an odd claim because here he has the distinction between truth-operations and truth-functions. It is this distinction that leads to the *Grundgedanke*. As the argument for the *Grundgedanke* is given in remarkably compressed style (even by the standards of the *Tractatus*), I shall go through the steps carefully in the remainder of this section before turning to what Wittgenstein himself says. What we have is an argument of remarkable force and generality. It allows us to see what *must* be said regarding the contributions made by logical constants to the truth-conditions of the sentences in which they occur in any compositional, truth-conditional theory of meaning.

Clearly, ~*p* and *p* do not mean the same.[11] They differ over the occurrence of negation in one but not the other. Going back to the definition of *expression* at 3.31, 'I call any part of a proposition that characterizes its sense an expression (or symbol)', it looks as if we should count the negation sign as an expression and a distinguishable part of '~*p*', thus a component of the picture that is the proposition '~*p*', and hence, more oddly perhaps, we should count negation as a component part of what '~*p*' represents. The thought that this is the role of logical constants leads not just to oddity but to incoherence. Here's why. We need the following three premises, coupled with a strong reading of 'contribution to truth-conditions' as 'component part'.

Premise 1 The meaning of a sentence is given by its truth-conditions, what has to be the case for the sentence to be true (and so to know the meaning of a sentence is to know its truth-conditions and to know that they are its truth-conditions).

[11] Wittgenstein says that they have opposite sense (4.0621). That they have *opposite* sense is a hangover from the doctrine of bipolarity that features prominently in the 1913 'Notes on Logic', as is the remark at 5.2341 that negation reverses the sense of a proposition. According to this doctrine, 'Every proposition is essentially true-false: to understand it, we must know both what must be the case if it is true, and what must be the case if it is false. Thus a proposition has two poles, corresponding to the case of its truth and the case of its falsehood. We call this the sense of a proposition' (NL 98–9). A proposition and its negation are said to have the same reference but opposite sense; the reference of both is the fact that makes one of *p* and ~*p* true and the other false (NL 95). Here Wittgenstein is trying to characterize negation in terms of truth and falsity. Bipolarity appears in the 'Notes dictated to G. E. Moore in Norway' of April 1914 and in some of the early entries in the 1914–16 *Notebooks*, but its death-knell is already sounded in the Moore notes when Wittgenstein gives negation priority over falsity: '*p* is false = ~(*p* is true) Def.' (NM 116).

Premise 2 (*Compositionality*) The meaning of a sentence is determined by the meanings of its subsentential constituents and its logico-grammatical structure. (To know these suffices for knowledge of its meaning in the sense that one *need* know nothing more).

Premise 3 The meaning of a subsentential expression is the (uniform) contribution it makes to the truth-conditions of the sentences in which it occurs.[12]

5.1 The Argument

Logically equivalent sentences are true in exactly the same ranges of (possible) circumstances, therefore have the same truth-conditions, therefore mean the same. Let *p* and *q* be any two declarative sentences. '*p.q*' and '~(~*p*∨~*q*)' are logically equivalent. Hence the truth-conditions of '*p.q*' are fully determined by the meanings of '*p*', '*q*', '~', and '∨' and the logical structure of '~(~*p*∨~*q*)'. These meanings, the uniform contributions '*p*', '*q*', '~', and '∨' make, are *component parts* of the truth-conditions of the sentences in which they can occur. But the meaning of '*p.q*', *i.e.* its truth-conditions, can be spelled out without mention of either of '~' and '∨'; and so one can know the meaning of '*p. q*', i.e. know its truth-conditions and that they are its truth-conditions, without recognizing any contribution from either of '~' and '∨'—compositionality guarantees this. Hence the component parts contributed by '~', and '∨' can make no essential contribution to the truth-conditions of '*p.q*', no contribution that one cannot but discern in identifying those truth-conditions. But '*p.q*' and '~(~*p*∨~*q*)' have the same truth-conditions. Therefore, the meanings of '~' and '∨' can make no essential contribution to the meaning of '~(~*p*∨~*q*)'.

An obvious move to make at this juncture is to ask, Could '~' and '∨' contribute component but idle parts to the truth-conditions of '*p.q*', parts that are no longer idle when these same truth-conditions are given as the truth-conditions of '~(~*p*∨~*q*)'? The answer is no, for if the component parts play different roles, these are just not the same truth-conditions. The component parts contributed by '~' and '∨' cannot be essential when *the very same truth-conditions* are the truth-conditions of '~(~*p*∨~*q*)' but not essential when these same truth-conditions are the truth-conditions of '*p.q*'. For, put another way, that would be to have different modes of presentation of their common truth-conditions in play when we are given the meanings of '~(~*p*∨~*q*)' and '*p.q*', respectively, and this is what must not be the case: the truth-conditions themselves constitute the sense (thought) expressed. (Or, perhaps better, as Frege put it, the sense is the thought that these conditions are fulfilled.) What one must know, when one understands the sentences, is the same in each case:

[12] All we need is that there is not more to the meaning of a subsentential expression than what it contributes to the meanings of the sentences in which it may occur, and that, in the spirit of compositionality, the meaning of a subsentential component is something 'uniform' or 'systematic', not something only available as a list ('In Sentence 1 *e* means __1, in Sentence 2 *e* means __2, . . . '). We do not need what I take to be—but see Platts 1979: 23–4—the stronger thesis found, e.g. in Quine (1981: 69) and in Davidson (1977: 220), that the meanings of words are 'abstractions' from the meanings of sentences.

'~(~p∨~q)' is true if and only if . . . ;
'p.q' is true if and only if . . . .

Here the dots must be filled by the same thing, for our starting point is that the two sentences have the same truth conditions and thereby mean the same. So, as said above, the meanings of '~' and '∨' can make no essential contribution to the meaning of '~(~p∨~q)'.

The conclusion that '~' and '∨' make no essential contribution to the truth-conditions of '~(~p∨~q)' is absurd. The Argument must point to a flaw underlying our reading of its premises. That flaw is readily discerned: it lies in the handling of contributions to truth-conditions on the model of component parts thereof—the building block model.[13]

From the claim that one can know the truth-conditions of '*p.q*' (and know that they are such) without recognizing any contribution from either of '~' and '∨', we are entitled to conclude only that any contribution made by '~' and '∨' is contained in the contribution made by '.' in the pattern of occurrence $\xi.\eta$. (Here 'ξ' and 'η' are 'place-holders' for sentences.[14]) Running The Argument with '*p.q*' and '~(~p∨~q)' swapped around, we find that any contribution made by '.' to the common truth-conditions of '*p. q*' and '~(~p∨~q)' is contained in the contributions made by '~' and '∨' in the pattern of occurrence $\sim(\sim\xi\vee\sim\eta)$. Thus '.' in the pattern of occurrence $\xi.\eta$ makes the same contribution as '~' and '∨' do in the pattern $\sim(\sim\xi\vee\sim\eta)$. And likewise for the connectives occurring in any truth-functionally equivalent pair of patterns of occurrence. What matters is the truth-function as map from *n*-tuples—in the example, pairs—of truth-values to truth-values, not how it is represented in the sentence. The patterns $\xi.\eta$ and $\sim(\sim\xi\vee\sim\eta)$ yield the same truth-function; they have the same truth-table. It is a constraint on what the advocate of a truth-conditional theory of meaning *must* mean by contribution to truth-conditions that this common (truth-) *function*, and only this, is what the patterns contribute to the truth-conditions of the sentences in which they occur.

A Tarskian recursive 'definition of truth' in Davidsonian employment as a meaning-theory meets this constraint, for it tells us that '~(~p∨~q)' is true if, and only if, '~p∨~q' is not true, if, and only if, it is not the case that ['~p' is true or '~q' is true], if, and only if, it is not the case that ['*p*' is not true or '*q*' is not true], which is to say ['*p*' is true and '*q*' is true].[15] In short, '~(~p∨~q)' is true if, and only if, ['*p*' is true and '*q*' is true]. Likewise,

[13] Strictly speaking, Klement's argument concerning the thought expressed by ⌈A ∨ A⌉ requires an extra assumption beyond my Premises 2 and 3, namely that the sense of the incomplete symbol '$\xi \vee \xi$' makes a *non-null* contribution to the sense of ⌈A ∨ A⌉. It is certainly in the spirit of Frege's composition principle that this be so. When *P* and *Q* are logically independent, '$\xi \vee \eta$' *cannot* make a null contribution to ⌈*P* ∨ *Q*⌉.

[14] The careful reader will realize that something odd is going on here as I have included the pattern itself in what purports to be a sentence of English. Still, readers of contemporary fiction understand such devices readily enough and Wittgenstein himself displays forms when he employs them at 6.1203. On quotation and display, see Read 1997.

[15] The 'which is to say' is the point where you, dear reader, and I have to engage in ordinary logical reasoning. Typically, after we have run through application of the relevant recursion clauses, we have to reason to obtain a simplification because application of the recursion clauses alone gets us something that reflects the surface logico-grammatical complexity of the sentence whose truth-conditions we are determin-

the recursive 'definition of truth' tells us that '*p.q*' is true if, and only if, ['*p*' is true and '*q*' is true]. Thus it tells us, whatever the meanings (truth-conditions) of *p* and *q* may be, that the pattern $\sim(\sim\xi \vee \sim\eta)$ makes the same contribution to the truth-conditions '$\sim(\sim p \vee \sim q)$' shares with '*p.q*' as $\xi.\eta$ makes (and tells us, thanks to the recursion clauses for '$\sim$' and '$\vee$', what that contribution is).

It is the *patterns of occurrence* of logical connectives that contribute to the truth-conditions; *what* they contribute is determined, in each case, by the recursion clauses for the connectives. Thus the contribution a connective makes to the truth-conditions of the sentences in which it occurs is at one remove from those truth-conditions; it depends on the pattern of occurrence in which it occurs. One might better say that the meaning of a logical connective is the (uniform) contribution it makes to the *determination* of the truth-conditions of the sentences in which it occurs. Indeed, this *is* how Michael Dummett puts the matter quite generally. For example, 'What is a Theory of Meaning? (II)' opens with these questions:

> Does the meaning of a sentence consist in its truth-conditions? Does the meaning of a word consist in the contribution it makes to determining the truth-conditions of any sentences in which it occurs? (Dummett 1976: 67)

In *Frege: Philosophy of Language* we find the same view attributed to Frege:

> For Frege, the sense of a word or any expression not a sentence can be understood only as consisting in the contribution which it makes to determining the sense of any sentence in which it may occur. (Dummett 1981a: 3)

However, on a number of occasions Dummett has followed Frege in saying that the sense of a part of a sentence is part of the sense expressed by the whole sentence in which it occurs (see e.g. Dummett 1981b: 43, 172, 251–3, 380). Dummett mostly says this when stressing a contrast between sense and reference: the reference of a complex expression is a function of the references of some of its parts, which function being determined by the other parts; the references of the parts are *not* parts of the reference of the whole. To use Dummett's own example, 'The capital of Denmark' refers to Copenhagen but Denmark is not readily thought of as a part of Copenhagen (Dummett 1981a: 159).[16] When not making that contrast between sense and reference, Dummett is more careful regarding the sense of parts of sentences:

ing. To see that two sentences with the same truth-conditions have the same truth-conditions may require considerable mental effort—and the availability of a complete proof procedure. The appeal to proof is in accord with what Wittgenstein says: 'But in fact all the propositions of logic say the same thing, to wit nothing' (5.43); 'The propositions of logic are tautologies' (6.1); and 'Proof in logic is merely a mechanical expedient to facilitate the recognition of tautologies in complicated cases' (6.1262). Presumably Wittgenstein does not see in an appeal to proof here any bar to the transparency of meaning, something to which he makes a strong commitment when he says, in 6.23222, that he cannot know the meaning of two expressions 'without knowing whether what they mean is the same or different'. But the absence of a complete proof procedure for second-order logic (with its standard semantics) should give him pause.

[16] In 'Notes for Ludwig Darmstaedter' Frege (1979: 255) says, 'We cannot say that Sweden is a part of the capital of Sweden'.

> [W]e have seen that, for Frege, the sense of a word or expression always consists in the contribution it makes to determining the thought expressed by a sentence in which it occurs. That is: the sense of a word consists in a rule which, taken together with the rules constitutive of the senses of the other words, determines the condition for the truth of a sentence in which the word occurs. (Dummett 1981a: 194)

Looked at this way, the senses expressed by parts of a sentence are not literally *parts* of the sense expressed by that sentence (contrary to what we saw Frege say).

At this point two observations are called for. First, although I have said that we have an argument of remarkable force and generality, we must be careful not to draw out too much from it. On the one hand, we will find that any two terms occurring in a *logically true* identity statement must make exactly the same contribution to the truth-conditions of pairs of sentences in which they are interchanged (which is not to say that *how* that contribution is determined for each expression is the same).[17] On the other, the sentences 'Vienna is larger than Klagenfurt' and 'Klagenfurt is smaller than Vienna' are true in exactly the same range of circumstances, they have the same truth-conditions, and compositionality entails that one may understand one but not the other; however, we cannot legitimately infer that one can grasp these common truth-conditions without recognizing the contribution of, say, 'ξ is larger than η'; one is entitled only to the claim that one can grasp these common truth-conditions without recognizing that contribution *if the contribution of* 'ξ is larger than η' *is distinct from the contribution of* 'η is smaller than ξ'—note carefully the correlation of argument-places. In contrast to the case above, in which we know that '.', '$\vee$' and '~' do not mean the same, hence cannot contribute the same component part, here we have no reason to believe that there is a difference of contribution.

Second, compositionality is stated in terms of what suffices to determine the meaning of a sentence. What suffices may not be necessary. For any pair of sentences p and q, p and $(p.q)\vee p$ are logically equivalent. The inferences from each to the other are *formally valid*, so on purely formal grounds $(p.q)\vee p$ has the same truth-conditions as p (provided q is meaningful), and hence what q means is irrelevant to this sentence's truth-conditions (*cf.* Dummett 1981a: 237).

[17] Frege maintains that '2 + 2' and '4' do not have the same sense (Frege 1967: §2). But if truths of arithmetic are truths of logic, as Frege maintains, and we are not to follow King and Klement in rejecting his claim that logical equivalents express the same sense, '2 + 2' and '4' must make exactly the same contribution to the truth-conditions of sentences in which they are interchanged. We can say that how that contribution is determined differs but we must say that the contribution determined by the senses expressed by '2' and by '$\xi + \xi$' in the pattern '2 + 2' is the same as the contribution made by '4'. And then it is hard to explain how '2 + 2 = 4' can differ in cognitive value from '4 = 4'. (For an argument that Frege's claims regarding the status of arithmetic truths in *Grundgesetze* are incoherent, see Milne 1989.)

6 Truth-functions, truth-operations, and Wittgenstein's *Grundgedanke*

I have laid out The Argument and its analysis carefully and at length. In the *Tractatus* the argument of our previous section occurs highly compressed at 5.41. The 5.2s and 5.3s tell us about truth-operations and truth-functions.[18] Wittgenstein says,

Truth-functions of elementary propositions are results of operations with elementary propositions as bases. (These operations I call truth-operations.) (5.234)

Negation, logical addition, logical multiplication, etc. etc. are truth-operations. (5.2341)

One operation can counteract the effect of another. Operations can cancel one another. (5.253)

An operation can vanish (e.g. negation in '$\sim\sim p$': $\sim\sim p = p$). (5.254)

All truth-functions are the results of successive applications to elementary propositions of a finite number of truth-operations. (5.32)

The occurrence of an operation does not characterize the sense of a proposition.

Indeed, no statement is made by an operation, but only by its result, and this depends on the bases of the operation. (5.25)

The sense of a truth-function of p is a function of the sense of p. (5.2341)

5.32 occurs immediately before 5.4. The latter is a restatement of Wittgenstein's fundamental idea, his *Grundgedanke*:

At this point it becomes manifest that there are no 'logical objects' or 'logical constants' (in Frege's and Russell's sense). (5.4)

5.41 explains why not:

The reason is that the results of truth-operations on truth-functions are always *identical* whenever they are one and the same truth-function of elementary propositions. (5.41, my emphasis)

And this, in condensed form, is *exactly* what we went through painstakingly in the previous section. From 5.41 it follows that, as they occur in sentences, (signs for) logical constants do not *stand for* anything, a consequence Wittgenstein draws out at 5.44:[19]

The proposition '$\sim\sim p$' is not about negation, as if negation were an object: on the other hand, the possibility of negation is already written into affirmation.

And if there were an object called '$\sim$', it would follow that '$\sim\sim p$' said something different from what 'p' said, just because the one proposition would be about $\sim$ and the other would not.

[18] Entries in the *Notebooks* suggest that Wittgenstein hit upon the distinction between functions and operations in mid-December 1914.

[19] Wittgenstein was to return to the non-representative character of the logical constants in his Cambridge lectures in 1931: 'To say that a word has meaning does not imply that it stands for or represents a thing. Such words as "and", "not", "or", etc. obviously do not stand for anything' (Wittgenstein 1980: 45).

What is a little puzzling is that this should occur such a long way after 4.0312 which states the point quite generally, that (signs for) logical constants are not representatives. 5.41 and the propositions immediately preceding it constitute a proof of Wittgenstein's *Grundgedanke*.[20]

In 5.44 the negation-sign is contrasted with other parts of language. The component-part model may work for expressions. In *Tractarian* terms, the flaw in The Argument is to treat (signs for) logical constants as expressions. Wittgenstein must and does deny that (signs for) logical constants are distinguishable parts of the propositions in which they occur. His denial comes in two parts.

First, we find that (signs for) logical operations are not expressions. This follows from what Wittgenstein says at 3.31 and 5.241:

> An expression is the mark of a form and a content. (3.31)

> An operation is not the mark of a form, but only of the difference between two forms. (5.241)

This in turn means that the occurrence of (a sign for) a logical operation in a proposition (sentence) cannot help characterize its sense. In view of what has just been reported regarding logical operations and truth-functions, this should come as no surprise. The occurrence of a negation as part of a double negation contributes *nothing* to what the proposition in which it occurs expresses. Thus there is no uniform contribution to the truth-conditions of the sentences in which they occur made by all occurrences of negation. And as for negation so for the other logical constants.

The second line of thinking that shows that (signs for) logical operations are not expressions—one that is already fully worked out in the 'Notes on Logic'—comes directly from consideration of negation.

> Can we not make ourselves understood with false propositions just as we have done up till now with true ones?—So long as it is known that they are meant to be false.—No! For a proposition is true if we use it to say that things stand in a certain way, and they do; and if by '*p*' we mean ~*p* and things stand as we mean that they do, then, construed in the new way, '*p*' is true and not false. (4.062)

> But it is important that the signs '*p*' and '~*p*' *can* say the same thing. For it shows that nothing in reality corresponds to the sign '~'. (4.0621)

This is the crux of the matter, although it is perhaps poorly expressed. (Why should one care how the same signs may be used to say things other than they do?) The thought, clearly, is that there is no part of reality that the negation sign 'picks up on' or 'latches on to'. We use it in saying one thing, rather than its opposite, in our language, but we can imagine a language that works equally well, and in which the propositional signs

[20] Where I and, as he notes, other commentators see only examples of the 'vanishing' of logical operations in 5.254 and 5.441, Leo Cheung (1999: §3) sees a proof, an invalid proof, of Wittgenstein's *Grundgedanke*. I'm inclined to think Cheung fails to appreciate the significance of 5.41 and of the distinction between truth-functions and operations (on which more in §7 below).

we use are put to a different use: the roles of the original sentence and the one to which the negation sign is added are turned about. In both languages the negation sign contributes to the sense expressed by the propositional sign (sentence) containing it, but it contributes differently. In both cases it marks a difference of content but it is not that one language achieves this in a better way than the other, that one better fits the world (reality) than the other.

All this calls for two comments. The first is that this seems a decidedly poor argument for (the sign for) a logical operation's not being an expression, for it seems that, within each of the languages, '~' does characterize the sense of the proposition (meaningful sentence) in which it occurs: it contributes to indicating what has to be the case in order for it to be true. Nonetheless, negation is not an object, not a feature of reality. The same point is made at 4.0621 when Wittgenstein says,

> The propositions '*p*' and '~*p*' have opposite sense, but there corresponds to them one and the same reality.

There is not a positive fact making one true and a negative fact making the other false. It is the same aspect of reality that accounts for the one's being true and the other's being false. That being so, the negation sign cannot itself latch on to a feature of reality.[21]

The second comment is that there is a change in Wittgenstein's way of thinking about pictures and propositions between November 1914 and the *Tractatus*. On 26 November Wittgenstein confided to his notebook,

> Can one negate a picture? No. And in this lies the difference between picture and proposition. The picture can serve as a proposition. But in that case something gets added to it which brings it about that now it says something. In short: I can only deny that the picture is right, but the picture I cannot deny. (NB 34)

This is very much at odds with what he was to say in the *Tractatus*. There we find him saying,

> Pictorial form is the possibility that things are related to one another in the same way as the elements of the picture. (2.151)

> *That* is how a picture is attached to reality; it reaches right out to it. (2.1511)

> A picture agrees with reality or fails to agree; it is correct or incorrect; it is true or false. (2.21)

> What a picture represents is its sense. (2.221)

> The agreement or disagreement of its sense with reality constitutes its truth or falsity. (2.222)

[21] Contrast how '*fa*', '*fb*', and '*ga*' are made true or false by different aspects of reality. The first two concern different objects but say the same thing about them; the first and third say different things about the same object. (Although *Wirklichkeit* admits a plural in German, Wittgenstein would not have been happy to say that 'different realities' correspond to these different realities.) Between '*p*' and '~*p*' the difference is much less. True, they say different things, but it's the same issue at stake; they affirm opposite sides of it, but it is the same issue.

There is not a further step of direction, just as there is not with language: a proposition shows how things stand, if it is true, says Wittgenstein, and *says that* they do so stand (4.022). We would do well to keep in mind that, just like assertion, picturing involves us, that it is something we do:

We picture facts to ourselves. (2.1)

Our normal practice is not such that we stop to think, 'Does this photograph purport to show how things are?—Or how they are not?' No more than do we normally stop to ponder whether an assertion has been made by the ordinary use of a declarative sentence (even if we may wonder whether the speaker knows what he says or whether he is lying). For all sorts of reasons a picture may be a misrepresentation, but if it is a misrepresentation it is ordinarily such because things are not as it indicates them to be; it is not ordinarily a *mis*representation because things are as it shows them to be. Pictures (and picturing) are in no different a position than assertions (and assertion).

Wittgenstein, I said, changed his mind regarding pictures and propositions. He changed his mind because he came to see how to accommodate logically complex propositions into the picture theory. From the perspective of the *Tractarian* account, one *can* negate a picture. We turn to this next.

7 Pictures and logically complex propositions

If, as they must not, logical constants do not represent, how does the picture theory accommodate logically complex propositions? Wittgenstein's answer is in the 4s of the *Tractatus*.

Logically equivalent sentences have the same truth-conditions, the same meaning. As we saw, the Frege of the *Grundgesetze* has a hard time accommodating this fact:

The names, whether simple or themselves composite, of which the name of a truth-value [a declarative sentence] consists, contribute to the expression of the thought, and this contribution of the individual [component] is its *sense*. If a name is part of the name of a truth-value, then the sense of the former name is part of the thought expressed by the latter name. (Frege 1967: 90).

In this context Frege uses the term 'name' broadly, broadly enough to count in logical constants (as follows from §§12, 28, and 30 of *Grundgesetze*). But that being so, his talk of the sense of a name being part of the thought expressed commits him to the component-part model of contributions to truth-conditions which falls foul of The Argument. Wittgenstein is right to reject this conception and to see the role of logical constants (logical operations) as one of contributing to the *determination* of truth-conditions. But whereas what is important from, say, Davidson's perspective, with its emphasis on learnable languages, is the *how* of that determination, the recursion conditions themselves, what is important to Wittgenstein is the *outcome*, what gets determined. In Davidson's case, what gets determined is an instance of Convention T, in which the object-language sentence and its metalinguistic translation appear; what is important is the way that the translation has been produced, the proof of this instance of the T-schema from the axioms of the truth-theory (see Davidson 1970: 61). In Wittgenstein's case, the product is a presentation of a sentence (proposition) as a truth-*function* of elementary propositions. The logical constants (operations) that have got us there are 'washed out'; the truth-function they determine is all. This is made manifest in the previously quoted 5,

A proposition is a truth-function of elementary propositions.

It is also manifested in Wittgenstein's one presentation of a propositional sign at 4.442:

For example, the following is a propositional sign:

' *p*	*q*	'
T	*T*	*T*
F	*T*	*T*
T	*F*	
F	*F*	*T*

Notice that the truth-table is *quoted*; it is *not* intended as a metalinguistic device for working out the properties of object-language sentences; rather—as the absence of a formula heading the right-most column hints—the table itself is an object-language expression: a propositional sign. We are told:

A proposition is an expression of agreement and disagreement with truth-possibilities of elementary propositions. (4.4)

We can express agreement with truth-possibilities by correlating the mark '*T*' (true) with them in the schema [truth-table]. (4.43)

The truth-table *does* express agreement with the truth-possibilities of the elementary propositions p and q, hence *is* a propositional sign, for it expresses a proposition: it makes clear the truth-conditions of the proposition whose sign it is; it makes clear which truth-function of the elementary propositions p and q it is. We see that it is the proposition for which '$p \supset q$' is a more familiar (propositional) sign.[22]

This exemplar is archetypical in this respect: it does all that, and no more than, a proposition has to, for it indicates, on an exhaustive list of the truth-possibilities of a set of elementary propositions, those whose obtaining suffices for the truth of the proposition itself. (These are the *truth-grounds* of the proposition (5.101).)

[22] Wittgenstein was to expand on the use of the truth-table as propositional sign in his lectures in the 1930s. See Wittgenstein (1980: 50; 1979b: 135–8; 1976: 177–81).

We need to go into this in a little more detail.

At 4.31 Wittgenstein says,

> We can represent truth-possibilities by schemata of the following kind ('*T*' means 'true', '*F*' means 'false'; the rows of '*T*'s' and '*F*'s' under the row of elementary propositions symbolize their truth-possibilities in a way that can easily be understood),

then goes on to draw up the now-standard columns displaying the eight, four, and two possibilities for, respectively, three, two, and one proposition. Each row of the truth table presents a truth-possibility of elementary propositions. What are elementary propositions?

> The simplest kind of proposition, an elementary proposition, asserts the existence of a state of affairs. (4.21)

For n elementary propositions, there are 2^n possibilities of truth and falsity, 2^n truth-possibilities, and consequently 2^{2^n} ways in which a proposition can agree and disagree with these truth-possibilities (4.27, 4.28, 4.42).

Now,

> If an elementary proposition is true, the state of affairs [it asserts to exist] exists: if an elementary proposition is false, the state of affairs does not exist. (4.25)

So, an elementary proposition is true or false as the state of affairs (*Sachverhalt*) it pictures/asserts to exist, exists or not, and hence

> Truth-possibilities of elementary propositions mean possibilities of existence and non-existence of states of affairs. (4.3)

When, in a truth-ground, an elementary proposition is flagged with an '*F*', the state of affairs it asserts to exist does not exist and the elementary proposition is false.

There are really no surprises here, but all this takes place within the picture theory of meaning. All propositions are pictures. The propositional sign from 4.442 is a picture when the '*p*' and '*q*' are fleshed out; they themselves are pictures. The flagging of pictures with combinations of '*T*'s and '*F*'s, and the further marking of some of these combinations with '*T*'s must itself constitute a picture. Taking the pictorial nature of elementary propositions as read, Robert Fogelin says,

> A proposition is just a set of pictures together with an assignment of truth values for the combinations of agreement and disagreement with reality. (Fogelin 1987: 49–50)

This, I suppose, is near enough what Wittgenstein has in mind provided one reads nothing more than a formal role into the expression 'truth-value'. Wittgenstein is at pains to point out,

> It is clear that a complex of signs '*F*' and '*T*' has no object (or complex of objects) corresponding to it, just as there is none corresponding to the horizontal and vertical lines or to the brackets.—There are no 'logical objects'. (4.441)

(A repudiation of Frege's The True and The False.)

There is a minor textual problem here. Propositions are pictures, but given n elementary propositions and the states of affairs they assert to exist, there are 2^{2^n} distinct propositions one can form, but only 2^n pictures. At least, that there are only 2^n pictures seems to be the upshot of 2.11 and 2.201. Whereas a proposition is an expression of agreement and disagreement with truth-possibilities of elementary propositions, and notice the plural there, truth-*possibilities*,

A picture presents a situation in logical space, the existence and non-existence of states of affairs. (2.11)

A picture depicts reality by representing a possibility of existence and non-existence of states of affairs. (2.201).

If we take this '*a* possibility' seriously, a picture can represent an individual truth-ground of a proposition but not the totality of its truth-grounds, save if it has only one truth-ground. Should we take it seriously? Probably not, for at 4.1 we read,

Propositions represent the existence and non-existence of states of affairs.

We may take it that this and the earlier characterization of pictures await the refinement supplied in the 4.2s and subsequently.[23]

An elementary proposition asserts the existence of a state of affairs. If it is false that state of affairs does not exist. If it is false its negation is true. Does the negation of an elementary proposition assert the non-existence of a state of affairs? According to the *Tractatus*, the answer is yes, provided that we read 'say' (*sagen*) as differing little from 'assert' (*behaupten*). Wittgenstein says,

The general form of a proposition is: This is how things stand. (4.5)

having previously said, at 4.022,

A proposition *shows* how things stand *if* it is true. And it *says that* they do so stand.

How do things stand if the proposition is true? In such a way that its truth-conditions are fulfilled. Reality *is* the existence and non-existence of states of affairs (2.06). Exactly one of the proposition's truth-grounds matches reality in this way: those states of affairs asserted to exist by the elementary propositions that are tagged with a T exist; those that are tagged with an F do not. In the case of the negation of an elementary proposition, what it says, what it expresses, is that the state of affairs pictured/asserted to exist by the

[23] A different but not unrelated matter also suggests a mismatch between pictures and propositions. Propositions are truth-functions of elementary propositions. Included amongst these are tautologies and contradictions (4.46, 5.101). But, 'Tautologies and contradictions are not pictures of reality' (4.462). It may not be worth pursuing this, for there is certainly a tension in what the *Tractatus* says. 'A proposition can be true or false only in virtue of being a picture of reality' (4.06); 'Among the possible groups of truth-conditions there are two extremes. In one of these cases the proposition is true for all truth-possibilities of the elementary propositions. We say that the truth-conditions are tautological' (4.46). 'A tautology has no truth-conditions, since it is unconditionally true' (4.461).

unnegated original does not exist. How is the non-existence of a state of affairs pictured? As Fogelin indicates, by attaching some conventional marker to a picture of the state of affairs.[24]

Before leaving this topic we should note that both before the *Tractatus* and after Wittgenstein would explain the role of negation in a way that works well for the negation of an elementary proposition but is at odds with the truth-table-as-propositional-sign archetype. Among a long series of entries in the *Notebooks* from 3 November 1914 we find:

> Negation refers to the finished sense of the negated proposition and not to its way of presenting.
> If a picture presents what-is-not-the-case in the forementioned way, this only happens through its presenting *that* which *is* not the case.
> For the picture says, as it were: '*This* is how it is *not*', and to the question '*How* is it not?' just the positive proposition is the answer. (NB 25)

Similarly, in his Cambridge lectures of 1931 Wittgenstein says of negation:

> And what corresponds to $\sim p$ [when it is true] is *p not being the case*. (Wittgenstein 1980: 52)

But if we take, in Russellian notation, the formula '$p \supset q$', say, and its negation '$\sim(p \supset q)$', written in the form Wittgenstein adopts in 4.442 and 5.101, we get '$(TTFT)(p,q)$' and '$(FFTF)(p,q)$', respectively. There seems to be no good reason to take the latter as presenting anything as 'how it is not', or, certainly, no good reason to read only the latter as doing so. It is the truth-*function* that matters, not the truth-*operations* that may occur in one or another expression of it.

8 Punctuation

We have seen that Wittgenstein conceives of the logical constants as non-representational elements that play a role in determining the sense of the propositions in which they occur. We have yet to see how he arrives at the claim that signs for logical constants function as punctuation marks. Ricketts, discussing Wittgenstein's object-language truth-table, says,

[24] David Shier suggests that 'we could devise a symbol, say a red circle with a slash through it, and superimpose it on the picture to indicate that the depicted state of affairs does not obtain' (Shier 1997: 71). Equally, one could leave the picture unadorned for non-existence, and attach a conventional marker for existence. That would be a different convention. Those, such as Shier, who maintain that a picture needs something extra in order to make an assertion *because* it could be used both to say what is the case and to say what is not, have not quite trained the spotlight on the right target. It is not that the picture already has a content and one needs a convention to determine what claim is made using that content (cf. 4.062). In the truth-tabular presentation of a complex proposition, however, one does need some convention, precisely to indicate which role belongs to the unadorned picture, and which to the picture sporting the conventional tag. This convention fixes whether it is the truth or the falsity of what is presented that is required in the truth-ground of a complex proposition.

The complex of 'T''s and 'F''s in one of Wittgenstein's tabular sentential signs—or Russell's sentential connectives with scope demarcating parentheses—are used to express a particular truth-function of the contained elementary sentences. Signs for logical operations do not then symbolize as the expressions in elementary sentences do. In particular, their use requires no coordinations of names and objects in addition to those that project elementary sentences onto reality. Signs for logical operations are thus a sort of punctuation. (Ricketts 1996: 82)

But this won't do, for it misses out what is essential about the signs for logical operations, and leaves us wondering in what respect they are akin to punctuation marks. The use of the truth-tabular propositional signs would seem to eliminate any need for signs for logical operations *as we normally conceive of them*, signs of the sort Wittgenstein himself uses throughout the *Tractatus*, '.', '∨', '⊃', '~'. However, the truth-table notation does make heavy use of other conventional markers, in this case '*T*' and '*F*' (*cf.* 4.441, quoted above). In what way are these akin to punctuation marks? Exactly in that they are non-representational elements that play a role in determining the senses of the propositions in which they occur.

Let us begin with Ian Proops's analysis:

The latter [the comma and the full-stop] are signs that, although they lack sense and reference, nonetheless do have a meaning in their own right.... Consider, for example, the following two sentences, which differ from each other only with respect to a comma.

4) The chef watched her assistant, stirring the soup.

5) The chef watched her assistant stirring the soup.

To my ear—or eye—(4) means that the chef watched her assistant while she (the chef) was stirring the soup, while (5) means that the chef watched her assistant perform the act of stirring the soup. In this example, the whole difference between the meanings of the two sentences is the difference indicated by the comma, for the meanings of the other constituents are the same in both sentences. But the comma, plainly, is far from being a semantically inert part of another sign. (Proops 2000: 15)

The comparison is apt and illuminating. Certainly the comma makes all the difference to what (4) says, as opposed to (5), and, for sure, no one thinks that answering to the comma there is a distinct feature of reality. (Although Proops says that the comma and the full-stop 'have meaning in their own right', we should not read too much into that: commas clearly do not make the same contribution to every sentence in which they occur.)

Proops goes on to say

Wittgenstein's point is that the logical connectives share with punctuation marks the feature of lacking sense and reference while nonetheless having a meaning in their own right. The point of the comparison is to bring out that the logical connectives make a *purely structural* contribution to the meanings of the sentences in which they figure. (Ibid.)

I am much less sure about the attribution of quite so clear-headed a view to Wittgenstein.

Certainly, the logical connectives are not semantically inert. The occurrence in a sentence of a logical connective can—but as double negation reminds us, only can, not must—make a difference to what is expressed by a sentence in which it occurs (its truth-conditions). Certainly, a logical connective is not best thought of as referring to some distinguishable component-part of reality. But in the lead-up to 5.4611 it is not obvious that this is what Wittgenstein is on about. His main complaint is that the usual logical connectives, for a variety of reasons, principal among which is their interdefinability (5.42, 5.451),[25] cannot be primitive signs. He then says,

> Though it seems unimportant, it is in fact significant that the pseudo-relations [*Scheinbeziehungen*] of logic, such as ∨ and ⊃, need brackets—unlike real relations.
>
> Indeed, the use of brackets with these apparently primitive signs is itself an indication that they are not the real primitive signs. And surely no-one is going to believe that brackets have an independent meaning. (5.461)

Now, brackets, like Proops's comma, are not semantically inert and, as Wittgenstein says, do not have an independent meaning. But notice that it is the use of brackets *along with* the logical connectives that is said to show us something significant about the connectives, the logical connectives are not themselves likened to brackets. And the possibility of bracket-free Polish notation means that whatever the use of brackets shows, the use of brackets had better not be essential to Wittgenstein's point.

What point that is comes more sharply into focus thanks to this passage in the 'Notes dictated to G. E. Moore in Norway' of April 1914:

> It is very important that the apparent logical relations ∨, ⊃, etc. need brackets, dots, etc., i.e. have 'ranges'; which by itself shews they are not relations. This fact has been overlooked, because it is so universal—the very thing which makes it so important. (NM 116)

In Polish notation, unlike the Russellian notation which Wittgenstein employs in the *Tractatus*, one does not need brackets or dots to mark off the scope ('range') of an occurrence of a connective, but occurrences of connectives still have scopes. In any

[25] At 5.42 Wittgenstein has already declared, 'It is self-evident that ∨, ⊃, etc. are not relations in the sense in which right and left etc. are relations,' something which, he says the 'Notes on Logic', is 'obvious to the plain man' (NL 101). He goes on in 5.42, again repeating the view of 'Notes on Logic', to say, 'The interdefinability of Frege's and Russell's "primitive signs" of logic is enough to show that they are not primitive signs, still less signs for relations.' Ricketts (1996: 81) says, 'Wittgenstein believes that Russell's notation misleads, for it notationally tempts us to think of the sentential connectives "¬" and "∨" as representing a property of or relation over items signified by sentences.' While this is certainly correct, given Wittgenstein's repeated indications that, *contra* Frege, sentences are not names (3.142, 3.143, 3.1432, 3.144, 3.3, 4.063, 5.02), it cannot be the whole story, for it sheds no light on why interdefinability is evidence against their being relations. Left (i.e. to the left of) and right (i.e. to the right of) would seem to be no less interdefinable yet they are not disparaged as pseudo-relations. What Wittgenstein had in mind, I suggest, is that, as in our example involving Vienna's being larger than Klagenfurt, 'ξ is to the left of η' and 'η is to right of ξ' can be viewed as picking out the same relation, so the interdefinability is harmless, but in the case of the logical connectives, they are distinct but interdefinable. Their interdefinability indicates at once that none is more fundamental than any other and that not all of them are fundamental: they do not all stand for irreducible features of reality—hence none does.

notation, the scope of an occurrence of a connective is the smallest (sub-)formula containing the occurrence. Now, it would be absurd to argue that addition and multiplication are not genuine functions because one needs brackets to disambiguate the two readings of

$$7 + 5 \times 12,$$

namely

$$(7 + 5) \times 12 \text{ and } 7 + (5 \times 12).$$

But that it would indeed be absurd so to argue is, in effect, Wittgenstein's point. That the signs for logical constants can be composed, and in composition have ranges, whether explicitly marked off by brackets and/or dots or taken care of by an economical notation, shows that they do not stand for relations. On Wittgenstein's way of thinking, when objects stand in a relation their so standing is a state of affairs, a thing, if one may put it this way, of quite a different kind from the objects related. But, continuing in this loose way of speaking, composability requires that, say, '$p \supset q$' be the same kind of thing as p and q, for it can occur as a component of, say, '$(p \supset q).{\sim}q$', just as '5 × 12' stands for a thing of the same kind as '5' and '12' themselves stand for, for only so can what it stands for be added, say, to what '7' stands for. (The point at issue is not undermined by the *Tractarian* thesis that propositional signs in their projective relations to the world do not stand for things in the way names do but rather represent their truth-conditions.)

In Russellian notation, brackets serve to disambiguate. Like the comma in Proops's example, their role is to determine what gets 'fed in' to the recursion clauses of a Tarskian 'definition of truth'; brackets and punctuation marks determine what the recursion clauses are to apply to.[26] The nature of their contribution to the determination of truth-conditions is, then, rather different from that of the logical constants; indeed, *it is not so much that they help determine truth-conditions as that they determine what it is whose truth-conditions are to be determined.* Wittgenstein's emphasizing the products of the recursion conditions, not the conditions themselves, may have helped him lose sight of this difference.

Thus, disappointingly, we have arrived at a point where we can see that Wittgenstein ought to have said something like he does at 5.4611, had indeed every reason to say something like what he does say there, but that his reasons for saying what he does in fact say are difficult to fathom; we have to face squarely indications that he may have said it for the wrong reasons.[27] In making the strong claim that the signs for the logical

[26] Davidson once suggested that 'the logical constants [of a language] may be identified as those iterative features of the language that require a recursive clause in the characterization of truth or satisfaction' (Davidson 1973: 71). (The special role of logical constants is maintained, *mutatis mutandis*, when, following Frege and Dedekind, one turns a recursive definition into an explicit definition.)

[27] In finding a second proof of 4.0312's *Grundgedanke* implicit in the *Tractatus*, one depending on 5.4611, Cheung (1999) offers a different route to a literal reading. As Wittgenstein's deployment of the operator N in

operations *are* punctuation marks he overstated matters, for there is a significant difference between the roles played by punctuation marks and those played by the logical constants in the context of a truth-conditional theory of meaning. By emphasizing the end-product of the application of the recursion clauses in the construction of a logically complex proposition, Wittgenstein overlooks the fact that punctuation marks out to what and in what order recursion clauses are to be applied, thus marking off the *inputs* to the recursion clauses, while the connectives single out which clauses apply and thus determine the *outputs* of the intermediate steps in the determination of truth-conditions.[28]

On a truth-conditional account of the meaning of sentences, to know the meaning of a sentence is:

(1) to know its truth-conditions;
(2) to know that they are its truth-conditions.

On this basis, one can know the meaning of sentences in isolation and of sentential idioms. For such knowledge (1) and (2) suffice. But if a speaker *of a language* is to understand previously unencountered sentences of that language, more is required. In addition to (1) and (2) the speaker must know:

(3) how the truth-conditions are determined by the meanings of the component parts of the sentence and their mode of composition in the sentence.

With the Davidsonian rereading of a Tarskian definition of truth as a meaning-theory in mind, Mark Platts has summed this up nicely:

> Giving the truth-conditions of a statement in the Tarskian manner determines its meaning by incorporating that statement, or the sentence used in making it, within a systematic account of the language as a whole, such that that sentence can be seen as but one of the indefinitely large number of sentences in the language. Less abstractly, the simple statement of the truth-conditions in the form of a T-sentence is indeed of little (though not of no) value, a 'snow-bound triviality' as Davidson has it; where the value is realised is in the construction of a device that serves to generate,

the early 5s and 6s shows, we can make do with a single logical operator. Now, in any context in which there is but one operation, no sign for the operation is needed as long as one clearly marks what it is to apply to and the order of application. For example, as long as it is understood that we are dealing with addition, we may express what is conventionally written as '4 + (7 + 3) = (4 + 7) + 3' thus: (4,(7,3)) = ((4,7),3). The general assertion of the associativity of addition may then take this form: $(x)(y)(z)[(x,(y,z)) = ((x,y),z)]$. Similarly, by using *N* alone, we can eliminate any explicit notational marker for a logical constant and make do with punctuation only, brackets and commas. This fails to establish 5.4611: it does not show that the familiar signs '.', '∨', '⊃' and '~' are punctuation marks. Nor, even if it did, would this show, as Cheung needs it to, that logical constants do not represent. One common notation for the exponentiation function merely juxtaposes the symbols for its arguments: x^y. This does not show that there is no exponentiation function.

[28] There is a hint of this contrast in what Hylton says: '[O]ne might wish to say that parentheses function not by having a sense themselves, but rather by indicating how other senses should be combined. In Wittgenstein's view, I think, something similar can be said of operations. An operation, as its name perhaps suggests, is less like an entity, that might be a constituent of a more complex entity, than it is like something we *do*' (Hylton 1997: 105). But Hylton, I think, fails to grasp that Wittgenstein, concerned with the product not the process, severely underplays the doing.

to prove, all the near trivial T-sentences. The journey, not the destination, matters. The journey that issues in the near-trivial destination shows *how* that destination is reached: it shows not just *what* the truth-conditions are but *how* the truth-conditions are determined. (Platts 1979: 57–8)

Wittgenstein saw, far more clearly than either Frege or Russell, that it is only the destination that matters for truth-conditions. One must arrive at the same destination with pairs of logically equivalent propositions; this common destination is the picture of the ways the world may be for the propositions to be true. This common picture is what is delivered by the two distinct propositional signs in their projective relation to the world. What Wittgenstein has forgotten by the time he gets to the elaboration of the picture-theory account of logically complex propositions is that we, the users of language, have to work out what a proposition pictures from the components of the propositional sign and their mode of composition in the sign. While neither punctuation marks nor logical constants stand for anything, both contribute to the sense expressed. In this they are alike. But *how* they contribute differs, and it is for this reason that it is misleading—I'm tempted to say, it is just plain wrong—to say that signs for logical operations *are* punctuation marks.

References

Baker, G. P and Hacker, P. M. S (1984) *Frege: Logical Excavations*. Oxford: Blackwell.

Cheung, Leo K. C. (1999) 'The Proofs of the Grundgedanke in Wittgenstein's *Tractatus*', *Synthese* 120, 395–410.

Davidson, Donald (1970) 'Semantics for Natural Language'. Reprinted in Davidson (1984), 55–64.

——(1973) 'In Defence of Convention T'. Reprinted in Davidson (1984), 65–75.

——(1977) 'Reality without Reference'. Reprinted in Davidson (1984), 215–25.

——(1984) *Inquiries into Truth and Interpretation*. Oxford: Clarendon Press.

Dummett, Michael (1969) 'The Reality of the Past'. Reprinted in his *Truth and Other Enigmas*. London: Duckworth, 1978, 358–74.

——(1976) 'What is a Theory of Meaning? (II), in G. Evans and J. McDowell, eds, *Truth and Meaning: Essays on Semantics*. Oxford: Oxford University Press, 67–137.

——(1981a) *Frege: Philosophy of Language*, 2nd edition. London: Duckworth.

——(1981b) *The Interpretation of Frege's Philosophy*. London: Duckworth.

——(1989) 'More about Thoughts', *Notre Dame Journal of Formal Logic* 30, 1–19.

Fogelin, Robert J. (1987) *Wittgenstein*, 2nd edition. London & New York: Routledge.

Frege, Gottlob (1967) *The Basic Laws of Arithmetic: Exposition of the System*, ed. and trans. M. Furth. Berkeley and Los Angeles: University of California Press.

——(1979) *Posthumous Writings*, ed. H. Hermes, F. Kambartel, and F. Kaulbach, trans. P. Long and R. White. Oxford: Basil Blackwell.

——(1980) *Philosophical and Mathematical Correspondence*, ed. B. F. McGuinness, trans. H. Kaal. Oxford: Basil Blackwell.

——(1984) *Collected Papers on Mathematics, Logic and Philosophy*, ed. B. F. McGuinness. Oxford: Basil Blackwell.

——(2004) *Frege's Lectures on Logic: Carnap's Student Notes, 1910–14*, ed. and trans. E. H. Reck and S. Awodey. Chicago and La Salle, Ill.: Open Court.

Hacker, P. M. S. (1981) 'The Rise and Fall of the Picture Theory', in I. Block, ed., *Perspectives on the Philosophy of Wittgenstein*. Oxford: Basil Blackwell, 85–109.

Hylton, Peter (1997) 'Functions, Operations and Sense in Wittgenstein's *Tractatus*', in W. W. Tait, ed., *Early Analytic Philosophy: Frege, Russell, Wittgenstein*. Chicago and La Salle, Ill.: Open Court, 91–106.

King, Jeffrey C. (2007) *The Nature and Structure of Content*. New York: Oxford University Press.

Klement, Kevin (2002) *Frege and the Logic of Sense and Reference*. New York and London: Routledge.

Milne, Peter (1989) 'Frege, Informative Identities, and Logicism', *British Journal for the Philosophy of Science* 40, 155–66.

Morris, Michael (2008) *Wittgentein and the Tractatus Logico-Philosophicus*. Abingdon: Routledge.

Platts, Mark (1979) *Ways of Meaning*. London, Boston, and Henley: Routledge and Kegan Paul.

Proops, Ian (2000) *Logic and Language in Wittgenstein's Tractatus*. New York and London: Garland.

Quine, Willard Van Orman (1981) 'Five Milestones of Empiricism', in his *Theories and Things*, Cambridge, Mass., and London: Belknap Press/Harvard University Press, 67–72.

Read, Stephen (1997) 'Quotation and Reach's Puzzle', *Acta Analytica* 19, 9–20.

Ricketts, Thomas (1996) 'Pictures, Logic, and the Limits of Sense in Wittgenstein's *Tractatus*', in H. Sluga and D. G. Stern, eds, *The Cambridge Companion to Wittgenstein*. Cambridge: Cambridge University Press, 59–99.

Russell, Bertrand (1903) *The Principles of Mathematics*. Cambridge: Cambridge University Press; 2nd edition, London: George Allen and Unwin, 1937.

Shier, David (1997) 'How Can Pictures Be Propositions?', *Ratio* 10, 65–75.

Stenius, Erik (1960) *Wittgenstein's Tractatus: A Critical Exposition of its Main Lines of Thought*. Oxford: Blackwell.

Stevens, Graham (2003) 'The Truth and Nothing but the Truth, yet never the Whole Truth: Frege, Russell and the Analysis of Unities', *History and Philosophy of Logic* 24, 221–40.

Whitehead, Alfred North and Russell, Bertrand (1910) *Principia Mathematica*, vol. 1. Cambridge: Cambridge University Press.

Wittgenstein, Ludwig (1913) 'Notes on Logic' (NL), Appendix I of Wittgenstein (1979a).

——(1914) 'Notes Dictated to G. E. Moore in Norway' (NM), Appendix II of Wittgenstein (1979a).

——(1922) *Logisch-Philosophische Abhandlung/Tractatus Logico-Philosophicus* (TLP), trans. C. K. Ogden, London: Routledge and Kegan Paul Ltd, 2nd edition 1933; trans. D. F. Pears and B. F. McGuinness, London: Routledge and Kegan Paul Ltd, 1961, 2nd edition 1974.

——(1976) *Wittgenstein's Lectures on the Foundations of Mathematics: Cambridge, 1939* (from the notes of R. G. Bosanquet, Norman Malcolm, Rush Rhees, and Yorick Smythies), ed. C. Diamond. Ithaca, NY: Cornell University Press.

——(1979a) *Notebooks 1914–1916* (NB), 2nd edition, ed. G. H. von Wright and G. E. M. Anscombe, trans. G. E. M. Anscombe. Oxford: Basil Blackwell.

——(1979b) *Wittgenstein's Lectures: Cambridge, 1932–1935* (from the notes of Alice Ambrose and Margaret Macdonald), ed. A. Ambrose. Oxford: Blackwell.

——(1980) *Wittgenstein's Lectures: Cambridge, 1930–1932* (from the notes taken by John King and Desmond Lee), ed. D. Lee. Oxford: Blackwell.

7

Logical segmentation and generality in Wittgenstein's *Tractatus*

Thomas Ricketts

How does Wittgenstein understand quantification in the *Tractatus*? In particular, what becomes of higher-order quantification? Higher-order quantification is central to Frege's and Russell's universalist conception of logic, the conception that Russell vividly encapsulates in *Introduction to Mathematical Philosophy*: 'logic is concerned with the real world just as truly as zoology, though with its more abstract and general features' (Russell 1919: 169). In particular, it is by means of quantification of predicate as well as singular term positions that the principles of logic abstract from the content that distinguishes the propositions of the special sciences. Logical principles get applied here in that their higher-order variables as well as their first-order variables may be instantiated by vocabulary from any science. This view of logic and its application implicates many features of the 'old' logic that Wittgenstein rejects in the *Tractatus*: a need for general principles to mediate specific inferences in argumentation (5.132), general validity as the mark of a logical law (6.1231), an important distinction between logical axioms and the truths derivable from them (6.127), to name a few.[1] The alternative the *Tractatus* offers to Frege's and Russell's universalist conception displaces higher-order quantification from the central role it had occupied. What happens to higher-order quantification itself in the *Tractatus* scheme?

Frege's and Russell's views of quantification are linked to their views of logical segmentation. On their views, there is no great divide as regards logical segmentation between atomic sentences and logically compound sentences. This comes out strikingly in Frege's view that every sentence is multiply analyzable as the completion of an incomplete expression by type-appropriate completers, and in his assimilation in logic of the grammatical predicates of colloquial language to his first-level incomplete

[1] References to the *Tractatus* and *Prototractatus* are by Wittgenstein's numbering system. In quotations I have variously used, combined, and emended the Ogden–Ramsey and Pears–McGuinness translations of the *Tractatus* and the McGuinness translation of the *Prototractatus*.

expressions, formed by removing occurrences of a proper name from sentences of arbitrary complexity. Frege, without hesitation or argument, views his predicates as a kind of name, and in so doing, permits quantification of his predicate positions, treating quantification of predicate positions as fully parallel to quantification of proper name positions. The uniformity in Frege's and Russell's view of logical segmentation comes out as well in their view of the truth-functional connectives. For example, both Frege and Russell view material conditionals as relational sentences, so that the structure of a material conditional is parallel to that of '*aRb*'.

Wittgenstein thinks that this overarching uniformity in Frege's and Russell's view of logical segmentation leads them to overlook a distinction between two kinds of generality, generality proper and formal generality, as Wittgenstein calls them in *Prototractatus* 5.00533 and 5.005341. I will present an account of this distinction and say something about its significance for Wittgenstein's philosophy of logic. The place to begin is with Wittgenstein's view of the structure of expressions of truth-functionally compound sentences.

1 Two kinds of structure

Early on, Wittgenstein sharply distinguishes the structure of his elementary sentences from the iterative structure of truth-functional compounds of elementary sentences. Quantification, we shall see, is the product of interaction between these two different structures.

The structure of elementary sentences is the representing structure of a model. Wittgenstein epitomizes this view in the 1913 'Notes on Logic:'

> In '*aRb*' it is not the complex that symbolizes but the fact that the symbol '*a*' stands in a certain relation to the symbol '*b*'. Thus facts are symbolized by facts, or more correctly: that a certain thing is the case in the symbol says that a certain thing is the case in the world. (NL 96[4]. See 4.0311, 3.21, 3.1432, and 2.15.)

The basic idea here is that in a sentence like 'Socrates teaches Plato', that 'Socrates' is related to 'Plato' in a particular way in the sentence—that 'Socrates' is the subject and 'Plato' the direct object of the verb 'teach' in the sentence—says that Socrates teaches Plato. It is the particular relation in which symbols stand in the sentence that expresses a sense, that represents a situation in logical space. In the material conditional

Socrates teaches Plato ⊃ Plato teaches Aristotle,

the antecedent and consequent express same sense as they do when they stand alone. Their role in the sentence is exhausted by each component sentence's expressing the sense it does. Hence, in the conditional, the antecedent and consequent do not go proxy for anything. The relation in which the antecedent and consequent stand in the material conditional does not say anything. In this conditional, we encounter a kind of logical structure very different from the modeling structure of its antecedent and consequent. How does Wittgenstein understand this kind of structure?

An elementary sentence models a state of things (Sachverhalt). The sentence is true, if the state of things it models obtains; it is false, if the represented possibility does not obtain. Wittgenstein's leading idea is that sentences generally are expressions of agreement and disagreement with the independent truth-possibilities of elementary sentences (4.4). We might form such an expression by taking several elementary sentence-models and surrounding them by indications of which of their truth-possibilities are the ones with which our new sentence agrees and which are the ones with which the new sentence disagrees. This may be done by the tabular notation Wittgenstein presents in the 4.3s and 4.4s. At 4.442 Wittgenstein presents a tabular expression for the material conditional '$p \supset q$'. The tabular notation removes the temptation to think of the arrangement of sentences around the hook as representing the holding of relation between items designated by 'p' and 'q' (4.441), and so makes it patent that the relationship of the elementary sentences indicated by 'p' and 'q' to this sentence is very different from the occurrence of a name in an elementary sentence.[2]

We can extract from the tabular sign at 4.442 a particular way of forming a truth-function of two elementary sentences, of forming that truth-function which disagrees with the truth-possibility that the first elementary sentence is true and the second false, and agrees with the others. This way of forming a truth-function of two elementary sentences can be generalized to forming truth-functions of truth-functions of elementary sentences. From two truth-functions of elementary sentences Φ and Ψ, we can form that truth-function of elementary sentences that disagrees with any truth-possibility of elementary sentences that verifies Φ and falsifies Ψ, and agrees with all the others (5.31). In this way, we obtain intrinsically iterative truth-operations (5.3).

The sentences to which a truth-operation is applied to produce a truth-function of elementary sentences are the bases of the operation. 5.21 tells us that we can '. . . represent [*darstellen*] a sentence as the result of an operation that produces the sentence from other sentences (the bases of the operation)' (cf. PTLP 5.001). This remark clues us into Wittgenstein's understanding of the familiar Frege–Russell notation for truth-functionally compound sentences. Using the bracket notation of 5.501, let us consider how Wittgenstein would represent the construction of a truth-functionally compound sentence by means of successive applications of his N-operation, the truth-operation which yields a truth-function that agrees only with those truth-possibilities of elementary sentences which falsify each sentence among the bases of the operation.

Let ξ_1 = elementary sentence p. Then $N(\bar{\xi}_1) = {\sim}p$.
Let ξ_2 have as values $N(\bar{\xi}_1)$ and elementary sentence q. Then $N(\bar{\xi}_2) = p \wedge {\sim}q$.
Let $\xi_3 = N(\bar{\xi}_2)$. Then $N(\bar{\xi}_3) = {\sim}(p \wedge {\sim}q) = (p \supset q)$.

[2] This point is emphasized in Sullivan (2000: 180–1).

This construction is summarized by the bracket expression:

$$N(N(N(p), q)).$$

This bracket notation is, of course, a variant on Frege–Russell notation. Wittgenstein views a formula in Frege–Russell notation as representing the construction of a sentence from given elementary sentences by successive applications of truth-operations. Wittgenstein thus thinks of Frege–Russell formulas somewhat as we do, when we use them in metamathematical reasoning as shorthand for structural descriptions of object-language formulas. That is, Wittgenstein uses a description of the construction of a sentence via the iterated application of truth-operations to elementary sentences as an expression of the sense of that sentence. Descriptions of the construction of sentences are used for those very sentences.

This construal of Frege–Russell notation for truth-functions yields a consequence that will prove central to Wittgenstein's account of quantification. A sentence is an expression of agreement and disagreement with the truth-possibilities of elementary sentences (4.4), and so an expression of a particular truth-function of given elementary sentences. Sentences that express the same truth-function of the same elementary sentences have the same sense. 3.34 distinguishes between 'the accidental features of sentences that arise from the particular way the sentence-sign is produced' and the 'essential features which are required in order to enable the sentence to express its sense'. As Wittgenstein understands them

$$p \supset q,$$
$$\sim p \vee q,$$

and

$$N(N(N(p), q))$$

all portray different successive applications of truth-operations to 'p' and 'q' that yield the same truth-function of 'p' and 'q'. When we use these formulas as expressions for this truth-function, we find ourselves with three sentences expressing the same sense.[3] The features of these sentences that characterize a particular construction of the truth-function they all express are accidental features of the sentences. In particular, the occurrence of a particular molecular subformula in a Frege–Russell sentence like the examples above is generally an accidental feature of the sentence, for the occurrence of a molecular subformula typically characterizes only an initial segment of a particular construction of a truth-function, not the truth-function constructed.

[3] 5.141 says, in effect, that sentences expressing the same sense are the same sentence. We might then say that we have here three different sentence-signs belonging to the same sentence, to the same symbol.

2 Sentence-functions and quantification

Wittgenstein says very little about quantification in *Tractatus*, and nothing explicitly about higher-order quantification. What little he says explicitly about quantification is concentrated in the 5.52s. There Wittgenstein tells us:

I separate the concept *all* from truth-functions.

Frege and Russell introduced generality [Allgemeinheit] in connection with logical product or logical sum. This made it difficult to understand the sentences '$(\exists x).fx$' and '$(x).fx$', in which both ideas are embedded.

By *Tractatus* standards, this remark is fairly straightforward. As Wittgenstein views matters, a Frege–Russell generalization expresses the result of applying a truth-operation to its instances. There are two facets to this expression: the collection together of the instances of the generalization—the concept *all*—and the indication of a particular truth-operation.

5.501 gives us Wittgenstein's view of the first task:

When a bracket-expression [Klammerausdruck] has sentences as its terms—and the order of the terms inside the brackets is indifferent—then I indicate it by a sign of the form '$(\bar{\xi})$'. 'ξ' is a variable whose values are terms of the bracket-expression and the bar over the variable indicates that in the brackets it goes proxy for (vertreten) all its values.

(Thus, if ξ has three values P, Q, and R, then

$$(\bar{\xi}) = (P, Q, R).)$$

The values of the variable is something to be stipulated.

The stipulation is a description of the sentences that the variable goes proxy for.

How the description of the terms of the bracket-expression is produced is not essential.

We *can* distinguish three kinds of description: 1. direct enumeration, in which case we can, instead of the variable, simply put in its constant values; 2. giving a function fx whose values for all values of x are the sentences being described; 3. giving a formal law that governs the construction of the sentences, in which case the bracket-expression has as its terms all the terms of a series of forms.

My earlier discussion of Wittgenstein's understanding of Frege–Russell notation for truth-functions illustrates the first way of stipulating the values of a variable. Quantification involves the second—the description of the bases of a truth-operation by means of a function, a propositional function or, as I shall say, sentence-function.[4]

The *Tractatus* offers no account of sentence-functions. I take the context for the notion to be provided in the 3.3s by Wittgenstein's notion of an expression. 3.3 is Wittgenstein's version of the context principle: 'Only sentences have sense; only in the nexus of a sentence does a name have meaning.'

[4] Two terminological points. First, to avoid confusion with Russell's pre-1910 notion of a proposition as well as with contemporary conceptions, I translate 'Satz' by 'sentence'. Second, as Peter Hylton (1997) has observed, Wittgenstein follows Whitehead and Russell's usage, and uses 'function' in the *Tractatus* to mean 'propositional function'. Like Russell, these are the only functions Wittgenstein recognizes.

3.31 I call any part of a sentence that characterizes its sense an expression (or a symbol).
(A sentence is itself a symbol.)
Everything essential to their sense that sentences can have in common with one another is an expression.
An expression is the mark of a form and a content.
3.318 Like Frege and Russell, I conceive of a sentence as a function of the expressions contained in it.

4.24 indicates the application of these ideas to elementary sentences:

I write elementary sentences as functions of names so that they have the form '*fx*', '$\varphi(x,y)$', etc.

So, let us begin with elementary sentence-functions.

To fix ideas, I will assume that elementary sentences have Russellian structures, asserting that individuals stand in assorted *n*-adic relations.[5] Something elementary sentences can have in common that characterizes their sense is the occurrence of a name. Names must then be expressions. Names are not the only expressions in elementary sentences. Two elementary sentences might be just alike except that one contains occurrences of one name where the second contains occurrences of another name:

$$aRb;$$
$$cRb.$$

Here the common expression properly contains a name. I'll call such expressions of elementary sentences 'predicates'. Wittgenstein's understanding of the expressions contained in elementary sentences is, then, in its way Fregean.[6] We can view the elementary sentence as a function of a name, holding the predicate fixed. We can also view it as a function of a predicate, holding a name fixed. Any elementary sentence is a value of various sentence-functions. These possibilities of analysis are intrinsic to the structure by virtue of which elementary sentences present states of things. It is this structure that makes elementary sentences functions of the expressions they contain.

We can then form a sentence-function from an elementary sentence by converting any name or predicate in it into a variable-expression. We can use these elementary sentence-functions to stipulate values of sentence-variables to serve as bases for an application of a truth-operation that yields a truth-function of those values.

For example, let ξ_1 have as values the values of the sentence-function '*x* teaches Plato'. Then $N(\bar{\xi}_1) = \sim(\exists x)(x$ teaches Plato). Let ξ_2 have as its value $N(\bar{\xi}_1)$. Then $N(\bar{\xi}_2) =$

[5] There has been a long-running debate whether, on the assumption that elementary sentences have Russellian structures, in e.g. 'Socrates teaches Plato,' the relation between 'Socrates' and 'Plato' is itself a name in the sentence. My purposes here do not require me to take a stand on this issue. Following Wittgenstein's usage in 4.24, I will restrict myself to examples involving names of individuals.

[6] See especially Frege (1879: §9), where Frege says that an expression of a judgment can be regarded as a function of the signs that occur in it. This is the only place in his writings where Frege applies the word 'function' to linguistic expressions.

$(\exists x)$(x teaches Plato). As before, we might collapse the series of stipulations of values for variables into an embedded bracketed expression prefixed by an indication of a truth-operation:

N(N(x teaches Plato)).

By means of these simple examples, Wittgenstein indicates his analysis of Frege's and Russell's quantifiers, at least their first-order quantifiers (5.52). These simple examples, however, lead directly to the issue of the Fogelin–Geach exchange, namely Wittgenstein's understanding of multiply embedded quantifiers. Here I side with Geach and those early readers of *Tractatus*—Russell, Ramsey, and Carnap—none of whom saw any difficulty here.

To begin, how might the truth-function expressed in Frege–Russell notation by '$(\forall x)$(x teaches Plato)' be portrayed using sentence-functions and the N-operation? We could obtain this truth-function by applying the N-operation to the totality consisting of the results of applying the N-operation individually to each value of the sentence-function 'x teaches Plato'. But how, Fogelin asks, are we to stipulate the values for a variable whose values are these individual negations? 5.32 states:

> All truth-functions are results of successive applications to elementary sentences of a finite number of truth-operations.[7]

Fogelin urges (1987: 80) that this thesis rules out the unordered applications of a truth-operation to potentially infinitely many sentences. This, however, is exactly what the description of the desired bases appears to require.

I maintain that Wittgenstein approaches this issue from a Fregean direction. To obtain the desired bases, we first apply the N-operation to a representative arbitrary value of our sentence-function, say 'Socrates teaches Plato', to get the truth-function

N(Socrates teaches Plato).

Conversion of the name 'Socrates' into a variable-name yields a sentence-function whose values are the results of applying the N-operation to each value of 'x teaches Plato'. A further application of the N-operation to the bases given by this sentence-function yields the truth-function expressed by our universal quantification. This Fregean approach both avoids the unordered applications of a truth-operation to potentially infinitely many sentences, and is continuous with the understanding of elementary sentence-functions just sketched. The form common to an elementary sentence and the state of things it represents makes it possible to remove a name from an elementary sentence and so to transform a picture into a proto-picture—an *Urbild*, a pattern or form—that collects together all those elementary sentences that flesh out this pattern. Suppose we replace a name in an elementary sentence that occurs within the representation of the construction of a particular truth-function of that elementary

[7] The idea is repeated at 6.001, a comment of the general form of sentences exhibited in 6.

sentence. The elementary sentence-function in the formula collects together its values so that the entire formula collects together expressions of that particular truth-function of values of the elementary sentence-function. In this way then, we have not only elementary-sentence functions, but also sentence-functions formed by conversion of occurrences of expressions in elementary sentences that occur in portrayals of truth-functions into variable expressions. Although 4.0411 demonstrates that Wittgenstein knew how the features of quantifier-variable notation suit it to the expression of multiple generality, he does not himself in the *Tractatus* introduce notation adequate to the use of these sentence-functions to stipulate values of sentence-variables. As Peter Geach observed, it is easy to supplement Wittgenstein's notational suggestions and usage in the 5.5s to provide such a notation.[8] I have noted how we can portray truth-functions of elementary sentences by embedded bracket-expressions. We need a device in these embedded bracket-expressions to indicate the stage at which a name in elementary sentences is converted into a variable-name to yield a sentence-function. To this end, in a bracket-expression I will write the variable-expression in boldface *outside* the nexus of an elementary sentence next to the representation of the sentence-function obtained by use of the variable-expression. So,

$$\mathrm{N}(\boldsymbol{x}\ x \text{ teaches Plato}) = \sim(\exists x)(x \text{ teaches Plato});$$
$$\mathrm{N}(\mathrm{N}(\boldsymbol{x}\ x \text{ teaches Plato})) = (\exists x)(x \text{ teaches Plato});$$

and

$$\mathrm{N}(\boldsymbol{x}\ \mathrm{N}(x \text{ teaches Plato})) = (\forall x)(x \text{ teaches Plato}).$$

With these sentence-functions, Wittgenstein can capture multiply embedded quantification over objects. For instance, '$(\forall y)(\exists x)(x$ teaches $y)$' goes over into

$$\mathrm{N}(\boldsymbol{y}\ \mathrm{N}(\boldsymbol{x}\ x \text{ teaches } y)).$$

This embedded bracket-expression portrayal encapsulates the following construction. We begin with an arbitrary instance of the elementary sentence-function 'x teaches y', say, 'Socrates teaches Plato'. Let ξ_1 have as its values the values of the sentence-function 'x teaches Plato'. So $\mathrm{N}(\bar{\xi}_1) = \mathrm{N}(\boldsymbol{x}\ x$ teaches Plato). Let ξ_2 have as its values the values of the sentence-function '$\mathrm{N}(\boldsymbol{x}\ x$ teaches $y)$'. The desired sentence is $\mathrm{N}(\bar{\xi}_2) = \mathrm{N}(\boldsymbol{y}\ \mathrm{N}(\boldsymbol{x}\ x$ teaches $y))$. This embedded bracket-expression notation is thus a variant on Frege's and Russell's first-order quantifier-variable notation. It shows us how to understand Frege–Russell quantifier-variable formulas as representations of truth-functions of elementary sentences.

There is a complication here that deserves mention. In the 5.53s, Wittgenstein rejects use of a sign for identity (as a primitive sign) in favor of expressing the identity of objects by means of the identity of names for those objects, difference of object by difference of

[8] See Geach (1981). Geach, however, brings in the notion of a class of sentences to construe his version of Klammerausdruck notation. In contrast, I believe that talk of classes of sentences is to be understood in terms of the stipulation of sentence-variables. The notion of a sentence-variable is prior to that of a class of sentences.

names. Identity and difference of variable-names is similarly treated. Multiply quantified sentences thus get a non-standard construal. For example, as Wittgenstein understands it, the above portrayal of '$(\forall y)(\exists x)(x$ teaches $y)$' says that everyone is taught by someone *else*. I won't be concerned here about the rules for interpreting bracket expressions in conformity with Wittgenstein's views on identity. Perhaps the complications that his views on identity introduce into the construal of multiply embedded quantifiers explain why Wittgenstein's examples in the 5.52s are all monadic quantifications.[9]

Wittgenstein's analysis of first-order quantification exploits the use of sentence-functions obtained from portrayals of truth-functions to stipulate the bases for further applications of truth-functions. We see here how the interactions between the picturing structure of elementary sentences and the iterative structure of portrayals of truth-functions add to the expressive power of language. As long as we are stipulating values of sentence-variables either by lists of sentences or lists of elementary sentence-functions in the same variables, the tabular notation of the 4s suffices, and the iterative possibilities of truth-operations are idle. The formation of sentence-functions from truth-functions and their use to describe the bases for further truth-operations opens the way to expressing truth-functions that are not expressible in the tabular notation. Of course, whether there are truth-functions not expressible in tabular notation depends on the forms of elementary sentences and the number of names.[10]

3 Form-series variables and higher-order quantification

I have been explaining Wittgenstein's analysis of first-order quantification, quantification of name positions, in terms of sentence-functions. We have seen how Tractarian sentence-functions gather together the instances of first-order generalizations to serve as bases of truth-operations. I have restricted my examples to sentence-functions whose arguments are names, indeed names of individuals, given my assumptions concerning the forms of elementary sentences. Everything I've said about the analysis of quantification of name-positions holds equally for quantification of predicate-positions in elementary sentences. The arguments for these sentence-functions are predicates from elementary sentences.[11] We will not then be able to capture Frege's or Russell's second-order quantification using these sentence-functions, because the substituends for these variables are predicates of arbitrary truth-functional and first-order complexity.

Wittgenstein conceives of generalizations as truth-functions of their instances. This conception rules out Frege's impredicative higher-order quantification on pain of

[9] For a thorough and instructive discussion of the *Tractatus* treatment of quantification and identity, see Brian Rogers and Kai F. Wehmeier (forthcoming). I agree with Rogers and Wehmeier that their weakly exclusive interpretation is the view best attributed to the *Tractatus*.

[10] In my discussion of embedded first-order quantification in *Tractatus*, I have benefited from the discussions in Kremer (1992) and Floyd (2001).

[11] In particular, the predicate-quantifier in 5.5261 can naturally be taken to be a quantification of a predicate-position in an elementary sentence.

vicious circularity.[12] What about Russell's predicative higher-order quantification? At this stage, it looks tempting to accommodate higher-order quantification by following Russell's example and admitting sentence-functions that take non-elementary sentence-functions as arguments. We then could form a sentence-function

$$(\forall x)(x \text{ teaches Plato} \supset \Psi(x)),$$

whose values are the results of replacing '$\Psi(x)$' by any first-order formula containing only 'x' free. So the values for this putative sentence-function would include:

$$(\forall x)(x \text{ teaches Plato} \supset \sim(\text{x is mortal}))$$
$$(\forall x)(x \text{ teaches Plato} \supset (\exists y)(x \text{ teaches } y \wedge y \text{ teaches } x)$$
$$(\forall x)(x \text{ teaches Plato} \supset (\forall y)(x \text{ loves } y \supset (\exists z)(y \text{ loves } z))).$$

I claim that Wittgenstein's view of expressions and sentence-functions does not permit sentence-functions that gather together the instances of Russell's predicative second-order generalizations. The arguments for sentence-functions are expressions, and expressions characterize the senses of the corresponding values of those functions. So far we have recognized as expressions the names and predicates of elementary sentences. It should be clear from the discussion in §2 how these expressions characterize both the senses of those elementary sentences as well as the senses of truth-functions of those elementary sentences. So, do the substituends for 'Ψ(x)' generally characterize the corresponding instances of our sample schema

$$(\forall x)(x \text{ teaches Plato} \supset \Psi(x))?$$

On the Tractarian construal of Frege–Russell first-order formulas, a molecular substituend for '$\Psi(x)$' indicates only an initial segment in the construction of the sentence which is portrayed by the entire formula. More precisely, each such substituend is the result of replacing a name with a variable-name in the portrayal of the construction of a sentence, a sentence which in turn is the basis for the application of a truth-operation in the construction of the corresponding instance of our schema. In general, an initial segment in a portrayal of a construction of a sentence does not characterize the sense of that sentence: it characterizes neither the elementary sentences of which that sentence is a truth-function, nor the specific truth-function of those elementary sentences that the sentence expresses.[13] Wittgenstein's discussion of the construction of sentences indicates as much. 5.25 tells us the occurrence of an operation in the construction of a sentence does not characterize its sense. 5.501 states that in portraying the application of a truth-operation to bases, how the bases are described is inessential (*unwesentlich*). Inessential for what? It is inessential as regards (the sense of)

[12] An impredicative predicate-variable over Fregean concepts may be instantiated by a predicate containing a quantified variable over those same concepts.

[13] In some cases, a subformula does characterize the sense expressed by a sentential-sign in which it occurs, for example, if the subformula is equivalent to an elementary sentence or if the entire sentence is equivalent to the subformula.

the sentence produced by the application of the truth-operation to those bases. All this is an elaboration of the discussion of Wittgenstein's construal of Frege–Russell notation for truth-functions at the end of §1.

There is one question that remains. I have said that the values of the sentence-function

$$\varphi(a)$$

are all the *elementary* sentences containing the name '*a*'. But according to 3.313, the sentence-function representing an expression has as values all the sentences containing the expression. Truth-functions whose truth-arguments include elementary sentences containing the name '*a*' themselves contain that name. Yet on my account these are not in general among the values of

$$\varphi(a).$$ [14]

4.23 observes that names—and so elementary sentence predicates—occur in sentences only in the context of elementary sentences. I earlier argued that it is the occurrence of expressions of elementary sentences in sentences generally that yield a sentence-function like

$$\sim(x \text{ teaches Plato}),$$

whose values are negations of the instances of the elementary sentence-function

$$x \text{ teaches Plato}.$$

5.442 states:

> If we are given a sentence, then *with it* we are also given the results of all truth-operations that have it as their base.

5.47 elaborates:

> An elementary sentence really contains all logical operations in itself. For '*fa*' says the same thing as '$(\exists x).fx \,.\, x = a$'.
>
> Where there is compositeness [*Zusammengesetztheit*], argument and function are present, and where these are present, we already have all the logical constants.

I suggest that Wittgenstein views the sentence-function whose values are the elementary sentences containing '*a*', to encompass virtually all sentences containing '*a*'. At the end of the chapter, I will return to say something about how 'virtually' should be understood here.

Let us return to the topic of higher-order predicative quantification. I have argued that Tractarian sentence-functions cannot be used to capture Russell's predicative second-order quantifications. Still, we can recognize in formal terms that exploit the

[14] I am grateful to Michael Potter for raising this issue.

iterative possibilities of truth-operations the Frege–Russell sentences that are instances of predicative second-order generalizations. It seems that we should be able to stipulate these sentences to be values of a sentence-variable and so to serve as the bases for a truth-operation. And so we can. The instances of a Russellian second-order generalization are not the values of a sentence-function. Rather, they are the members of a form-series (*Formenreihe*). Wittgenstein's third way from 5.501 for describing the values of variables gives him the resources to simulate higher-order quantification.

5.501 says that we can describe the sentences that are to be the values of a variable, the terms of a bracket-expression, by 'giving a formal law that governs the construction of the sentences, in which case the bracket-expression has as its terms all the terms of a series of forms'. I shall call variables whose values are stipulated by formal laws, form-series variables. Wittgenstein says very little about this third means for stipulating the values of a variable. It is linked to his notion of an operation. An operation is a logically significant recursive notational procedure for generating sentences from sentences (see the 5.23s). At 4.1273, we find an informal presentation of Wittgenstein's only example of a form-series variable:

If we want to express in begriffsschrift the general sentence, '*b* is a successor of *a*', then we require an expression for the general term of the form-series:

$$aRb,\ (\exists x){:}aRx\ .\ xRb,\ (\exists x,y){:}aRx\ .\ xRy\ .\ yRb, \ldots$$

The general term of a form-series can only be expressed by means of a variable, for the concept: member of this form-series, is a *formal* concept. (Frege and Russell overlooked this. The way they want to express general sentences like the above is therefore incorrect; it contains a vicious circle.)

We can determine the general series of forms by giving its first term and the general form of the operation that produces the next term out of the sentence that precedes it.[15]

I'm not going to be concerned here with the formal details of specifying form-series. Göran Sundholm, after raising assorted difficulties in interpreting Wittgenstein's theses, examples, and notational scraps on this topic, concludes:

The author of the *Tractatus*... constitutes the finest example of a philosopher whose technical formal capacities do not reach the outstanding level of his logico-philosophical thinking. (Sundholm 1992: 76)

I will assume that wherever there is a procedure for recognizing members of a class of Russellian first-order sentences in terms of logically significant features of their construction, there is a formal law that generates a form-series whose members are these sentences.[16]

[15] In *Prototractatus*, a version of this remark (PTLP 5.005342, 5.00535, and 5.005351) follows very closely the introduction of formal generality in 5.00531.

[16] Two points. First, specification of the form-series may, as in 4.1273, require imposing a logically arbitrary alphabetic ordering on signs. This is permitted so long as any alphabetic ordering of the same signs could be employed to specify a form-series of the same terms. Second, my talk of a procedure to recognize the members of a class of sentences is ambiguous between decision procedures (recursive) and search procedures (semi-recursive). For my purposes here, all I need is the weaker reading. Nothing I say excludes the stronger reading.

Suppose we have, then, a form-series whose members are the first-order sentences formed from some countable stock of names. There will then be a form-series of sentence-functions of a single variable individual-name formed from these sentences. These terms are the substituends for Russell's second-order variables ranging over, as Russell would put it, first-order propositional functions of a single individual. We can now exploit this form-series to construct that series of sentences we get when we replace '$\Psi(x)$' in

$$(\forall x)(x \text{ teaches Plato} \supset \Psi(x)),$$

first by the first term in our enumeration of sentence-functions, then by the second term, then by the third term. That is, we have a form-series constructed from the form-series enumerating first-order sentence-functions which enumerates the Russellian instances of '$(\forall x)(x$ teaches Plato $\supset \Psi(x))$'. We can stipulate a variable with the members of the form-series as values. Application of a truth-operation to the bases given by this variable thus simulates Russellian second-order quantification.[17]

In general, to represent the construction of the application of a truth-operation to the terms of a form-series, first construct the first term of the form-series. At the next stage, use a form-series variable to gather the members of the form-series and apply the desired truth-operation to it. Understood in this way, the simulation of embedded Russellian second-order quantifications poses no problems. Consider a second-order formula of the form

$$(\forall \Psi)(F[\Psi] \supset (\exists \Phi)G[\Psi, \Phi]),$$

where '$F[\ ]$' and '$G[\ ,\ ]$' are first-order matrices. Let us suppose that 'x is mortal' is the first sentence-function in our form-series of first-order sentence-functions. So, first, we construct the first-order sentence

$$G[x \text{ is mortal}, x \text{ is mortal}].$$

Now introduce a form-series variable whose first term is our sentence and whose n^{th} term is the result of substituting for the second occurrence of 'x is mortal' the nth sentence-function in our enumeration, and apply generalized disjunction to the bases given by this variable. This gives us the truth-function portrayed by:

$$(\exists \Phi)G[x \text{ is mortal}, \Phi].$$

This sentence can now be a basis for further truth-operations. Moreover, sentence-functions can be obtained from it via the replacement of expressions within elementary sentences by variable-expressions. We can then go on to construct the sentence:

$$F[x \text{ is mortal}] \supset (\exists \Phi)G[x \text{ is mortal}, \Phi].$$

Consider the form-series whose n^{th} term is the result of substituting the nth sentence-function in our enumeration in this frame for 'x is mortal'. An application of generalized conjunction to the terms of this series gives us the desired truth-function. Our

[17] In this use of form-series to simulate higher-order quantification, I benefited from conversations with Michael Potter and Peter Sullivan.

truth-function is then a generalized conjunction, whose conjuncts are conditionals. The antecedent of each conjunct is a first-order sentence. The consequent is a generalized disjunction, each disjunct of which is a first-order sentence.

Our form-series enumerating first-order sentence-functions thus gives us simulations of Russellian second-order quantifications over first-order propositional functions of a single individual. We must, however, remember that the second-order variables are not Tractarian variable-expressions used in the portrayal of sentence-functions in stipulations of the values of a sentence-variable. They are, as I shall say, pseudo-variables, to be understood in terms of the form-series determined by the form-series enumerating first-order sentence-functions and the construction of the formula, the bracket expression, that fills the scope of the pseudo-variable. We may, nevertheless, incorporate these pseudo-variables into the bracket notation for portraying truth-functions of elementary sentences, if we wish. We can now formulate a formal law that generates all sentence-functions that take a single individual name as argument in this extended notation. These expressions are the substituends for Russell's third-order variables generalizing over, as he puts it, second-order propositional functions of a single individual. We can then repeat the preceding construction to simulate Russellian fourth-order quantifications by introducing a new type of pseudo-variable. And so on. In this way, modulo the original form-series generating a class of first-order sentences, we can simulate full predicative second-order logic. I see no bar to the construction of the entire ramified hierarchy in Tractarian terms.[18] I would like to think that this procedure is what Wittgenstein has in mind, when he speaks in 5.252 of advancing from type to type in the hierarchies of Russell and Whitehead.

I spoke of the *Tractatus analysis* of first-order quantification, but only of its *simulation* of Russellian second-order quantification. I approached Wittgenstein's treatment of quantification by searching for Tractarian resources for stipulating variables whose values are the instances of the desired quantifications. In the case of first-order quantifications, sentence-functions gathered together the desired instances. Sentence-functions are not, however, guaranteed to gather together sentences that are the result of arbitrarily many applications of truth-operations to elementary sentences. So, to accommodate higher-order quantification, I turned to form-series. A Russellian variable over second-order propositional functions of a single individual may be instantiated by any first-order monadic predicate of arbitrary logical complexity. My Tractarian simulation of second-order quantification restricts the substituends for my second-order pseudo-variables to the sentence-functions present in a given form-series. There is no guarantee that there is a form-series whose members include all first-order sentences. That will depend on the forms of elementary sentences and the number of names. The generality present in my Tractarian simulation of higher-order

[18] Here in addition to pseudo-variable sentence-functions, we will need to introduce notation for pseudo-higher-order sentence-functions that take pseudo-lower-order sentence-functions as pseudo-arguments.

quantification is formal generality, not quantificational generality with its source in the representing structure of elementary sentences.

This difference between proper quantificational generality and formal generality has a further consequence. For Russell, all generality is captured by quantification of positions defined within the ramified theory of types. While the hierarchy of the ramified theory can be simulated by the use of form-series, form-series can also be used to express truth-functions that are not expressible with the ramified theory. Indeed, the use of form-series to define the ancestral of a relation over countably many objects is an example. The same sort of construction used to simulate Russellian second-order quantification can be applied to *any* form-series whose members are truth-functions of elementary sentences obtained by arbitrarily many applications of truth-operations and so arbitrarily many embeddings of scopes of variable-expressions in elementary sentences. Nothing logically distinguishes the simulation of higher-order quantification from other uses of form-series.

4 The general form of sentences

The account I have given of the sources of generality in the representing structure of elementary sentences and the iterative possibilities of truth-operations sheds light on the proper understanding of the general form of sentences. The general form of sentences is presented at 6:

> The general form of a truth-function is: $[\bar{p}, \bar{\xi}, \mathrm{N}(\bar{\xi})]$.
> This is the general form of sentences.

This specification of the general sentence-form presents several problems. The one most salient in the present context is that the general form of sentences says nothing about how the bases for later applications of truth-operations depend on previously constructed sentences.

I have given several examples of constructions of sentences, constructions of truth-functions of elementary sentences from elementary sentences. Each construction begins with elementary sentences, given individually or by elementary sentence-functions. Each step in the construction has two stages: stipulation of a sentence-variable and the application of a truth-operation to the terms the variable goes proxy for. For the construction to be a stepwise construction, we must at each stage stipulate the values of that stage's sentence-variable in terms of what has gone before. This does not mean that the sentences represented by the variable stipulated at stage *n* must have been constructed at an earlier stage. This will be the case only if the variable is stipulated by means of a list. We may stipulate a variable at stage *n* by means of a sentence-function obtained from a sentence constructed at an earlier stage. To have constructed a sentence is to have available the sentence-functions obtainable from it. After all, the instances of the sentence-function are constructible in the same way as the value of it that appears at an earlier stage in the construction. Here then we have a potentially infinitary step in the construction of sentences. Wittgenstein treats form-series stipulations in a parallel way. Having constructed a sentence at an early stage, we may at a later stage stipulate a sentence-variable by means of a formal law for a form-series whose first member is that sentence. The application of a truth-operation to the values of a form-series variable counts as a single step, even though arbitrarily many steps may be required for the construction of the individual terms of the form-series. Form-series then introduce another potentially infinitary dimension into sentence construction.

These examples suggest that the general form of sentences has an intrinsically schematic character.[19] Wittgenstein does say that the general sentence-form is a variable (4.53), and uses the notation for form-series introduced at 5.2522 to specify it at 6. Nevertheless, I do not take this specification itself to be the specification of sentences via a form-series, at least not in the way that my earlier examples of the use of form-series variables in the construction of sentences are. The general sentence-form is rather a schema for the construction of any sentence: it is the most general form for the construction of truth-functions of elementary sentences. I take 6.001 to indicate as much in commenting on the notation for the general sentence-form that 6 presents:

> This says nothing other than that every sentence is the result of successive applications of the operation $N(\bar{\xi})$ to elementary sentences.

So, the opening term of the general form is an indication that the construction of sentences begins with some elementary sentences. The ways in which elementary sentences may be specified, apart from lists of individual sentences, depends on their forms, the forms of their constituting names, and the number of names of each form. Hence, the general form is properly silent here (see 5.55–5.551). The general form is also silent as to how a later specification of the bases for an application of the N-operation must depend on sentences constructed earlier. The bases are specified by stipulating the values of a sentence-variable. The only requirement here is that the stipulation of the values of the variable '*is a description of the symbols and states nothing about what is signified*' (3.317).[20]

3.317 requires that the description be in terms of symbols and so pertain to the senses expressed by the values of the stipulated variable.[21] It is clear how the stipulation of values for a variable by means of a sentence-function satisfies this condition: not only the arguments for a sentence-function but the sentence-function itself characterizes the senses of the values of that sentence-function. What about a stipulation by means of a form-series? The sentences of a form-series are the sentences produced by successive applications of an operation (5.2521–5.2522). An operation, 5.241 says, does not characterize a form, but a

[19] I'm indebted here to Juliet Floyd, who years ago in an unpublished paper urged the schematic character of the general sentence-form.

[20] The antecedent of this passage in PTLP is 5.0052, and is a part of the material from 5.003–5.005341 that contains the antecedents of TLP 5.501.

[21] This requirement is consistent with my claim in §3 that the way in which the bases for an application of a truth-operation are described does not characterize the sense of the sentence thereby produced.

difference of forms: it is a formal relation between the bases of the operation and its result (5.22 and 4.122). I suggest that two sentences share the same logical form, if both are the same truth-function of elementary sentences of the same form. The logical form of an elementary sentence corresponds to the sentence-function obtained by replacing each name in the elementary sentence with a variable-name (see 3.315). Consider the N-operation. It yields a sentence which is true just in case the bases to which it is applied are all false. This is the difference it marks—or the relation it expresses—between its bases and its results. Consider now the operation that generates the bases for the truth-function which says that some object b R-follows a. This operation, applied to the conjunction of disjunctions which says, so to speak, that n distinct steps along the relation R suffice to get from a to b, yields the conjunction of disjunctions which says that it takes $n + 1$ steps. The difference between n steps and $n + 1$ steps along a relation is the difference in form, the difference in sense, marked by this operation. Differences in logical form are differences in sense. Since operations mark a difference in form as between bases and result, the specification of a form-series in terms of the iterated application of an operation does involve a description of symbols, and not just signs.

I can now redeem an earlier promise. I noted how the values of the elementary sentence function

$$\varphi(a)$$

are only the elementary sentences containing '*a*', and not the truth-functions of those elementary sentences, whereas, according to 3.313, the sentence-function representing an expression has as values all the sentences containing the expression. I suggested that the elementary sentence-function virtually encompasses all the sentences containing '*a*'. I can now explain what 'virtually' comes to here. The variable whose values are all the sentences containing '*a*' will be the restriction of the general form of sentences to bases which include some elementary sentences that are values of the elementary sentence-function '$\varphi(a)$'.[22]

If my suggestion about the schematic character of the general sentence-form is correct, the notation of 6 is not an example—or not merely an example—of Wittgenstein's limitations as a designer of notation. Rather, in light of the exposition of the 5s on the theme, 'A sentence is a truth-function of elementary sentences,' 6 crystallizes the open-ended presentation of the general form from 4.5: 'Es verhält sich so und so.' That's the way it is.[23]

[22] I take Anscombe (1959: 97) to voice a similar idea.

[23] I have benefited from conversations on the topics of this chapter with Enzo DePellegrin, Juliet Floyd, Peter Hylton, Michael Potter, Peter Sullivan, and especially Warren Goldfarb. I presented some of the ideas in this paper at a *Tractatus* conference at Utrecht in 2000 and at an earlier Stirling *Tractatus* workshop in 2004. I have presented versions of this paper at the 2005 Stirling *Tractatus* conference, at the 2007 Leipzig Wittgenstein und Wissenschaft conference, and at colloquia at Carnegie Mellon University and Stanford University. Discussions on these occasions prompted improvements in the paper. I am especially grateful to Peter Sullivan for extensive comments on the penultimate version of this chapter.

References

Anscombe, G. E. M. (1959) *An Introduction to Wittgenstein's Tractatus*. Philadelphia: University of Pennsylvania Press.

Floyd, Juliet (2001) 'Number and Ascription of Number in Wittgenstein's *Tractatus*', in Floyd and Shieh (2001), 145–92.

Floyd, Juliet and Shieh, Sanford, eds (2001) *Future Pasts: The Analytic Tradition in Twentieth Century Philosophy*. Oxford: Oxford University Press.

Fogelin, Robert (1987) *Wittgenstein*, 2nd edition. London: Routledge.

Frege, Gottlob (1879) *Begriffsschrift, eine der arithmetischen nachgebildete Formelsprache des reinen Denkens*. Halle: Verlag von Louis Nebert.

Geach, Peter (1981) 'Wittgenstein's Operator N', *Analysis* 41, 168–71.

Hylton, Peter (1997) 'Functions, Operations, and Sense in Wittgenstein's *Tractatus*', in Tait (1997), 91–105.

Kremer, Michael (1992) 'The Multiplicity of General Propositions', *Nous* 26, 409–26.

Rogers, Brian and Wehmeier, Kai F. (*Review of Symbolic Logic*, forthcoming) 'Tractarian First-Order Logic: Identity and the N-Operator'.

Russell, Bertrand (1919) *Introduction to Mathematical Philosophy*. London: Allen and Unwin.

Sullivan, Peter (2000) 'The Totality of Facts', *Proceedings of the Aristotelian Society* 100, 175–92.

Sundholm, Göran (1992) 'The General Form of the Operation in Wittgenstein's *Tractatus*', *Grazer Philosophische Studien* 42, 57–76.

Tait, W. W., ed. (1997) *Early Analytic Philosophy*. La Salle, Ill.: Open Court Press.

Wittgenstein, Ludwig (1913) 'Notes on Logic' (NL), in Wittgenstein (1979).

—— (1979) *Notebooks: 1914–1916*, 2nd edition, ed. G. H. von Wright and G. E. M. Anscombe, trans. G. E. M. Anscombe. Chicago: University of Chicago Press.

—— (1989) *Logisch-philosophische Abhandlung/Tractatus logico-philosophicus: kritische Edition*, ed. B. F. McGuinness and J. Schulte. Frankfurt am Main: Suhrkamp.

8

Does the *Tractatus* contain a private language argument?

William Child

It is a central feature of so-called 'new' readings of Wittgenstein that they find in the *Tractatus* an absence of positive philosophical doctrines, a kind of quietism, and an explicitly therapeutic approach that have traditionally been associated with Wittgenstein's later philosophy. Seen from the perspective of a 'new' reader, this reveals 'a novel kind of continuity' in Wittgenstein's thought (Crary 2000: 4). Seen from the perspective of a 'traditional' reader, it involves reading back into the *Tractatus* elements that properly belong to Wittgenstein's later work. I want to explore the justice of this complaint in connection with a recent argument of Cora Diamond's, that the *Tractatus* contains a private language argument: an argument to the effect that private objects in other people's minds can play no role in the language I use for talking about their sensations.[1] The argument Diamond finds in the *Tractatus* is not the private language argument of *Philosophical Investigations*. But the appearance of such an argument in Wittgenstein's early writings, she thinks, brings out important elements of continuity in his work: it challenges the orthodox idea that 'the topic of privacy' is distinctively 'a topic of Wittgenstein's *later* philosophy', that its appearance in the later work is 'indicative of a shift in Wittgenstein's philosophical interests to topics within the philosophy of mind, not of interest to him in the *Tractatus*', and that 'his treatment of the topic [is] an illustration of the fundamental shifts in his overall philosophical position' (262). And the '*Tractatus* private language argument', Diamond thinks, is important not only for our understanding of the development of Wittgenstein's philosophy but also for the discussion of realism and anti-realism more generally, 'especially as that discussion has been shaped by Michael Dummett' (284).

My chapter has three parts. In section 1, I challenge Diamond's interpretation on internal grounds. To find a private language argument in the *Tractatus*, I argue, we

[1] This argument is presented in 'Does Bismarck Have a Beetle in his Box? The Private Language Argument in the *Tractatus*' (Diamond 2000). All page references to Diamond in the text and notes refer to this paper.

would have to read Tractarian doctrines about naming and use in ways that might, indeed, be suggested by Wittgenstein's later writings but which are entirely absent from the earlier work. In section 2, I discuss the account of sensation language that Wittgenstein offered in 1929, and argue that it poses a prima facie challenge to 'new' readings of the *Tractatus*. Section 3 explores the relation between the *Tractatus* and Dummettian realism. I defend Dummett's suggestion that the *Tractatus* embodies a form of semantic realism against Diamond's arguments, and show the helpfulness of Dummett's framework in reflecting on the nature of Tractarian analysis.

1 The '*Tractatus* private language argument'

Diamond argues that the *Tractatus* contains a private language argument, the immediate target of which is a set of views advanced by Russell in the period 1905–13. Russell held that 'Every proposition which we can understand must be composed wholly of constituents with which we are acquainted' (Russell 1912: 32). He also held that each of us is acquainted with objects that no one else can be acquainted with: our own experiences and, possibly, our selves.[2] So there are propositions about a subject's self and experiences that only she can understand, because they contain constituents with which only she can be acquainted. But even though I cannot understand such a proposition about another person's self or experiences, I can identify it by description and may know that it is true: for I can understand a descriptive proposition, composed wholly of constituents with which I am acquainted, which is true or false in virtue of the truth or falsity of the proposition I do not understand. In that way, Russell thought, someone else's private objects can play a role in one's language. For example:

> when we say anything about Bismarck, we should like, if we could, to make the judgement which Bismarck alone can make, namely, the judgement of which he himself is a constituent. In this we are necessarily defeated, since the actual Bismarck is unknown to us. But we know that there is an object B, called Bismarck, and that B was an astute diplomatist. We can thus *describe* the proposition we should like to affirm, namely, 'B was an astute diplomatist', where B is the object which was Bismarck . . . This proposition, which is described and is known to be true, is

[2] Sense-data, for Russell, are not themselves private objects. A sense-datum is the object that a subject is aware of—and Russell regarded it as 'a theoretic possibility' for the sense-data perceived by one person also to be perceived by another person (1913: 34). But he also held that, as well as experiencing a sense-datum, I can experience my experiencing of that sense-datum: 'my seeing the sun', or 'my desiring food', he writes, are 'object[s] with which I am acquainted' (Russell 1912: 27). And, he thought, 'it is obvious' that it is only our *own* experiencings that we can experience in this way, not other people's (Russell 1912: 27). Now for Russell, these experiencings are not private objects in exactly the sense involved in the private language argument of *Philosophical Investigations*. For the target of that private language argument is the view that sensations are '*logically*' private objects. And Russell regarded it as an *empirical* fact, rather than 'a matter of a priori necessity', that the only experiencings we can experience are our own (1913: 34–5). So there is a case for thinking that it is just a mistake to see Russell's view as one on which others' sensations are private objects in any recognizably Wittgensteinian sense. But for present purposes I shall accept this part of Diamond's interpretation in order to focus on what she says about the *Tractatus*.

what interests us; but we are not acquainted with the proposition itself, and do not know *it*, though we know it is true. (Russell 1912: 31)

In this case the private object is Bismarck himself: Bismarck's ego. But, according to Diamond, Russell gives the same treatment to propositions about anything else with which only Bismarck can be acquainted: Bismarck's toothache, for example.[3] It is this set of views that is challenged by the private language argument Diamond finds in the *Tractatus*.

In this first part of the chapter, I do two main things. First, I examine the case Diamond offers in support of her claim that the *Tractatus* contains a private language argument. She highlights an important difference between Russell's account of quantification and the treatment of general sentences that Wittgenstein offers in the *Tractatus*. But, I shall argue, in claiming that the *Tractatus* treatment of generality entails that private objects in other people's minds can play no role in our language for talking about others' sensations, Diamond imputes to the *Tractatus* a specific conception of use, with a definite epistemic content, that is not actually there; she reads back into the *Tractatus* a conception of use that only surfaces in Wittgenstein much later on. Second, I examine Diamond's account of the kind of positive treatment of claims about others' sensations that, she thinks, is implicit in the *Tractatus*; Tractarian analysis of such propositions, she says, 'involves an early version of the later insistence that where there are symptoms (inductively based justifications for a kind of statement) there must also be criteria (something that would count, non-inductively, as justification)' (281). Here, again, I argue, Diamond interprets the *Tractatus* in a way that makes it seem far more similar to Wittgenstein's later work than it really is—by reading into the *Tractatus* epistemic concerns that have no place there.

I should emphasize at the outset that I am not arguing that, at the time of writing the *Tractatus*, Wittgenstein actually envisaged analysing statements about others' sensations in a way that treated those sensations as private objects. In the first place, there is plainly no explicit account of sensation language in the *Tractatus*. That is a particular instance of a more general point: that Wittgenstein simply had no views about the specific kinds of analysis that would be appropriate for particular parts of ordinary language. In the second place, insofar as it is reasonable to speculate about the kind of analysis of sensation language that Wittgenstein might have favoured, had he been pressed to provide one, it seems plausible that he would have opted for something along the lines of the account of thoughts suggested by *Tractatus* 5.542 and the remarks in a much-quoted 1919 letter to Russell (WIC 98–9). On this view, thoughts are composed of psychical objects: objects that are straightforwardly open to public, empirical investigation, just as we can investigate the worldly objects with which those psychical objects are correlated. There is no reason why Wittgenstein should not have envisaged

[3] Diamond writes: 'I want to emphasize that, although Russell's example is the self, his discussion of it is meant to apply to everything with which other people are directly acquainted, and with which we ourselves cannot be acquainted' (266).

extending this treatment from the case of others' thoughts to the case of others' sensations—regarding them not as private objects knowable only by their possessors but as psychical objects whose nature is discoverable by empirical investigation. My case against Diamond, therefore, is not that the *Tractatus* actually envisages an account of sensation language that treats others' sensations as private objects. It is, rather, that there is nothing in the *Tractatus* to rule out such an account. Diamond sees a private language argument in the *Tractatus*. I argue that her interpretation depends on projecting onto the *Tractatus* views of Wittgenstein's that only appear later on.

1.1 Quantification, naming, acquaintance, and use

On Diamond's interpretation, the key to the '*Tractatus* private language argument' lies in the difference between Russell's and Wittgenstein's treatments of generality. A crucial element in Russell's view is the idea that I can grasp a general proposition that is true or false in virtue of the truth or falsity of singular propositions that I cannot understand. For example, I understand the proposition, 'There is something that is a toothache and is had by Bismarck'. That proposition is true (we may suppose) in virtue of the truth of the proposition '*t* is a toothache', where '*t*' is a name for some token toothache of Bismarck's. But I cannot understand the proposition '*t* is a toothache', since I cannot be acquainted with the toothache that is named by '*t*'. What allows me, nonetheless, to understand the general proposition is the fact that I am acquainted with the universal, *toothache*, and that I have 'a *general* grasp of what it is for a property or relation to be instantiated in *some* or *all* cases' (271). Those resources, Russell thinks, allow me to understand the sentence 'there is an object that is a toothache and is had by Bismarck', even though I could never be acquainted with such an object or give it a name. Diamond stresses that on Wittgenstein's account, by contrast, a general proposition cannot have as one of its instances a singular proposition that I cannot understand. To see why, we need to appreciate Wittgenstein's treatment of general sentences.[4]

On Wittgenstein's view, all propositions are truth-functions of elementary propositions. And every truth-function is representable in terms of a single truth-function, joint negation, which Wittgenstein represents with the symbol, N: '$N(\bar{\xi})$', he explains, 'is the negation of all the values of the propositional variable ξ' (TLP 5.502). And, he says, there are three ways in which we can specify the propositions that are represented by the variable ξ—the propositions that are jointly negated by the N operation: '1. direct enumeration, in which case we can simply substitute for the variable the constants that are its values; 2. giving a function fx whose values for all values of x are the propositions to be described; 3. giving a formal law that governs the construction of the propositions' (TLP 5.501). To understand how Wittgenstein proposed to see general propositions as truth-functions of elementary propositions, we need to take the second case. Consider the general proposition, $\forall x \neg fx$. Wittgenstein's idea is in

[4] The summary account that follows is mine, rather than Diamond's.

effect that the construction of that proposition has two stages. The first stage is to describe the propositions to which the N operation is to be applied: in this case, by specifying that the relevant propositions are *all* values of the function fx. That gives us the propositions $(fa, fb, fc \ldots)$. The second stage is to apply the operation of joint negation to those propositions, to yield the complex proposition $(\neg fa \,\&\, \neg fb \,\&\, \neg fc \ldots)$. That complex proposition, on Wittgenstein's account, captures the content of the general proposition, $\forall x\ \neg fx$, or equivalently, $\neg\exists x\ fx$. A second application of the N operation, to the complex proposition that was produced by the first application, will yield the general proposition $\neg\forall x\ \neg fx$, or $\exists x\ fx$.

The feature of this account that Diamond highlights is that, according to Wittgenstein, 'a quantified sentence is a construction from singular sentences' (270). The idea of construction needs to be handled with care—though the idiom of 'construction' is drawn from Wittgenstein himself (e.g. 5.503). In the case of the proposition $(\neg fa \,\&\, \neg fb)$, the complex proposition is constructed from elementary propositions in a straightforward sense; we take two elementary propositions that are, as it were, already to hand and construct a complex proposition by conjoining them. But in the case of a general proposition, we need not be in a position to specify the relevant propositions in any way other than as all values of the function fx. So the manner in which a general proposition is a construction from elementary propositions is different from the manner in which a proposition like $(\neg fa \,\&\, \neg fb)$ is such a construction—in a way that corresponds to the difference between Wittgenstein's first and second ways of describing the propositions to which the N operation is to be applied. But as long as we bear that difference in mind, there is no harm in employing the 'construction' idiom, as Diamond does. Now Diamond emphasizes that a crucial feature of Wittgenstein's account of the construction of complex propositions from elementary propositions is that, if I am to understand a non-elementary proposition, then the elementary propositions that go into its construction must themselves be propositions that I understand: they must be made up of names that I understand, or could understand. That is obvious in the case where a complex proposition is constructed from elementary propositions that are given by direct enumeration: I cannot understand $(\neg fa \,\&\, \neg fb)$ unless I understand fa and fb. But it is equally true, Diamond thinks, for the case where a general proposition is constructed by successive applications of the N operator to *all* values of a given propositional function fx: the instances of a general proposition that I understand must be elementary propositions that I can in principle understand. And that rules out the possibility that I could understand a general proposition about toothaches that had, as one of its instances, a singular proposition about Bismarck's toothache that I could not understand. On Wittgenstein's account, therefore, I cannot use general sentences to identify by description an object that Bismarck can name and I cannot. So, *pace* Russell, private objects in others' minds can play no role in my language—by figuring in the truth-conditions of my claims about others' sensations.

I agree with Diamond that, in the *Tractatus*, if there are objects that I cannot name, they can play no role in my language. But that will only show that private objects in others' minds can play no role in my language if we assume that I cannot have names for such objects. If we reject that assumption and allow that I *can* give names to others' private objects, there will be no barrier to allowing that propositions about others' experiences are true or false in virtue of the presence or absence of private objects in their minds. So if there is a private language argument in the *Tractatus*, it depends on the principle that it is impossible for one person to have names for other people's private sensations. What might explain or justify that principle? There are two obvious possibilities. First, one might claim that someone can only name an object if she is acquainted with it—in Russell's sense of acquaintance. Since one cannot be acquainted with others' private objects, it would follow immediately that one cannot have names for others' private objects. Second, one might claim that which objects are referred to by the names in one's language is determined by our use of those names, and then argue that we cannot use a name in a way that makes it function as a representative of another person's private sensation. Some interpretations of the *Tractatus* suggest the first route. Diamond's reading pursues the second route. But, I shall argue, neither is supported by the text of the *Tractatus*.

In early readings of the *Tractatus*, it was often assumed that objects were sense-data and consequently that naming an object would require Russellian acquaintance with it. On this interpretation, the *Tractatus* will have the same reason as Russell for holding that one cannot give a name to a private object in someone else's mind: that one cannot be acquainted with such an object. The idea that Tractarian objects are sense-data is not as popular as it once was. But we can still ask whether there is anything to be said for the idea that Tractarian naming requires something like Russellian acquaintance. An argument for that idea would involve two claims: that understanding a name requires knowing (*kennen*) its meaning (*Bedeutung*); and that knowing (*kennen*) the *Bedeutung* of a name is or involves being acquainted with it. But against that argument, I would respond that, while the *Tractatus* does say that understanding a name involves knowing its referent, it does not say or imply that knowing the referent of a name requires acquaintance with the referent. I will expand on both points.

TLP 3.263 tells us that 'The meanings of primitive signs can be explained by means of elucidations. Elucidations are propositions that contain the primitive signs. So they can only be understood if the meanings [*Bedeutungen*] of those signs are already known [*kennen*].' Since the primitive signs are names, that gives firm support to the claim that understanding a Tractarian name requires knowing (*kennen*) its referent. But does knowing (*kennen*) the referent of a name require acquaintance with it? Nothing in the *Tractatus*, I think, suggests that it does.

In the first place, we know that Wittgenstein himself preferred to translate 'kennen' as 'know' rather than 'be acquainted with' in the passages that talk about knowing (*kennen*) an object.[5] Second, 2.0123 tells us that 'if I know [*kennen*] an object I also

[5] See his comments on the translation of 2.0123 (LO 59).

know [*kennen*] all its possible occurrences in states of affairs'; 2.01231 adds that 'if I am to know [*kennen*] an object, though I need not know [*kennen*] its external properties, I must know [*kennen*] all its internal properties'. But both remarks are hard to make sense of if knowing an object is conceived in terms of Russellian acquaintance. How could mere Russellian acquaintance with an object automatically bring with it knowledge of all its possible occurrences in states of affairs—knowledge of its internal properties? And how could one be acquainted with an object without knowing any of its external properties? Third, and relatedly, it is worth remarking that, in Russell, the idea that a subject is *acquainted with* an object has, as its converse, the idea that the object is *given to* the subject.[6] Now in the *Tractatus*, the remarks just quoted about knowing (*kennen*) an object are immediately followed by this: 'If all objects are given, then at the same time all *possible* states of affairs are also given' (2.0124). That juxtaposition makes it plausible to think that, for Wittgenstein too, knowing an object goes hand in hand with the object's being given to one. In that respect, his view parallels Russell's. But now consider 5.524: 'If objects are given, then at the same time we are given *all* objects.'[7] If Wittgenstein had Russell's conception, on which an object's being given is a matter of acquaintance, it would be impossible to make sense of that remark: Russellian acquaintance with objects does not and could not bring with it acquaintance with *all* objects. That strongly suggests, again, that Wittgenstein has a quite unRussellian conception of what it is to know an object: knowing (*kennen*) an object is not acquaintance.[8] There is, I conclude, no reason to think that the *Tractatus* holds that one can have a name for an object only if one is acquainted (in Russell's sense) with that object.

Diamond herself explicitly rejects the idea that the connection between Tractarian names and objects is effected by a mental act in favour of the view that 'what objects we are thinking about is something that is shown in the language we use' (274).[9] So, when she takes it for granted that I cannot name others' private sensations, she is not depending on the impossibility of my being acquainted with such objects. What, then, is she depending on? Why can't the language I use show that the objects I am thinking about include others' private objects—just as, presumably, it might show that the objects I am thinking about include ones that are unknowably distant in time or

[6] See e.g. 'The Relation of Sense-data to Physics' (1914: 7), where the question whether *sensibilia* can exist without being *given* is treated as equivalent to the question, whether *sensibilia* can exist without anyone being *acquainted with* them.

[7] 'Given' (*gegeben*) appears eight times in the index of the *Tractatus*. In each of those occurrences, something's being *given* is associated with generality.

[8] Peter Sullivan drew my attention to the contrast between Russell's association between acquaintance with an object and the object's being given and Wittgenstein's association between knowing an object and its being given. A similar point is made by Hidé Ishiguro: 'Such claims [as TLP 2.0124 and 5.524] would be nonsense if we were to understand the given as something that we are immediately aware of, or anything like Moore or Russell's sense-data, or even objects of phenomenological reflection' (Ishiguro 2001: 40).

[9] For Diamond's rejection of the mental act model, see 290, n. 33. See also the comments about understanding at 273.

space, or that they are infinitely many in number?[10] The answer is that Diamond has a particular conception of the *use* of language, from which it follows that other people's private objects are irrelevant to the use I make of my language: 'Objects known only to Bismarck', she writes, 'play no role in the language which I use in everyday life in talking about Bismarck and things in his mental life' (275). But what reason is there to think that the conceptions of *use* and of *playing a role* that are implicit in that claim are conceptions to which Wittgenstein was already committed in the *Tractatus*?

When Wittgenstein later turned his attention to the question, how we communicate with one another about sensations, he certainly did construe 'use' in a way that led him to conclude that, if we construe sensations as private objects, then others' sensations will drop out as irrelevant to our use of sensation language (PI §293). But that is not the only way of construing 'use'. Someone who disagrees with the arguments Wittgenstein offers in *Philosophical Investigations* may agree with the principle that words have the meanings they do because of the use we make of them; she may simply disagree about how we should construe 'use'.[11] Now the *Tractatus* clearly maintains that there is a close association between meaning and use: 'In philosophy', it tells us, 'the question "What do we actually use this word or this proposition for?" repeatedly leads to valuable insights' (6.211); 'In order to recognize a symbol by its sign we must observe how it is used with a sense' (3.336); and 'If a sign is *useless*, it is meaningless' (3.328). But those are very general claims. By themselves, they tell us very little about the kinds of meanings our words can or do have. If we are to accept Diamond's interpretation, we need to be convinced that, at the time of the *Tractatus*, Wittgenstein already had the conception of use that is required to make a private language argument go through: a conception on which a *state of affairs* can be relevant to my use of a sentence only if I can recognize whether or not that state of affairs obtains, and on which an *object* can be relevant to my use of a word only if I can recognize the presence or absence of that object. But for that to be true, the *Tractatus* would need to forge a much closer tie between the notion of use and epistemic notions such as justification, evidence, and recognition than it actually does. Without that, there is nothing in the *Tractatus* to show that we cannot use names that refer to others' private objects. My complaint, then, is that Diamond's reading imputes to the *Tractatus* a notion of use that carries a much greater and more definite epistemic weight than anything that is actually to be found there.

There is a general lesson. It is agreed on all sides that the *Tractatus* is not concerned with epistemology: it contains no theory of perception, for example, and no account of ostensive definition; and it says explicitly that theory of knowledge has no special or foundational place in philosophy (4.1121). Those who think that the *Tractatus* bases a

[10] For Wittgenstein's acknowledgement that there may be infinitely many objects, see TLP 4.2211: 'Even if the world is infinitely complex, so that every fact consists of infinitely many states of affairs and every state of affairs is composed of infinitely many objects, there would still have to be objects and states of affairs.'

[11] For discussion of this point, and other relevant issues, see Craig (1982).

theory of thought and language on a realist ontology see this lack of interest in epistemology as going hand in hand with Wittgenstein's agnosticism about the nature of objects. On this view, an account of the epistemological underpinnings that link our thought and language to objects would be needed for a complete understanding of the relation of thought and reality. But since Wittgenstein has no idea what kinds of things objects will turn out to be, he is not in a position to give such an account; and since he does not regard the nature of objects as a *philosophical* question at all, he does not think it is philosophy's business to do so. Those who read the *Tractatus* in less realist ways offer a different account.[12] On their view, the absence from the *Tractatus* of any account of acquaintance or ostensive definition is not a gap in Wittgenstein's overall view; it is, instead, a positive feature of his understanding of the relation between thought and reality. For, on this interpretation, the link between names and objects is determined not by acquaintance but by our use of propositions. But those who offer such an account cannot stop at that point. For we do not understand the suggestion that meaning is determined by use unless we have some account of what use is: of what aspects of our use of propositions go to determining their meaning. If it is claimed that specific consequences flow from this account of meaning (for example, that we cannot use words in such a way that they refer to private objects in others' minds), that claim needs to be justified by an account of use that shows how those consequences follow. And where the consequences have an epistemic dimension (as they do in the case of Diamond's private language argument), the account of use must have an epistemic dimension too. The surprising implication would then be that the *Tractatus* does, after all, contain or imply a definite epistemology—albeit an epistemology that is allied to an account of use, rather than to an account of acquaintance or ostensive definition.[13] An alternative would be to hold that the *Tractatus* gives us only the outline of an account of meaning—an outline that needs to be filled in with a specific account of use before we can draw any conclusions about the meanings of the words of our ordinary language. Either way, though, we face a challenge if we accept this kind of use-based account of the relation between names and objects in the *Tractatus*: How exactly is use to be understood? In what ways, if any, is use constrained by epistemology? And what is the textual justification for these claims about use?

[12] For accounts of this sort, see for example Ishiguro (1969), McGuinness (1981), Goldfarb (unpublished).

[13] These remarks are directly relevant to McGuinness's reading, as well as to Diamond's. He writes that in the *Tractatus* 'the *use* of language shows that something is taken as a true elementary proposition, in that the other things that are then unhesitatingly said reflect its being so taken. Put mentalistically this means that we base inferences on it.' And, he says, 'what a proposition means and whether it is true depends (a) on its multiplicity . . . (b) on the relation that it has to the propositions that are or would be unhesitatingly accepted as true in the language' (1985: 101). McGuinness finds an 'implicit verificationism' (1985: 102) in this account. But that claim needs justification; for there is no reason why 'unhesitating acceptance' must be understood in a way that involves some kind of verificationism.

1.2 Criteria, symptoms, and truth-conditions

I have urged that we should not read the concerns of Wittgenstein's later philosophy back into the relation between use and meaning in the *Tractatus*. The same point applies when we turn to consider what positive account of propositions about others' sensations would be consistent with the *Tractatus*.[14] Here, Diamond argues that Tractarian analysis of such propositions 'involves an early version of the later insistence that where there are symptoms (inductively based justifications for a kind of statement) there must also be criteria (something that would count, non-inductively, as justification)' (281). Consider the proposition 'Bismarck's toothache is getting worse'. This is not itself an elementary proposition, since it stands in logical relations to other propositions that do not contain it: for example, the proposition 'Bismarck does not have toothache'. So it must be analysable. And 'what shows us how to analyze the sentence will be pursuing further the sentence's inferential relations'. Diamond proceeds:

> Suppose we think that Bismarck's behaviour gives us grounds, but merely inductive grounds, for inferring that his toothache is getting worse. . . . Well, what is it that the behaviour gives us inductive evidence for? What would show us that *THAT* is the case?. . . . If such-and-such behaviour is mere inductive evidence, then *something else* has to be what it is evidence *for*: there has to be something else that would, if it were established, constitute grounds for inferring that the toothache is getting worse. (280)

'The stage is set' she thinks, for a view on which propositions about others' mental states are analysable in terms of non-inductive behavioural grounds for asserting them.

It is true that, if something is inductive evidence for the claim that Bismarck's toothache is getting worse, there must be something else that the evidence is evidence for. But how should we interpret this point? The basic idea concerns truth-conditions: if there is inductive evidence that Bismarck's toothache is getting worse, there must be something that would *constitute* Bismarck's toothache's getting worse—something that would *make it true* that Bismarck's toothache is getting worse. But Diamond reads the point in an epistemic way; 'there has to be something', she writes, 'which would, if established, count as non-inductive *grounds for holding* that it is getting worse' (280, my emphasis).[15] It is this epistemic reading of the point that leads Diamond to say that the *Tractatus* contains 'the view that a process *in someone else's mind* stands in need of outward criteria' (281), and that the *Tractatus*' 'conception of logical analysis' embodies a critique of philosophical realism because it 'involves a tie between what our non-

[14] As I said above, I think it plausible that, had Wittgenstein considered how to treat claims about others' sensations within the framework of the *Tractatus*, he would have opted for an account similar to that which he suggests for thoughts. But I put that aside for the purposes of investigating Diamond's suggestion.

[15] Note the slippage, in the displayed quotation in the previous paragraph, from the constitutive formulation, 'if such-and-such behaviour is mere inductive evidence [that Bismarck's toothache is getting worse], then something else has to be *what it is evidence for*', to the epistemological formulation, 'there has to be something else that would, if it were established, constitute *grounds for inferring* that the toothache is getting worse'.

elementary sentences say is so, and sentences capable of serving as grounds for taking those sentences to be true or false' (286). But there is no basis in the text or context of the *Tractatus* for this interpretation.

The *Tractatus* does use the word 'grounds' in giving an account of the relation between a non-elementary proposition and the elementary propositions that figure in its analysis: in the terminology of the *Tractatus*, the '*truth-grounds* [*Wahrheitsgründe*] of a proposition [are] those truth-possibilities of its truth-arguments that make it true' (5.101). And Wittgenstein applies the notion of truth-grounds in explaining inferential relations (5.12) and in explaining probability (5.15). But that notion of a ground has nothing to do with evidence. Giving the truth-grounds of an ordinary proposition is a matter of spelling out what has to be the case for it to be true. There is no suggestion that the truth-grounds of a proposition function as evidence for asserting the ordinary proposition.[16]

Contrary to Diamond's account, therefore, there is nothing in the *Tractatus* conception of analysis that suggests any particular way of analysing propositions about others' sensations. In particular, there is nothing to suggest an account that exploits anything like Wittgenstein's later conception of criteria: non-inductive grounds for assertion that are built into the meanings of the relevant terms. The lesson, once more, is that we should not read Wittgenstein's later views back into the doctrines of the *Tractatus*.

2 The *Tractatus* and Wittgenstein's 1929 theory of sensation language

Whatever may be implicit in the *Tractatus*, the *Tractatus* itself contains no actual account of sensation language. But there is an account in the writings and conversations from 1929. Does reflection on the 1929 account tell us anything about the *Tractatus*? I shall argue that the 1929 account of sensation language appears to give a role to others' private objects that is prima facie puzzling if we read the *Tractatus* in the way advocated by 'new' or 'resolute' readings of the sort championed by Diamond and others. If the *Tractatus* already contains the views that 'new' readings attribute to it, how well can we make sense of what Wittgenstein says about sensations in 1929? The discussion of that question certainly does not lead to a knock-down argument against Diamond, or anyone else. But it does pose a puzzle that 'new' readings need to address.

I start with a statement of the 1929 theory:

[16] Having advanced her positive suggestion about the treatment of sensation language in the *Tractatus*, Diamond herself goes on to make the point that the kind of 'grounds' that are revealed by Tractarian analysis of a sentence 'need not... have any connection with what, in our actual practice, we really do count as grounds for accepting that sentence as true' (281). As far as I can see, that acknowledgement completely undercuts the suggestion that the *Tractatus* contains something that is in any way at all a precursor of the later treatment of mental language in terms of symptoms and criteria.

> We could adopt the following way of representing matters: if I, LW, have toothache, then that is expressed by means of the proposition 'There is toothache'. But if that is so, what we now express by the proposition 'A has toothache', is put as follows: 'A is behaving as LW does when there is toothache'. Similarly we shall say 'It is thinking' and 'A is behaving as LW does when it is thinking'. (You could imagine a despotic oriental state where the language is formed with the despot as its centre and his name instead of LW.) It's evident that this way of speaking is equivalent to ours when it comes to questions of intelligibility and freedom from ambiguity. But it's equally clear that this language could have anyone at all as its centre.
>
> Now, among all the languages with different people as their centres, each of which I can understand, the one with me as its centre has a privileged status. This language is particularly adequate. How am I to express that? That is, how can I rightly represent its special advantage in words? This can't be done. For, if I do it in the language with me as its centre, then the exceptional status of the description of this language in its own terms is nothing very remarkable, and in the terms of another language my language occupies no privileged status whatever.—The privileged status lies in the application, and if I describe this application, the privileged status again doesn't find expression, since the description depends on the language in which it's couched. And now, which description gives just that which I have in mind depends again on the application. (PR 88–9)

The point made in the second paragraph is repeated in the parallel account in *Wittgenstein and the Vienna Circle*:

> All these languages can be translated into one another. Only what they have in common mirrors anything.
>
> Now it is noteworthy that *one* of these languages has a distinctive status, namely that one in which I can as it were say that I feel *real* pain.
>
> If I am A, then I can, to be sure, say 'B is behaving as A does when he feels pain', but also 'A is behaving as B does when he feels pain'. One of these languages has a distinctive status, namely the one whose centre I am. The distinctiveness of this language lies in its application. It is not expressed. (WVC 49–50)

Wittgenstein says that the language with me as its centre has a 'privileged', 'distinctive', or 'exceptional' status; it has a 'special advantage'; it is 'particularly adequate' for describing or expressing my own experience. But the privileged status of this language cannot itself be expressed in language. Instead, it 'lies in the application' of language. Wittgenstein is here committed to the idea that there really is something—a special advantage to the language with me as its centre—which cannot be expressed in language but lies in the application of the language. Though he does not use the words 'show' and 'say', he could just as well have put the point he is making in those terms: there is a special advantage to the language with me as its centre; we cannot *say* what the special advantage is—it cannot be 'rightly represent[ed] . . . in words'; but its advantage is *shown* by the application of the language. And this use of the say/show distinction seems entirely serious. Wittgenstein is saying that there really is something privileged or distinctive about the language with me as its centre, which we cannot express in language but which comes out in the application of language. There is no hint that these remarks are not meant literally; so that, for example, Wittgenstein means only that we find it seductive to think that the language with me as its centre has a privileged status that lies in its application, but that, on further reflection, we can see that the seductive thought is just a mistake. On the contrary, all the indications are that, in these passages from 1929, when Wittgenstein is explaining, applying, and developing his earlier views, he is making a straightforward use of the distinction between saying and showing. Those who argue that the *Tractatus* intends no serious use of that distinction owe us an account of these passages; if Wittgenstein never thought that there was anything that could be shown but not said, what is he doing here with the distinction between what can be 'represented in words' and what 'lies in the application of the language'?

That is one way in which the 1929 account of sensation language poses a challenge for 'new' readings of the *Tractatus*. I turn now to the detail of the 1929 account.

2.1 The role of private objects in the 1929 account

What exactly is the account of sensation language that Wittgenstein proposes in 1929? It is natural to read the passages I have quoted as starting with the idea that sensations are private objects: objects that are accessible only to the subject, with identity conditions that are independent of any links to behaviour or external context. On this interpretation, the issue addressed in the 1929 discussion is how, on that assumption, we should account for the meanings of sensation words. And the proposal is in effect that each of us has two sensation languages: a purely introspective, private sensation language for our own use; and a public sensation language with which we can communicate about sensations, our own and other people's. So my word 'toothache', say, has two different meanings. In its communicative use, 'toothache' has a purely behavioural meaning; in this sense, 'the word "toothache" means the same in "I have toothache" and "He has toothache"' (PR 91). But my word 'toothache' can also be used with another sense, to pick out 'what is *primary*' in my experience (PR 91): the experiential datum itself. Used in that sense, my word cannot intelligibly be applied to anyone else; nor can its meaning be understood by anyone else. Thus, when Wittgenstein says that I understand all the different languages with different people as their centres, he is referring only to the public meanings of words in these languages: when Jones says 'There is toothache' or 'Brown is behaving as Jones does when there is toothache', I understand the public meaning of what he says; but his words also have a private meaning which I cannot understand. On this interpretation of the 1929 account, the solipsist who tries to capture the distinctiveness of his own pains by saying 'only I feel *real* pain' (see WVC 50), or 'only I really have *sense-data*' has a genuine insight: the words 'real pain' and 'sense-data' on his lips really do have meanings on which they are applicable to him and no one else. But, on Wittgenstein's view, that very fact makes the insight inexpressible. For if the proposition 'Other people have sense-data' is unintelligible, he thinks, then the propositions 'Other people do not have sense-data' and 'Only I have sense-data' are unintelligible, too: 'In the sense of the phrase "sense data" in which it is inconceivable that someone else should have them, it cannot, for this very reason, be said that someone else does not have them' (PR 90). So the solipsist's genuine insight is something that can only be shown, not said.[17]

If this two-level reading of the 1929 account is correct, then private objects in other people's minds have an important role to play in Wittgenstein's account. For the account depends on the idea that each of us has her own private sensations, sensations with which other people cannot be acquainted. Admittedly, others' private sensations play no role in the meanings of terms in my own sensation language: for my word 'toothache', in its third-person application, gets its meaning not by attachment to others' private sensations but by association with behavioural dispositions. Nonetheless, other people's private sensations do play an essential role in the philosophical account of sensation language: it is a central part of that account that each person has a private language in which they can talk about their own private sensations; and if I can understand the philosophical account, I must understand the idea that other people have private sensations. But how can I understand that idea; how am I to make sense of the claim, 'each person refers directly to her own private sensations'? If I use the term 'private sensation' with the meaning it has in my own private language, then a private sensation is something that only I can have; in that sense, the phrase 'other people's private sensations' makes no sense. If I use the term 'private sensation' with its communicative meaning, on the other hand, its meaning is explained in terms of behaviour. In that case, what I mean when I say 'other people refer to their own private sensations' is, more or less, that other people use sensation words in ways that are tied up with their own behavioural dispositions. But (on the current interpretation) that is not at all what Wittgenstein is proposing.

We face a dilemma. If the two-level interpretation of Wittgenstein's 1929 account is right, then Wittgenstein has a definite philosophical view of the workings of sensation

[17] It is interesting to compare this 'two-level' reading of Wittgenstein's 1929 view with the accounts offered by Schlick and Carnap in writings of a similar period (see Schlick 1949, originally published in 1935; Carnap 1995, originally published in 1932). Something like the two-level view figures in both Schlick and Carnap. But, as I read them, the two-level view figures only as something to be rejected. Thus, Schlick thinks there are in fact 'uniform one-to-one correspondence[s]' between subjective experiences on the one hand and physical states or events on the other. Because of those correspondences, he thinks, we can translate first-person claims about experience into physical language without loss. But suppose there were no such uniform connections. On that supposition, 'there would be a world of feeling which could not be talked about in the physical language . . . [A]ll that I could communicate would be expressible in [the physical] language. . . . [But] in addition to it there would be a private language in which I could reflect about the world of feeling' (1949: 405). But in Schlick's view, that is only how things would be given counterfactual suppositions; it is not how things actually are. Similarly, Carnap describes a view from which it would follow that protocol language (the 'language of direct experience' or 'phenomenal language' (1995: 44)) 'could be applied only solipsistically [and] there would be no intersubjective protocol language' (1995: 80). But he says explicitly that that is not his view. (If I understand him correctly, I differ here from Pears. He writes that 'it is a corollary of [Schlick's and Carnap's] theory that the content of experience remains necessarily incommunicable, but [that] this does not matter, because if the content of experience does vary from one person to another, the variation will necessarily remain undetectable' (Pears 1988: 303). As I read Schlick and Carnap, they think that, given the correlations that actually obtain between subjective experiences and physical states, the content of experience is not in fact incommunicable).

language. But if sensation language does work in the way that he is suggesting, that philosophical view will itself be impossible to state. One way of reacting to that dilemma is to regard it as a *reductio* of the two-level interpretation. On this view, the idea that we can understand a philosophical theory, even though we can find no intelligible way of stating it, is a delusion: if our attempts to state a theory turn out to issue in nonsense, then we cannot rescue that theory by claiming that the nonsense is illuminating nonsense and that, though the attempt to state the theory makes no literal sense, it succeeds in pointing to a deeper 'truth'. That will be the reaction of someone who reads the 1929 discussions in the spirit of 'resolute' readings of the *Tractatus*. On this view, however the 1929 writings and conversations are to be understood, they cannot be advancing a philosophical view on which each person has a private sensation language in which she uses words to refer to her own private objects.[18]

A different way of reacting to the dilemma is to hold that something like the two-level interpretation really does make best sense of the 1929 writings and to insist, therefore, that others' private objects do play a crucial role in Wittgenstein's account of sensation language. If we take this line, we must explain how we are supposed to understand the account, given that it cannot be stated. And the obvious suggestion will be that, when we attempt to state the account, we are trying to say something that can really only be shown in the language. In support of that suggestion, we can note the apparently serious use of the notion of showing in the statement of the 1929 theory, which I highlighted above. We can also note how plausible the interpretation is in its own right. Wittgenstein says explicitly that there is a use of the term 'sense-datum' in which I can talk about what is *primary* in experience, and in which it is inconceivable that someone else should have sense-data (PR 90–1). There seems no plausibility in the suggestion that Wittgenstein's point is supposed to apply only to himself; on the contrary, the idea seems to be that for each person there will be a use of the term 'sense-datum' which that person can employ in talking about what is primary in her experience, and in which it is inconceivable that anyone else should have sense-data. And if that is right, it follows immediately that Wittgenstein is committed to recognizing the existence of senses of the term 'sense-datum' that he himself cannot understand.

I acknowledge that the bearing of these considerations on the *Tractatus* is indirect. In particular, I am not claiming that Wittgenstein's 1929 discussions contain an account of sensation language of the kind which Russell had advocated some twenty years earlier and which, according to Diamond, was rejected in the *Tractatus*. But I do claim that the most plausible reading of the 1929 account is strikingly 'irresolute' in at least two respects. First, it makes serious use of the idea that something that cannot be represented in language (namely, the privileged status of the language with oneself as its centre) can nonetheless be shown by the application of language. Second, it promotes a philosophical view of sensation language that assigns a crucial role to others' private

[18] I consider in the next section what positive interpretation of the 1929 discussions we might offer if we take this 'resolute' view.

objects, a role that it is impossible to articulate in language. If that is right, then it puts some pressure on the kind of 'resolute' reading of the *Tractatus* that has been promoted by Diamond and others. For if Wittgenstein was steadfastly 'resolute' in the *Tractatus*, is it credible that he would in 1929 have offered a view of sensation language that is 'irresolute' in the way I have described?

In fact, Diamond's own remarks about the *Tractatus*' treatment of sensation language face a direct version of the same challenge; for they attribute to the *Tractatus* a view of sensation language that is 'irresolute' in exactly the same way as the account that I have attributed to *Philosophical Remarks*. According to Diamond, the '*Tractatus* private language argument' rules out construing talk about *other people's sensations* in terms of reference to private objects in their minds. But it leaves in place the idea that we should construe a person's talk of *her own sensations* in terms of her reference to private objects in her own mind (see Diamond 268, 283). On Diamond's reading of the *Tractatus*, however, the suggestion that each person's first-person uses of sensation words refer to private objects that cannot be referred to by anyone else—including us, the philosophers offering this account—ought to be rejected as unintelligible. For, *ex hypothesi*, we can make no sense of this talk of others' private objects. And if cannot understand this talk of others' private objects, we cannot understand the view of the first-person use of sensation words that Diamond says Wittgenstein accepted in the *Tractatus*. But a guiding principle of Diamond's reading of the *Tractatus* is that we should not attribute to Wittgenstein views which, by the standards of the *Tractatus*, would make no literal sense. Diamond's own interpretation of Wittgenstein's view of sensation language, however, seems to do precisely that.

2.2 A deflationary interpretation of the 1929 account?

Suppose someone rejects the interpretation of the 1929 account that was explored in the previous section. What alternative might she offer?

It might be suggested that we should read the 1929 account of sensation language in essentially the same deflationary way as we read certain comments in *Philosophical Investigations*. In PI §403, Wittgenstein writes:

> If I were to reserve the word 'pain' solely for what I had hitherto called 'my pain', and others 'LW's pain', I should do other people no injustice, so long as a notation were provided in which the loss of the word 'pain' in other connexions were somehow supplied. Other people would still be pitied, treated by doctors and so on. It would, of course, be *no* objection to this mode of expression to say: 'But look here, other people have just the same as you!'[19]

We can fill out this suggestion in a way that emphasizes the parallel with Wittgenstein's 1929 discussion. Suppose each of us uses the word 'pain' only in the first-person case; in the third-person case, we replace the word 'pain' with the phrase 'A is behaving as

[19] See also the discussions at BB 59 and 66 of the suggestion that other people might give me a special place in the notation.

I behave when there is pain'. Each of these languages has the same expressive power as our ordinary language; we understand them all equally well. In *Philosophical Investigations*, the point of this thought-experiment is that the difference between the different languages is purely notational; as long as the non-standard notation suggested in §403 is used in a way that preserves all the ordinary links with external circumstances, bodily injury, behavioural manifestations, sympathetic reactions, medical treatment, and so on, there will be no substantive disagreement between someone who uses the non-standard notation and someone who uses our ordinary notation.[20] And when he makes this point in *Philosophical Investigations*, Wittgenstein is obviously not committing himself to treating anyone's sensations as private objects: the argument starting at §243 has ruled out treating one's own sensations as private objects; the argument of §293 has ruled out treating others' sensations as private objects.

Now on the deflationary interpretation of Wittgenstein's 1929 discussion, we should read the 1929 account in exactly the same way as we read the discussion in *Philosophical Investigations*. On this view, the account does not start from the idea that each of us makes direct, introspective reference to our own sensations; and the point of introducing the notation 'A is behaving as LW behaves when there is toothache' is not to offer an explanation of how, given that starting point, it is possible for us to talk about others' sensations—an explanation that works by analysing or explicating the meaning of 'A has toothache' in terms of behaviour. Rather, Wittgenstein is simply accepting our sensation language as given, taking it for granted without explanation that we do succeed in communicating with one another about the character of our sensations, and arguing, as he does in PI §403, that we could use a different notation without loss. Indeed (on this interpretation) the point Wittgenstein makes in PI §403 is precisely anticipated in *Philosophical Remarks*:

> The two hypotheses that other people have toothache and that they behave just as I do but don't have toothache, possibly have identical senses. That is, if I had, for example, learnt the second form of expression, I would talk in a pitying tone of voice about people who don't have toothache, but are behaving as I do when I have. (PR 93)

But this is an implausible interpretation. In the first place, even if we want to maximize the closeness of Wittgenstein's 1929 account to the mature view contained in *Philosophical Investigations*, it is hard to believe that the alternative notation he suggests—'A is behaving as LW behaves when there is toothache', and so on—is intended *merely* as a variant way of talking about sensations. On the contrary, the alternative notation seems evidently intended as an attempt to articulate the use of sensation language in a particularly revealing way; it is an attempt to show what is 'logically essential' to the

[20] In *Philosophical Investigations*, this point is used to argue that the differences between solipsists, idealists, and realists are, similarly, merely notational. (I disagree here with the reading offered by David Bell (1996), which sees PI §403 as expressing Wittgenstein's own adherence to a kind of solipsism.)

representation of toothache—which, on the 1929 theory, is a particular connection to behaviour.[21]

The deflationary interpretation reads back into the 1929 discussion features of Wittgenstein's mature treatment of sensation language that had still to be worked out. But it seems implausible that Wittgenstein had really freed himself, as early as this, of the idea that sensation terms get their meanings in the first-person case by direct attachment to the introspectible character of sensations. The 1929 material certainly contains early versions of some of the views Wittgenstein worked out later on. For example, there is a critique of the claim that I can grasp what it is for someone else to have toothache by starting with my knowledge of what it is for me to have toothache and appealing to the formulation, 'when I say he has toothache, I mean he now has what I once had' (PR 91; cf. PI §§350–1). There is a critique of the tendency to treat the proposition 'I cannot feel your toothache' as a statement of empirical fact rather than a logical truth (PR 90; cf. PI §253). There is a critique of the tendency to model mental phenomena on physical phenomena, and to treat having an experience on the model of awareness of an object (PR 88, 94). And there is an attempt to make use of the link between having a sensation and behaving in certain ways, and of the resemblances between different people's sensation-behaviour, to understand our capacity to communicate about sensations. But this is evidently a first shot. There is no sign in 1929 of many of the key elements of the later view: for example, the suggestion that sensation terms are acquired in the first-person case not by introspection but as a learned addition to natural expressive behaviour. And there is certainly no sign in 1929 of the argument against private ostensive definition, which was developed only in 1936. For the deflationary interpretation to be plausible, the 1929 account would need to contain not just early signs of Wittgenstein's later views, but the later views themselves. But it clearly does not do that.

If the deflationary reading were right, that would show that the 1929 account of sensation language made no use at all of the idea of sensations as private objects. That would remove the sense of tension I have suggested between the 1929 account and the 'resoluteness' that 'new' readers ascribe to the *Tractatus*. But if I am correct in rejecting the deflationary reading, the tension remains.[22]

[21] For the idea that one way of representing the facts may be more perspicuous than another, because it 'show[s] clearly what [is] logically essential to the representation', see PR 88.

[22] In a recent paper (published after this chapter was written), David Stern has offered an account of the 1929 account of sensation language that is broadly in line with the 'deflationary reading' I have described (see Stern 2010). In particular, he stresses the continuity between the strategy Wittgenstein pursues in exploring an alternative form of sensation language in PR 88–9, and the strategy pursued in PI §403. (It should be stressed that the focus of Stern's deflationary reading of the 1929 account is the development of Wittgenstein's view of sensation language; Stern is no defender of a 'new' or 'resolute' reading of the *Tractatus* (on which, see Stern 2004, 40–55).) I hope to respond to Stern's discussion in future work.

3 The *Tractatus* and Dummettian realism

Fifty years ago, Michael Dummett famously suggested that *Philosophical Investigations*, with its 'doctrine that the meaning is the use', 'contains implicitly a rejection of the classical (realist) Frege-*Tractatus* view that the general form of explanation of meaning is a statement of the truth-conditions' (1959: 185). On Dummett's view, a central element in the development of Wittgenstein's philosophy is the transition from a realist view of meaning in the *Tractatus*, where the meaning of a proposition is explained in terms of its truth-conditions, to a form of anti-realism in Wittgenstein's later philosophy, where the meaning of a proposition is explained in terms of the conditions that would justify its assertion. That view of Wittgenstein's development has received occasional endorsement.[23] But those who have quoted Dummett's view have more often taken issue with it. What has attracted most disagreement is the suggestion that the later Wittgenstein advances some form of anti-realism; there has been less dissent from Dummett's suggestion that the *Tractatus* advances a form of semantic realism.[24] So it is a notable feature of 'new' readings of Wittgenstein that they challenge Dummett's realist interpretation of the *Tractatus*, as well as his anti-realist interpretation of the later Wittgenstein.[25] But while many critics have castigated Dummett's account of Wittgenstein's development, few have paused to consider the details of what Dummett actually says. Diamond is a welcome exception to that tendency. So it is worth examining her discussion to see how convincing a case has been offered against Dummett's perspective on the *Tractatus*.

As we have seen, Diamond argues that there is a private language argument in the *Tractatus*. And, she suggests, that fact helps to show that Dummett has mislocated the *Tractatus* in relation to realism and anti-realism: 'Dummett reads Wittgenstein as having put forward global realism in the *Tractatus*' (284), she writes; but reflection on the private language argument in the *Tractatus*, and on the Tractarian conception of logical analysis, shows that in fact the *Tractatus* is significantly anti-realist. I shall argue that Diamond's criticisms are misplaced; they misunderstand Dummett's discussion of realism and its relation to the *Tractatus*. Furthermore, I show, Dummett's discussion of realism and anti-realism still provides an illuminating tool with which to reflect on the *Tractatus* and to understand its conception of meaning and analysis.

3.1 The Tractatus *and bivalence as a criterion for realism*

Diamond starts with what she calls 'a very simple point'. Russell's 'metaphysical and epistemological views . . . make him a realist in a familiar philosophical sense . . . He is a realist about other people's private objects' (284). Wittgenstein is opposed to Russell's

[23] Notably from Saul Kripke (1982: 71–8).

[24] Though some critics have explicitly disagreed with this part of Dummett's account, too; see e.g. Hacker 1986: 62–4.

[25] See e.g. Crary, who emphasizes that 'the standard narrative about the development of Wittgenstein's thought' (2000: 2), of which Dummett is regarded as a prime exponent, is to be rejected in its entirety.

treatment of talk about others' sensations. 'In a straightforward sense', she says, 'the *Tractatus* is therefore anti-realist, at any rate about other people's private objects' (284). And, she thinks, it is anti-realist *in Dummett's sense*. In support of this last claim, she quotes Dummett's observation that 'the colourless term "anti-realism".... denotes not a specific philosophical doctrine but the rejection of a doctrine' (Dummett 1991: 4, quoted by Diamond at 284). Since the *Tractatus* rejects Russell's realism about others' private objects, Diamond argues, it counts as anti-realist by that Dummettian standard. But of course, when Dummett says that the term 'anti-realism' denotes the rejection of a doctrine, he is not talking about the rejection of just *anything* that might be called 'realism'; he means the rejection of the specific doctrine that he calls 'realism'. So to see whether the view of sensation language that Diamond finds in the *Tractatus* really is a form of anti-realism in Dummett's sense, we need to be clear about what it takes for a view to be a form of realism in Dummett's sense.

For Dummett, a dispute between realism and anti-realism can arise whenever, as we usually conceive things, statements of some class may be true or false even though we have no way of establishing their truth or falsity. The realist takes the notion of truth as fundamental for semantics and explains the meanings of statements of the disputed class in terms of the conditions that must obtain in order for them to be true. The anti-realist takes some aspect of the use of a statement as fundamental—standardly, the notion of what justifies or warrants asserting that statement—and explains the meanings of statements of the disputed class in terms of that.[26] This way of characterizing the dispute between realist and anti-realist leads naturally to another: that realists affirm the principle of bivalence for statements of the disputed class (the principle that every statement is determinately either true or false) whilst anti-realists repudiate the principle.[27]

Now on this way of seeing things there is, Dummett insists, a crucial distinction between *anti-realism* and *reductionism*.[28] Behaviourism, for instance, is a reductionist view; it reduces the truth of statements about sensations and mental states to the truth of statements about behaviour.[29] But a position can be reductionist without being anti-

[26] As Dummett explains, an anti-realist may *present* her theory of meaning as a theory of truth-conditions. For, having defined the meanings of statements of the disputed class in terms of the conditions that would warrant asserting them, she may use the equivalence between 'P' and 'P is true' to define a notion of truth in terms of the notion of warranted assertability. She can then present her theory as one that explains meaning in terms of the conditions for a statement to be true, in the sense that she has defined. What is distinctive of a realist theory of meaning, therefore, is not merely that it presents meaning in terms of truth-conditions; it is that it takes the notion of truth as primitive in explaining meaning. (For discussion of this point, see Dummett 1991: 318; 1978: xxii–xxiii.)

[27] Diamond notes one way in which Dummett's more recent writings modify the bivalence criterion for realism. On the modified view, what makes a position anti-realistic is not that its adherents actually do repudiate bivalence but that they ought to do so, given the nature of their position: 'the criterion for having an anti-realist position becomes that of occupying a position that undercuts the ground for accepting bivalence' (1993a: 467). We shall see below a second way in which the recent writings modify the bivalence criterion.

[28] For what follows, see Dummett (1982; 1993a; 1991: ch. 15).

[29] Reductionism in this sense does not require an equivalence in meaning between the original statement and a statement, or class of statements, of the reductive class; see e.g. Dummett (1991: 322–3).

realist; as long as the reductionist accepts the principle of bivalence for statements of the disputed class, her position will be both reductionist and realist. Thus, a behaviourist will be a realist if she maintains that, for any significant statement about a person's mental state, there is some class of true statements about the person's actual or possible behaviour that makes the original statement true or false.

We are now in a position to evaluate Diamond's claim that the *Tractatus* contains an important element of anti-realism, in Dummett's sense. Suppose Diamond is right that the *Tractatus* would analyse propositions about others' toothaches in terms of propositions about their behaviour. Whether this position is a form of realism or a form of anti-realism will depend on whether or not Wittgenstein accepts bivalence for propositions about others' toothaches. That, in turn, will depend on the question, whether he accepts bivalence for the propositions about behaviour in terms of which the original propositions are analysed. And he surely does accept bivalence for those propositions; for it is a fundamental feature of the *Tractatus* that bivalence holds for every proposition whatsoever. That is explicit at many places: for example, in the claims that 'A proposition must restrict reality to two alternatives: yes or no' (4.023) and that 'A picture agrees with reality or fails to agree; it is correct or incorrect, true or false' (2.21). But Wittgenstein's commitment to bivalence can be spelled out in more detail. An elementary proposition 'asserts the existence of a state of affairs' (4.21). And 'a state of affairs . . . is a combination of objects' (2.01). So the state of affairs exists, and the elementary proposition is true, if objects are combined in the relevant way; the state of affairs does not exist, and the elementary proposition is false, if objects are not combined in the relevant way. Given that, how could bivalence fail for an elementary proposition?[30] It cannot fail because of vagueness or indeterminacy; the sense of an elementary proposition is determinate. Nor can an elementary proposition be meaningful yet fail to be true or false because it contains an empty name: an 'elementary proposition' that contained an empty 'name' would have no sense at all. Bivalence, then, holds for elementary propositions. But every proposition is a truth-function of elementary propositions. And a proposition that is composed by truth-functional combination from elementary propositions each of which is either true or false must itself be true or false also. So bivalence holds for every proposition, whether elementary or not. Thus, even if the *Tractatus* does repudiate Russell's particular kind of realism about statements concerning others' sensations, that has no tendency to show that it embodies any kind of Dummettian anti-realism. If bivalence is the touchstone of realism, then the *Tractatus* is resolutely realist.

[30] This puts the case for bivalence in a way that emphasizes the ontological comments at the beginning of the *Tractatus*. It does not follow that we must see Wittgenstein's commitment to bivalence as being *grounded in* a substantial ontology. We could equally well see it as an independently motivated commitment, which Wittgenstein then expresses in ontological terms. For this view of things, see e.g. McGuinness: 'Wittgenstein's remarks about facts are . . . merely a way of asserting the principle of bivalence, *for which no ground can be given*. The ontological or realist myth of *TLP* is apparently an attempt to give such a ground but it suffers from the *Selbstaufhebung*—the self-cancelling character—of the whole of *TLP*' (1985: 102).

There is another respect in which Diamond thinks the *Tractatus* is anti-realistic by Dummett's own standards. Realists hold that understanding a statement consists in knowing what has to be the case for it to be true. Dummett challenges the realist to explain how we can know what has to be the case for a given statement to be true in a case where we are incapable of recognizing that condition as obtaining or not. He considers the suggestion that our grasp of truth-conditions in such a case can be explained by analogy with our grasp of truth-conditions in simpler cases, where we can directly recognize a statement's truth or falsity. On this view:

> We come to understand the condition for the truth of [statements of the disputed class] via a conception of an ability to determine their truth or falsity effectively and directly, an ability which we do not ourselves possess, but of which we conceive by analogy with those abilities we do have . . . [For example] we understand statements about other people's mental states by appeal to the conception of a capacity to inspect the content of a mind, arrived at by analogy with our capacity to inspect the contents of our own minds. (1991: 344–5)

Diamond suggests that Russell's account of our talk of others' sensations involves a manoeuvre of just this sort.[31] On Russell's view, I cannot directly recognize the truth or falsity of the statement that Jones is in pain. But I can recognize the truth or falsity of the claim that I am in pain. By analogy with that, I can form a conception of Jones's capacity to recognize the truth or falsity of the statement that he is in pain. And that, in turn, gives me a conception of what makes the statement that Jones is in pain true. On Diamond's interpretation, the *Tractatus* rejects this Russellian use of analogy—for it rejects the idea that one can generalize from an object that one can name (one's own pain) to an object that one cannot name (Jones's pain). Since the use of analogy, she thinks, is integral to Dummett's account of 'what realism is' (285), of 'what a realistic theory of meaning involves' (284), Diamond again concludes that the *Tractatus* is, by Dummett's own lights, a form of anti-realism. But her argument misstates the place of analogy in Dummett's discussion. Dummett does not say that the appeal to analogy is part of 'what a realistic theory of meaning involves'. He offers it, rather, only as one of three possible realist responses to the challenge of explaining our grasp of what it is for a statement to be true.[32] So even if Diamond is right to say that the *Tractatus* rejects this use of analogy, that does nothing to show that the *Tractatus* is significantly anti-realist in Dummett's sense.

3.2 The Tractatus *and global realism*

According to Diamond, 'Dummett reads Wittgenstein as having put forward global realism in the *Tractatus*' (284). It is true that Dummett regularly cites the *Tractatus*,

[31] Indeed, Dummett himself mentions Russell in connection with this use of analogy (1991: 346).

[32] The other two responses that Dummett considers are: (i) that 'the practice of reasoning in accordance with the canons of classical logic' in the disputed area itself '*constitute[s]* a grasp of a notion of truth satisfying the principle of bivalence' (1991: 342), and (ii) that the practice of reasoning classically '*warrants the ascription* to a speaker of a grasp of that notion, without the need for further explanation or justification' (1991: 343).

alongside the works of Frege and Davidson, as a clear case of a theory on which 'the general form of explanation of meaning is a statement of... truth-conditions', realistic-ally conceived (1959: 185; cf. 1991: 304; 1976: 34). But we should be careful about the idea that Dummett sees *anyone* as a *global* realist or anti-realist. Pursuing this point in the light of some of Dummett's more recent formulations focuses attention on an interesting issue about realism and the *Tractatus* conception of analysis.

In Dummett's view, even Frege, 'an archetypal realist', is not a realist about absolutely every class of statement. For, rejecting the ultra-realism of Meinong, Frege gives an anti-realist account of statements containing empty terms: 'According to this, such terms do not denote anything; the sentences in which they occur express intelligible propositions (thoughts in Frege's terminology), but these propositions are neither true nor false' (1991: 325; cf. 1982: 269–70; 1993a: 468). Dummett's reason for classifying Frege as an anti-realist concerning statements about non-actual objects is straightforward: Frege rejects bivalence for such statements. Now Russell also rejects Meinongian realism. But he does so in a way that maintains bivalence for statements of the disputed class by interpreting the apparently singular terms in such statements not as empty singular terms but instead as descriptions. So if we are to classify Russell as an anti-realist concerning statements about non-actual objects—as Dummett thinks we should—we cannot count bivalence alone as the criterion for an account's being realistic; we must also look at how the account succeeds in respecting bivalence. Thus: 'Integral to any given version of realism are both the principle of bivalence for statements of the disputed class, and the interpretation of those statements at face value, that is to say, as genuinely having the semantic form that they appear on their surface to have' (1991: 325).[33]

What does this imply for the *Tractatus*? Wittgenstein praises Russell for 'perform[ing] the service of showing that the apparent logical form of a proposition need not be its real

[33] See also his (1993a: 468): 'the true criterion for a realist interpretation of any given class of statements... includes construing apparent singular terms occurring in them at face-value, to be explained in terms of their referring to elements of the domain of quantification'. This revision, as Dummett observes, has the effect of bringing his criterion for realism into closer contact with 'the intuition that realism has to do with the existence of objects' (ibid.). It might be thought that, if we adopt this revised criterion for realism, we give up the idea that there could be a position that was both reductionist and realist. For (to take one example) it is natural to think that phenomenalism, which reduces statements about physical objects to statements about experiences, cannot treat what appears to be a singular term referring to a physical object as a genuine singular term; the only genuinely singular terms, it may seem, will be those that refer to individual experiences. So, according to this line of thought, if we adopt Dummett's revised criterion for realism, phenomenalism inevitably becomes a form of anti-realism. But Dummett disagrees. In his view, whether a phenomenalist is a realist or an anti-realist depends on whether or not 'he denie[s] that the notion of reference, as applied to names of material objects, play[s] any role in the account of how a statement containing such names [is] to be determined as true or as false. Whether he could be said to deny this would depend upon the details of his translation into the sense-datum language...: specifically, on whether, in the process of translation, a name for a material object would be replaced by a term for some complex of sense-data, or whether it would be dissolved altogether so that no corresponding term remained in the sense-datum sentence' (1982: 253). So in Dummett's view, it remains possible for a position to be both reductionist and realist, even when we strengthen the criterion for realism in the way discussed.

one' (4.0031; cf. 4.002). And he echoes a central point of Russell's theory of descriptions: when a proposition of ordinary language contains a term that seems to function by referring to a complex, and the complex does not exist, the proposition will not be nonsensical but simply false (3.24; cf. 2.0201; NB 93, 101). Accordingly, the *Tractatus* agrees with Russell that, at the level of complete analysis, the apparently singular term for the complex is analysed away; bivalence is preserved by giving an analysis on which the apparently singular term turns out not to be a genuinely singular term. By Dummett's revised criterion, therefore, we must count Wittgenstein, like Russell, as an anti-realist concerning statements about complexes that do not exist. So it is wrong to think, as Diamond does, that Dummett sees the *Tractatus* as embodying a form of global realism.

It might be argued, however, that by the revised criterion the *Tractatus* is not simply anti-realistic about this limited class of statements but comes close to advancing a *global* form of anti-realism. The argument starts from the working assumption that no singular term of ordinary language is a Tractarian name—a term that, at the level of complete analysis, directly designates an object.[34] Suppose that assumption is correct. And suppose that an analysis only meets Dummett's condition of 'construing apparent singular terms at face value' if it treats the singular terms of ordinary language as singular terms at the level of complete analysis. Then the *Tractatus* will be profoundly anti-realist; for, on these suppositions, it will fail to construe *any* apparently singular term at face value. But that is surely not what Dummett intended; even if he does not think that the *Tractatus* is globally realist, he surely does not think it is globally anti-realist. Can we apply the revised criterion for realism in a way that does not have this surprising result?

One form of analysis envisaged in the *Tractatus* is the analysis of propositions about complexes into propositions about their constituents. In *Philosophical Investigations* §§60–3 Wittgenstein uses that form of analysis as a model in his critical discussion of philosophical analysis. His example there is a proposition that includes a description: 'The broom is in the corner'. But suppose I give my broom a proper name: 'Broom'. And suppose that Tractarian analysis reveals that propositions containing the ordinary proper name 'Broom' are to be analysed in terms of propositions containing the Tractarian names 'A' and 'B', which denote, respectively, the brush and the broomstick. And suppose analysis stops there. So our analysis of the proposition 'Broom is in the corner' will be this: 'A is in the corner & B is in the corner & A is attached to B'. Part of the point of this analysis is to show how the original proposition can have a sense even in a situation where the complex that the proper name 'Broom' seemed to denote does not exist. In that respect, this Tractarian analysis shares the aim of a Russellian analysis. But Russell's analysis proceeds differently: it does not invoke singular terms standing for parts of the broom; instead it relies on quantification. Suppose the ordinary proper name 'Broom' is equivalent to the description, 'The

[34] This can be no more than a working assumption since Wittgenstein gives no examples of names or objects. But there are clear indications that he expects the logical form and the simple elements revealed by analysis to be remote from those that ordinary language seems to have (see e.g. WVC 41–2).

broom whose name is "Broom".'[35] Then Russell's analysis will come out like this: 'There is a broom whose name is "Broom" & There is only one broom whose name is "Broom" & That broom is in the corner.' Is the difference between these two kinds of analysis significant when we come to apply Dummett's revised criterion for realism?

A hard-line application of the criterion would suggest not. On the face of it, the name 'Broom' is a singular term. So treating the sentence 'Broom is in the corner' at face value requires treating 'Broom' as a singular term. And neither Wittgenstein's analysis nor Russell's does that. From this point of view, any difference between the two analyses is irrelevant. But we might argue that respecting surface structure is a matter of degree, and that the Tractarian analysis stays closer to the surface of the original proposition. The name 'Broom' does not occur at the level of complete analysis. But there is a sense in which, at least in the case where the broom exists, the *Tractatus* does represent the ordinary proper name 'Broom' as functioning by denoting an ordinary object, the broom. The analysis reveals how the name 'Broom' is attached to a complex thing by way of having an analysis into simpler elements that denote the parts of that thing. On Russell's view, by contrast, there is no referential relation at all between the term 'Broom', or any part of its analysis, and the object, Broom, or any part of it. So, in cases where Tractarian analysis proceeds in the way we have been discussing, it is arguably in keeping with the spirit of Dummett's revised criterion to classify the *Tractatus* account as a form of realism—or at least, as closer to realism than Russell's position.

We cannot conclude that the result of Tractarian analysis will in general be something that qualifies as a form of realism by Dummett's revised criterion. For there may be cases where analysis proceeds differently: not by anatomizing the referential relations between apparently singular terms of ordinary language and their referents but by revealing that there are no such referential relations at all. If Tractarian analysis does turn out to include cases like that, then for those classes of propositions, the *Tractatus* should be counted anti-realistic. But, as Wittgenstein emphasizes, it is impossible to tell in advance how analysis will proceed. So if we adopt Dummett's revised criterion, the question, how far the *Tractatus* implies a realist account of the propositions of our ordinary language and how far it implies an anti-realist account, can be answered only by considering in detail, for each area of discourse, what would be involved in a Tractarian analysis of that area. Given Dummett's insistence that the issue between realism and anti-realism can only be settled piecemeal, that is just what we should expect.[36]

[35] For the suggestion that we might analyse ordinary proper names in this way, see Russell (1912: 32), where he offers 'the man whose name was *Julius Caesar*' as a possible analysis of the name 'Julius Caesar'.

[36] Earlier versions of this chapter were presented at the Stirling conference on 'The *Tractatus* and its History' in September 2005, and at the Universities of Oxford and Warwick. I am grateful to the participants on those occasions, to the Editors of this volume, and to OUP's anonymous reader, for very helpful comments. I am also grateful to the Arts and Humanities Research Council for an award under the Research Leave Scheme, which enabled me to complete the chapter.

References

Bell, David (1996) 'Solipsism and Subjectivity', *European Journal of Philosophy* 4, 155–74.

Carnap, Rudolf (1995) *The Unity of Science*, trans. Max Black. Bristol: Thoemmes Press.

Craig, Edward (1982) 'Meaning, Use and Privacy', *Mind* 91, 541–64.

Crary, Alice (2000) 'Introduction' to Crary and Read (2000).

Crary, Alice and Read, Rupert, eds (2000) *The New Wittgenstein*. London and New York: Routledge.

Diamond, Cora (2000) 'Does Bismarck Have a Beetle in his Box? The Private Language Argument in the *Tractatus*', in Crary and Read (2000), 262–92.

Dummett, Michael (1959) 'Wittgenstein's Philosophy of Mathematics'. Reprinted in his (1978).

—— (1976) 'What is a Theory of Meaning ? (II)'. Reprinted in his (1993b).

—— (1978) *Truth and Other Enigmas*. London: Duckworth.

—— (1982) 'Realism'. Reprinted in his (1993b).

—— (1991) *The Logical Basis of Metaphysics*. London: Duckworth.

—— (1993a) 'Realism and Anti-realism', in his (1993b).

—— (1993b) *The Seas of Language*. Oxford: Clarendon Press.

Goldfarb, Warren (unpublished) 'Objects, Names, and Realism in the *Tractatus*'.

Hacker, P. M. S. (1986) *Insight and Illusion: Themes in the Philosophy of Wittgenstein*, revised edition. Oxford: Oxford University Press.

Ishiguro, Hidé (1969) 'Use and Reference of Names', in Peter Winch, ed., *Studies in the Philosophy of Wittgenstein*. London: Routledge, 20–50.

—— (2001) 'The So-Called Picture Theory: Language and the World in *Tractatus Logico-Philosophicus*', in H-J. Glock, ed., *Wittgenstein: A Critical Reader*. Oxford: Blackwell.

Kripke, Saul (1982) *Wittgenstein on Rules and Private Language*. Oxford: Blackwell.

McGuinness, Brian (1981) 'The So-called Realism of the Tractatus'. Reprinted in his (2002).

—— (1985) 'Language and Reality'. Reprinted in his (2002).

—— (2002) *Approaches to Wittgenstein*. London and New York: Routledge.

Pears, David (1988) *The False Prison*, vol. 2. Oxford: Clarendon Press.

Russell, Bertrand (1912) *The Problems of Philosophy*. (Page references are to the 1976 Oxford University Press edition.)

—— (1913) *Theory of Knowledge: The 1913 Manuscript*, ed. Elizabeth Eames. London: George Allen and Unwin, 1984.

—— (1914) 'The Relation of Sense-Data to Physics'. Reprinted in J. Slater, ed., *The Collected Papers of Bertrand Russell*, vol. 8. London and New York: Routledge, 1986, 3–26.

Schlick, Moritz (1949) 'On the Relation between Psychological and Physical Concepts', in H. Feigl and W. Sellars, eds, *Readings in Philosophical Analysis*. New York: Appleton-Century-Crofts.

Stern, David (2004) *Wittgenstein's Philosophical Investigations*. Cambridge: Cambridge University Press.

—— (2010) 'Another Strand in the Private Language Argument', in A. Ahmed, ed., *Wittgenstein's Philosophical Investigations: A Critical Guide*. Cambridge: Cambridge University Press.

Wittgenstein, Ludwig (1922) *Logisch–Philosophische Abhandlung/Tractatus Logico–Philosophicus* (TLP). Trans. C. K. Ogden, London: Kegan Paul Ltd, 2nd edition 1933. Trans. D. F. Pears and B. F. McGuinness, London: Routledge and Kegan Paul Ltd, 1961, 2nd edition 1974.

—— (1958) *The Blue and Brown Books* (BB). Oxford: Blackwell.

—— (1973) *Letters to C. K. Ogden* (LO), ed. G. H. von Wright. Oxford: Blackwell.

—— (1974) *Philosophical Investigations* (PI), trans. G. E. M. Anscombe. Oxford: Blackwell.

—— (1975) *Philosophical Remarks* (PR), ed. R. Rhees, trans. R. Hargreaves and R. White. Oxford: Blackwell.

—— (1979a) *Wittgenstein and the Vienna Circle* (WVC), conversations recorded by Friedrich Waismann, ed. B. F. McGuinness, trans. J. Schulte and B. F McGuinness. Oxford: Blackwell.

—— (1979b) *Notebooks 1914–1916* (NB), 2nd edition, ed. G. H. von Wright and G. E. M. Anscombe, trans. G. E. M. Anscombe. Oxford: Blackwell.

——(2008) *Wittgenstein in Cambridge: Letters and Documents 1911–1951* (WIC), ed. B. F. McGuinness. Oxford: Blackwell.

9

Logic and solipsism

James Levine

A familiar puzzle regarding conceiving the inconceivable runs as follows: Suppose someone holds that there are examples of that which is inconceivable, or which cannot be conceived of, perhaps, for instance, the infinite or God. What, then, are we to understand by a sentence purporting to provide such an example? What, for example, are we to understand by the sentence 'The infinite cannot be conceived of'? Insofar as understanding the sentence requires thinking, or conceiving, of the infinite, then if the sentence can be understood, it is false; but if the sentence cannot be understood, it would not seem to be meaningful at all. By this reasoning, no sentence of the form '*X* cannot be thought of' can be both understood and true: it will be either false or meaningless, so that we cannot coherently provide an example of that which cannot be thought of.[1]

My main purpose in this chapter is to show that understanding different responses to this sort of puzzle helps clarify differences between how Russell and the early Wittgenstein regard the issue of solipsism. I begin, however, by considering some aspects of one of the most historically prominent uses of this sort of puzzle, namely, Berkeley's 'master argument' for idealism; or, more accurately, by considering some issues raised by A. N. Prior's analysis of the master argument. I then use that discussion as a means to examine how Russell and Wittgenstein regard the issue of solipsism. Finally, I argue that a more recent debate between Thomas Nagel and Donald Davidson involving the possibility of alternative conceptual schemes recaptures features of the debate between Russell and the early Wittgenstein regarding solipsism.

1 Berkeley's master argument and Prior's analysis

My discussion here is in three sections. First, I present the master argument as involving a version of the sort of argument I outlined above and discuss what Prior regards as correct in the argument. Second, I discuss two sorts of criticisms Prior makes of Berkeley's argument as an argument for idealism. Third, I discuss a further criticism

[1] Puzzles of this sort play a prominent role in Moore (1990 (e.g. 11–12)) and Priest (2002 (e.g. 22–3, 57–9)).

made of Berkeley by J. L. Mackie, who otherwise follows Prior's analysis of the argument.

1.1 *The master argument: what Prior regards as 'impeccable' in it*

Berkeley presents his master argument[2] both in his *Principles of Human Knowledge* (1710/34) and in his *Three Dialogues between Hylas and Philonous* (1713). In particular, in *Three Dialogues*, Berkeley writes:

Philonous . . . I am content to put the whole upon this issue. If you can conceive it possible for . . . any sensible object whatever, to exist without the mind, then I will grant it actually to be so.

Hylas If it comes to that the point will soon be decided. What more easy than to conceive a tree or house existing by itself, independent of, and unperceived by, any mind whatsoever? I do at present time conceive them existing after that manner.

Philonous How say you, Hylas, can you see a thing which is at the same time unseen?

Hylas No, that were a contradiction.

Philonous Is it not as great a contradiction to talk of *conceiving* a thing which is *unconceived*?

Hylas It is.

Philonous The tree or house therefore which you think of is conceived by you?

Hylas How should it be otherwise?

Philonous And what is conceived is surely in the mind?

Hylas Without question, that which is conceived is in the mind.

Philonous How then came you to say, you conceived a house or tree existing independent and out of all minds whatsoever?

Hylas That was I own an oversight; but stay, let me consider what led me into it.—It is a pleasant mistake enough. As I was thinking of a tree in solitary place, where no one was present to see it, methought that was to conceive a tree as existing unperceived or unthought of; not considering that I myself conceived it all the while. . . . (1713: 163–4; here, as throughout, emphasis is in the original)

I am not concerned here with Berkeley's apparent assimilation of conceiving and perceiving,[3] nor with his view that 'what is conceived is surely in the mind';[4] rather, I am concerned only with his argument to the effect that no one 'can conceive it possible' for a 'sensible object' to exist 'unthought of'. The argument (at least in the analysis developed by Prior that I discuss below) proceeds by assuming that, in order to conceive such a possibility, one will have to apprehend the propositional content expressed by a sentence of the form '*A* exists unthought of' and coherently conceive that content to be true; and it supposes further that apprehending such a propositional content requires thinking of the entity designated by the expression replacing '*A*', in which case apprehending that content precludes it from being true. Given that anyone

[2] So labeled by André Gallois (1974).

[3] For some discussion on this point, see Armstrong (1965: 10).

[4] For some discussion relevant to this point, see Russell (1912: ch. 4).

apprehending such a content should be able to recognize that the act of apprehending it precludes it from being true, apprehending such a content is incompatible with coherently conceiving it to be true.

Thus, for Berkeley, Hylas initially thinks he may conceive it possible that sensible objects exist unthought of, because he thinks that he can coherently conceive as true that a given tree exists unthought of. However, in doing so, Hylas failed to notice that in apprehending that content, he was thinking of that tree; and once he recognizes this, he recognizes that he cannot coherently conceive that content to be true. As Berkeley writes in the *Principles* (§23), in order to

> conceive it possible [that] the objects of your thought may exist without the mind . . . it is necessary that you conceive them existing unconceived or unthought of; which is a manifest repugnancy. . . . [T]he mind, taking no notice of itself, is deluded to think it can and does conceive bodies existing unthought of, or without the mind, though at the same time they are apprehended by, or exist in, itself. (1710/34: 70)

Further, in the *Principles* (§24), Berkeley concludes his presentation of the master argument by writing:

> To me it is evident those words ['absolute existence of sensible objects in themselves' or 'without the mind'] mark out either a direct contradiction, or else nothing at all. And to convince others of this, I know no readier or fairer way than to entreat they would calmly attend to their own thoughts; and if by this attention the emptiness or repugnancy of those expressions does appear, surely nothing more is requisite for their conviction. It is on this therefore that I insist, to wit, that the *absolute existence of unthinking things* are words without a meaning, or which include a contradiction. (Ibid.)

Here Berkeley may be suggesting the sort of dilemma I introduced at the outset. For, by the argument above, a sentence of the form '*A* exists unthought of' may be said to 'mark out either a direct contradiction, or else nothing at all'. To one who understands such a sentence, it will 'mark out' the 'direct contradiction' or 'repugnancy' of thinking of a given entity as unthought of;[5] while to one who fails to understand such a sentence, it will 'mark out . . . nothing at all'—it will consist of words that are 'empty' or 'without a meaning'. Thus, insofar as Berkeley has a reason for holding that the status of such phrases as 'the absolute existence of sensible objects in themselves' or of 'unthinking things'—or the status of the claim 'Sensible objects exist unthought of'—depends on the status of sentences of the form '*A* exists unthought of', he has a reason for holding, as he indicates here, that such expressions are either 'words without a meaning, or which include a contradiction'. For Berkeley, the 'emptiness or repugnancy' of such expressions will become clear to those who 'calmly

[5] Although see §1.3 below, where I discuss Mackie's criticism that Berkeley fails to distinguish what is contradictory from what is 'operationally self-refuting'.

attend to their own thoughts' and recognize—what Hylas at first fails to notice—that a sentence of the form '*A* exists unthought of' cannot be both understood and true.

In his 1955 analysis of Berkeley's argument, Prior abbreviates 'person *s* is thinking (entertaining the proposition) that *p*' as

$$T(s, p),$$

and he defines 'person *s* is thinking truly that *p*' as

$$T(t)(s, p) =_{\mathrm{df}} T(s, p) \& p.$$

Further, he analyzes 'person *s* is thinking about (or of) *a*' as 'for some φ, *s* is thinking that ϕa', which he thus abbreviates as

$$(\exists\varphi)T(s, \varphi a).^{6}$$

For Prior, that is, for a proposition to be about entity *a* is for it to be of the form '*Fa*', and for a person to think about (or of) *a* is for that person to entertain a proposition that is about *a*.[7] He then defines '*a* is thought-of (by someone)' as

$$\tau a =_{\mathrm{df}} (\exists x)(\exists\varphi)T(x, \varphi a),$$

in which case, '$\sim\tau a$' abbreviates '$\sim(\exists x)(\exists\varphi)\ T(x, \varphi a)$' or '*a* is unthought-of (by anyone)'.

Given these definitions, if

(1) Person *s* is thinking that *a* is unthought-of by anyone [$T(s, \sim\tau a)$],

then

(2) *a* is thought-of,

since there is an *x*, namely *s*, and there is a φ, namely $\sim\tau$, such that $T(x, \varphi a)$. Thus, if *s* is thinking that *a* is unthought-of by anyone, then *s* is not thinking that truly. And, more generally, by this reasoning, given any object *y*, no one can ever think truly that *y* is unthought-of. For anyone who thinks that *y* is unthought-of will be thinking of *y*, in which case it will be false that *y* is unthought-of. That is to say, this reasoning establishes that

(3) $(\exists y)\ T(t)(s, \sim\tau y)$

cannot be true. And this is Prior's reconstruction of what he finds correct in Berkeley's master argument. Prior deduces the negation of (3) as a 'logical law' (1955: 38); and, since he interprets Berkeley in the master argument as concerned to establish that law, Prior takes his formal derivation as substantiating this aspect of Berkeley's argument. As Prior writes, 'Formally, Berkeley's argument so far is impeccable' (1955: 37).

[6] Note that in using variables in predicate (as well as sentential) positions, Prior does not thereby take himself to be committing himself to properties (or propositions); see e.g. his (1971: ch. 3).

[7] By this definition, only singular sentences are about the sorts of entities designated by proper names, so that general sentences that contain no proper names cannot be about such entities. And this seems, in fact, to have been Prior's view; see e.g. his (1971: 112–13).

Given that understanding a sentence requires entertaining or thinking the proposition expressed by that sentence, then, by this argument, if an instance of

$$\sim \tau y \quad [y \text{ is unthought-of}]$$

is understood, then it is false. When it is put this way, it is obvious that one possible conclusion to draw at this point is that we are not able to understand instances of '$\sim\tau y$'. That is, the argument so far is meant to establish only that no instance of '$\sim\tau y$' can be both understood and true, and this is compatible with our not being able to understand any instance of '$\sim\tau y$' at all, a point, I will argue below, that is relevant for interpreting the position of the early Wittgenstein.[8]

1.2 The master argument: Prior's criticisms

While Prior thus regards Berkeley as recognizing that (3) is logically false, he also holds that the master argument fails 'as a defence of the Berkeleyan philosophy'. In particular, he claims that the argument 'proves' both 'too much' and 'too little' (1955: 37).

In claiming that the argument proves 'too much', Prior is claiming that, if successful, the argument would establish conclusions that are more extreme than the idealism that Berkeley wishes to defend. For, if the argument establishes that no 'sensible objects' can exist unthought-of, then it also establishes that no entities whatsoever exist unthought-of; for the logic of the argument does not depend upon any features unique to 'sensible objects' as distinct, for example, from minds. As Prior indicates, the variable 'y' in (3) need not range only over 'sensible objects', 'so that whatever the argument proves about "sensible objects" it proves about minds also' (1955: 37); but Prior presumes that Berkeley would not wish to hold of minds, as he holds of 'sensible objects', that they cannot exist unthought-of.[9]

Another version of the charge that the master argument 'proves too much' is made by those who note that the same style of argument may be used to establish not only that no ('sensible') object can be thought truly as being unthought-of, but also that no object can be thought truly as being unthought-of by me now.[10] For, insofar as my now entertaining the proposition φa requires my thinking about a now, then if I think that a is not thought-

[8] Thus, I am regarding a statement of the form 'That p cannot be thought truly' as a 'material mode' expression of the corresponding 'formal mode' statement of the form "p" cannot be both understood and true', so that both would be correct if what replaces 'p' is either a genuine statement that is necessarily false or is not a sentence at all. I recognize that in cases in which 'p' is replaced by something that is not a sentence at all, the 'material mode' expression is more likely to be regarded as itself nonsensical than the 'formal mode' expression. However, from the point of view of the *Tractatus*, either form of expression will be nonsensical (whether or not 'p' is replaced by a genuine sentence). For if we cannot say of a sentence with sense that it is a sentence with sense (see e.g. 5.525, 5.5351), neither can we say of what is nonsensical that it is nonsensical or cannot be understood. And thus from the point of view of the *Tractatus*, both forms of expression are ultimately to be rejected and serve, at best, as rungs on the ladder that are eventually to be thrown away.

[9] By the second (1734) edition of the *Principles*, Berkeley indicates that, while we can have no ideas of minds (or 'spirits'), we can nevertheless have 'notions' of them (see e.g. 1710/34: 72, 98); but then the question arises as to why he could not hold that, while we have no 'ideas' of sensible objects existing unthought-of, we may nevertheless have 'notions' of such objects. See also n. 11 below.

[10] See e.g. Armstrong (1965: 10); Mackie (1976: 53); Priest (2002: 65).

about by me now, I will be thinking about *a* now, in which case, for any object *y*, I will be unable to think truly that *y* is not-thought-about by me now. Hence, if the master argument establishes the idealist view that no object can exist unthought-of, the same style of argument also establishes the solipsistic view, which Berkeley clearly did not intend to embrace, that no object can exist unthought-of by me now.[11]

However, in claiming that the argument proves 'too little', Prior is claiming that the master argument does not succeed in establishing the idealist conclusion. According to Prior, while Berkeley is correct in arguing that (3) cannot be true, he has not thereby established that

(4) $T(t)(s, (\exists y)\sim\tau y)$

cannot be true. For Prior, while Berkeley correctly recognizes that there is no object of which one can think truly that it is unthought-of, he does not thereby establish that no one can think truly the general proposition that something is unthought-of;[12] but for Prior, it is this latter claim that Berkeley would need to establish in order to demonstrate his idealism. As Prior writes: 'It is [(4)], however, which Berkeley must argue to be impossible if he is to establish his idealism; in thinking that he has done so, he has confused two possible orders of his operators' (1955: 38; cf. Prior 1971: 91).[13]

[11] Thus, as Peter Sullivan has pointed out to me, in *Principles* §81, Berkeley writes that he 'see[s] no reason to deny' that there may be 'a great variety of spirits of different orders and capacities, whose faculties, both in number and extent, are far exceeding' his own, in which case 'there may be, for aught that I know, innumerable sorts of ideas or sensations, as different from one another, and from all that I have perceived, as colours are from sounds' (1710/34: 94), thereby allowing that there may be 'sensible objects' that are not, and cannot be, thought-of by himself. But then he goes on to claim that even granting 'the scantiness of my comprehension, with regard to the endless variety of spirits and ideas that may possibly exist, yet for any one to pretend to a *notion* of Entity or Existence, *abstracted* from *spirit* and *idea*, from perceiving and being perceived, is, I suspect, a downright repugnancy and trifling with words' (ibid.). From this it would appear that the crucial issue for Berkeley is not what we can and cannot have 'ideas' of, but rather what we can and cannot have 'notions' of. That is, it seems that while Berkeley may hold that he can have a 'notion', if not an 'idea', of sensible objects that he is not thinking of, he also holds that no one can have either an 'idea' or a 'notion' of a sensible object existing unthought-of by absolutely anyone. But if so, then it would seem that Berkeley should not be 'content to put the whole' case for idealism 'upon' his master argument; for there is nothing in that argument that addresses this sort of distinction between 'ideas' and 'notions'.

[12] Thus, Prior's criticism of Berkeley depends upon his view that entertaining (truly or falsely) a general proposition '$(\exists y)Fy$' does not require thinking of, or about, any object that is F; see note 7. See Armstrong (1965: 10) for an opposing view as to what general sentences are 'about'.

[13] In the same year that his paper on the master argument was published, Prior gave his John Locke Lectures, subsequently published as *Time and Modality* (1957). Central to those lectures is Prior's rejection of the Barcan formula in quantified modal logic (see especially chapters 3 and 4), which, for Prior, involves a structurally similar sort of confusion concerning 'possible orders of operators' as he holds is involved in Berkeley's master argument. The Barcan formula is the schema '$\Diamond(\exists x)\varphi x \rightarrow (\exists x)\Diamond\varphi x$', where '$\Diamond$' abbreviates 'it is possible that' (see Prior 1957: 26). Put in its contrapositive form, it is the schema '$\sim(\exists x)\Diamond\varphi x \rightarrow \sim\Diamond(\exists x)\varphi x$'; and in arguing against the Barcan formula, Prior holds that in establishing that the antecedent of an instance of this schema is true one does not thereby prove that the corresponding consequent is true. In arguing against Berkeley, Prior claims that in establishing that the negation of (3) is a logical law one does not thereby establish that the negation of (4) is a logical logic; and in doing so, he is, in effect, claiming that '$\sim(\exists x) T(t)(s, \varphi x) \rightarrow \sim T(t)(s, (\exists x)\varphi x)$', which is structurally similar to the Barcan formula (in its contrapositive form), is not a logically valid schema.

For Prior, that is, in the context of the master argument (and again ignoring the distinction between conceiving or thinking and perceiving), the idealist conclusion is

$$\text{(Id)} \qquad \sim (\exists y)(y \text{ is unthought-of}) \quad [\sim (\exists y) \sim \tau y],$$

the claim that nothing is unthought-of, while the negation of idealism is

$$\sim \text{(Id)} \qquad (\exists y)(y \text{ is unthought-of}) \quad [(\exists y) \sim \tau y],$$

the claim that something is unthought-of. And, on Prior's analysis, while Berkeley argues correctly that understanding an instance of '$\sim\tau y$' precludes its truth, so that no instance of '$\sim\tau y$' can be both understood and true and no one can provide an example of an object that is unthought-of, he does not thereby establish that understanding '$(\exists y)\sim\tau y$'—that is, the general sentence expressing the negation of idealism—precludes its truth, and thus does not thereby establish that '$(\exists y)\sim\tau y$' cannot be both understood and true, or that no one may coherently suppose that there are objects that are unthought-of. But, for Prior, this is what Berkeley would have to establish if he is to use his master argument to establish that (Id) is true.[14]

In defense of Berkeley, one might recall at this point that he regarded his objections to 'abstract general ideas' as central to his argument for idealism.[15] Thus, for example, as against Locke, Berkeley denies that we can have an 'abstract' idea of a triangle that is 'neither oblique nor rectangle, neither equilateral, equicrural nor scalenon' (1710/34: 52); and if, in doing so, Berkeley holds that no one can think that something is *F* without thinking of some specific entity that it is *F*, then he would have a reason for holding that, in establishing that (3) cannot be true, he had thereby established that (4) cannot be true. For he would have a reason for holding that thinking truly that there are entities that are unthought-of requires thinking truly of some specific object that it is unthought-of—something that Prior agrees Berkeley has shown cannot be done. Understood in this way, the disagreement between Prior and Berkeley may turn on a disagreement regarding the relation between what is required to understand a general sentence and what is required to understand its instances. Prior's criticism of Berkeley depends upon the view that there can be cases in which, even though no instance of '*Fx*' can both be understood and true, the corresponding general sentence '$(\exists x)Fx$' can be both understood and true, a view which, given his rejection of 'abstract' ideas, Berkeley may have reasons for rejecting.[16]

[14] Prior's central criticism here of Berkeley would seem to differ from the criticism presented by Bernard Williams in his (1966). Unlike Prior's, Williams's criticism depends upon Berkeley's assimilation of conceiving with having an image (or, in the case of sight, visualizing), and accuses Berkeley of failing to recognize that we can imagine (or visualize) a thing without thinking of it as perceived (or seen). Suffice to say, my purpose here is not to compare or assess these different analyses and criticisms of Berkeley's argument; rather, it is to use Prior's analysis as a means to introduce issues at stake between Russell and Wittgenstein.

[15] See Berkeley's 'Introduction' to the *Principles*, from §6 forward (1710/34: 46–60).

[16] Hay (1953) argues for this sort of connection between Berkeley's master argument and his rejection of 'abstract ideas'.

1.3 A further criticism of Berkeley by Mackie

J. L. Mackie accepts Prior's account, not only of what is correct in Berkeley's argument, but also of the ways in which Berkeley's argument both 'proves' both 'too much' and 'too little' (1964: 200–2; cf. 1976: 52–4). However, Mackie also makes a further criticism of Berkeley. In particular, he claims that Berkeley confuses establishing that no instance of '$\sim\tau y$' can be thought truly with establishing that each instance of '$\sim\tau y$' is 'internally self-contradictory'. For Mackie, what is correct in Berkeley's reasoning does not establish that no instance of '$\sim\tau y$' can *be* true; it only establishes that no such instance can be both *thought and true*. In Mackie's terminology, while Berkeley establishes that each instance of '$\sim\tau y$' is 'operationally self-refuting'—that is, cannot be 'coherently asserted'—he 'has confused operational self-refutation with outright logical impossibility' (1964: 201).

Given that he agrees with Prior that, in establishing that no instance of '$\sim\tau y$' can be thought truly, Berkeley does not succeed in establishing that '$(\exists y)(y$ is unthought-of)'—that is, ~(Id)—cannot be thought truly, Mackie holds that Berkeley fails to establish even that ~(Id) is 'operationally self-refuting', let alone that it is logically impossible. However, given Mackie's distinction between 'operational self-refutation' and 'logical impossibility', the question arises as to whether Berkeley's master argument would establish the truth of (Id), even if Berkeley had succeeded in establishing that ~(Id) cannot be thought truly. Suppose (as against Prior and Mackie) that Berkeley is right to hold that if no instance of '$\sim\tau y$' can be thought truly then ~(Id) cannot be thought truly; would he thereby be right to hold that ~(Id) is 'logically impossible', and hence that (Id) is 'logically necessary'? And would he then be right to regard his master argument as establishing (Id) as a necessary metaphysical truth?[17]

Mackie broaches these sorts of questions in the final paragraph of his discussion, where he writes:

> We might be tempted to believe that there is a special form of philosophical argument which enables us to establish positive conclusions by showing that certain contrary statements would be self-refuting. This would go against empiricism, for if any view would literally refute *itself*, its denial would be a necessary truth. However, our analysis shows that this challenge to empiricism evaporates on closer inspection.... The items that would have been troublesome are those that we have now classified under operational self-refutation. Empiricism would have been seriously undermined if they had enabled us to establish as necessary truths such propositions as 'I am (essentially) a thinking being', 'I know something', or 'Material objects do not exist unconceived'. But since the items opposed to these are at most operationally self-refuting, the detection of them does not lead to any such necessary truths. (1964: 203)

[17] Thus, we might distinguish here two sorts of issue. First, would establishing that ~(Id) cannot be thought truly, thereby establish that ~(Id) is a contradiction and hence that (Id) is the negation of a contradiction or, in Wittgenstein's terminology, a tautology? Second, whether or not (Id) is a tautology, would establishing that ~(Id) cannot be thought truly establish that (Id) is a necessary metaphysical truth? See in this connection n. 19 below.

Thus, even though Mackie (like Prior) denies that the general claim 'Material objects exist unconceived' is 'operationally self-refuting', he indicates here that, even if it were, that would not suffice for establishing its negation as a necessary truth. For Mackie, that is, even if successful Berkeley's master argument can only show that ~(Id) cannot be both true and thought; it cannot establish that ~(Id) cannot *be* true *simpliciter*, and so cannot establish (Id) as a necessary truth. For Mackie, Berkeley's master argument involves a kind of 'self-refutation' which cannot establish a positive necessary truth regarding ultimate metaphysics; it can at most establish a result as to what we cannot coherently think as true.

On the analysis presented here, then, there are three steps involved in Berkeley's master argument. The first is the argument that (3) cannot be true, so that no instance of '$\sim\tau y$' can be both understood and true; here, Prior agrees with Berkeley. The second is the view if (3) cannot be true, then neither can (4), or alternatively, that if no instance of '$\sim\tau y$' can be both understood and true, then neither can '$(\exists y)\sim\tau y$'—that is, ~(Id)—be both understood and true; here, Prior claims that Berkeley is guilty of a logical fallacy. The third is the view that if (4) cannot be true, so that '$(\exists y)\sim\tau y$' cannot be both understood and true, then '$(\exists y)\sim\tau y$' is 'internally self-contradictory', so that its negation—that is, (Id)—is necessarily true; here, Mackie claims that, even if Berkeley had established that ~(Id) 'operationally self-refuting', he would not have thereby succeeded in establishing that ~(Id) cannot be true or hence that (Id) must be true.[18]

In the next two sections, I argue that presenting Berkeley's master argument as involving these steps helps to clarify the different ways in which Russell and the early Wittgenstein regard the issue of solipsism. For I argue that their different approaches to solipsism may be understood as different responses to an argument for solipsism that is structurally similar to Berkeley's master argument for idealism understood as involving these three steps. In section 2, I argue that while Russell accepts the first step in the argument for solipsism, he also accepts a view of generality that enables him to reject the second step, so that Russell's response to solipsism has the same form as Prior's response to the master argument. In section 3, I argue that Wittgenstein not only follows Russell in accepting the first step in the argument for solipsism but also accepts a view of generality that, unlike Russell's, commits him to the second step; however, I argue that, like Mackie, Wittgenstein does not take the kind of 'operational self-refutation' that he reaches in this second step as establishing solipsism as a necessary

[18] Unlike Prior and Mackie, Armstrong accepts a view of generality according to which establishing that no instance of '$\sim\tau y$' can be both understood and true suffices for establishing that ~(Id) cannot be both understood and true (see n. 12); but, like Mackie, he then argues that reaching this latter conclusion does not suffice for establishing Berkeley's intended metaphysical conclusion:

[T]here is no contradiction in the statement 'There are things that nobody has thought of.' It is only *the making of the statement* that involves absurdity: it is only the fact that is has been put forward that entails that it is false. But if this proposition, 'There are things that nobody has thought of' is not self-contradictory, then Berkeley has quite failed to prove that the physical world must be thought of in order to exist. (1965: 11)

metaphysical truth.[19] In order to defend this interpretation of Wittgenstein, I present an interpretation of his show/say distinction and discuss how it bears on 'the problem, how much truth there is in solipsism' (5.62).

2. Russell on solipsism

My discussion here is in three subsections. First, I introduce, in general terms, both what Russell regards as correct and as incorrect in arguments for solipsism. Second, I substantiate these points by examining Russell's discussion of solipsism in his 1913 manuscript *Theory of Knowledge*. Third, I show briefly that the points that Russell makes in his 1913 discussion of solipsism are fundamental to him both before and after that discussion.

2.1 The structure of Russell's response to solipsism

Russell's understanding of the issue of solipsism is closely linked to his views regarding the constitution and apprehension of propositions. For Russell, a (genuine or logically) proper name serves to designate a particular without any intervening Fregean sense, so that a sentence containing a proper name expresses a proposition that contains among its constituents the particular designated by that name. Hence, given his 'Principle of Acquaintance', according to which apprehending a proposition requires being acquainted with each of its constituents, Russell holds that apprehending a proposition expressed by a sentence containing a proper name requires being acquainted with the particular designated by that name. That is to say, Russell accepts:

(PoA⋆) $(\forall x)(\forall \varphi)(\forall y)(T(x, \varphi y) \rightarrow \mathrm{Acq}(x, y))$,

according to which if a person x thinks (entertains, apprehends) the proposition that ϕy, then x is acquainted with y. (While Russell denies during this period that there are entities that are propositions, and so denies that apprehending a proposition is ultimately to be analyzed as a two-place relation between a thinking subject and a proposition, he nevertheless uses his 'multiple–relation' theory of judgment so that (PoA⋆) may be interpreted as being correct. See, for example, 1911a: 219–21.)

Given (PoA⋆), if

(1⋆) Person s thinks (entertains, apprehends) the proposition: I am not acquainted with a $[T(s, \sim\mathrm{Acq}(s, a)]$,

then

(2⋆) $\mathrm{Acq}(s, a)$.

[19] In particular, I argue that in being unable to think truly the negation of solipsism Wittgenstein takes solipsism to be neither tautologous (§3.3) nor a necessary, if unstateable, metaphysical truth (§3.4).

For if s thinks the proposition expressed by 'I am not acquainted with a', then, by (PoA*), s will be acquainted with a, in which case s is not thinking that proposition truly. More generally, by this reasoning, given any named particular y, no one can ever think truly: I am not acquainted with y, in which case

(3*) $(\exists y)\ T(t)(s, \sim Acq(s, y))$

cannot be true.

Put another way, and given his view that understanding a sentence requires apprehending the proposition it expresses, then, in accepting (PoA*), Russell is committed to accepting

(Und*) Understanding a sentence containing a proper name requires being acquainted with the entity designated by that name.

And in accepting (Und*), he is also committed to holding that no instance of

$\sim Ay$ [I am not now acquainted with y],

where 'y' is replaced by a proper name, can be both understood and true. For by (Und*), anyone understanding an instance of '$\sim Ay$' will be acquainted with the entity designated by the name replacing 'y', in which case that sentence will be false.

However, Russell is committed to holding that there are entities with which he is not acquainted. That is to say, he wishes to maintain that

(4*) $T(t)(s, (\exists y)\sim Acq(s, y))$

can be (and is) true, so that he can think truly that there are particulars with which he is not acquainted, even though he cannot think truly any proposition that says of a named particular that he is not acquainted with it. Thus, while Russell holds that no instance of '$\sim Ay$' can be both understood and true, he does not wish to hold that the general sentence '$(\exists y)\sim Ay$' cannot be both understood and true; on the contrary, he wants to maintain that it can be understood and known (or at least reasonably believed) to be true.

In fact, one of his original reasons for introducing his theory of denoting concepts in his 1903 *Principles of Mathematics* was to explain how we can 'deal with' entities—specifically, infinite classes—with which we are not acquainted;[20] and although he came to argue in his 1905 paper 'On Denoting' that the theory of denoting concepts is incoherent, the theory he advances in 'On Denoting' is also intended to explain how we can 'reach' entities with which we are not acquainted.[21] For Russell, what is at stake is providing a coherent account of general propositions that enables him to hold, among other things, that there are cases in which a sentence of the form '$(\exists x)Fx$' can be

[20] See e.g. Russell (1903: 145).

[21] See e.g. the second and penultimate paragraphs of 'On Denoting'; also Russell (1911a: 230). For some discussion of the differences between the theory of denoting concepts and the theory of 'On Denoting', and Russell's reasons for rejecting the former in favour of the latter, see my (2001: esp. 84–106), and my (2005).

understood and true even though no instance of '*Fx*' can be both understood and true. According to Russell, this is a problem of 'logic' that the theory of denoting concepts and the theory of 'On Denoting' both intend to address, though only the latter does so successfully.

By 1911, Russell holds that the only particulars with which we are acquainted are sense-data.[22] Thus he equates the acquaintance involved in understanding (genuine) proper names with the act of sensing, which he sharply separates from the sort of acquaintance—which he calls 'conceiving'—that he holds is required to understand parts of speech other than proper names, which he holds serve to stand for universals.[23] Hence, for Russell, the logical problem of providing an account of general propositions that explains how we can 'reach' in thought particulars with which we are not acquainted becomes the problem of explaining how I can 'reach' particulars other than the sense-data I am currently sensing, and thereby becomes closely linked for him to the problem of solipsism. For Russell, to hold, by (PoA⋆), that (3⋆) cannot be true is to hold that no one can think truly of any named particular that it lies outside of his or her present experience; and to infer from this that (4⋆) cannot be true is to hold that

~(Solip) $(\exists y)$(I am not now acquainted with y) $[(\exists y)(\sim Ay)]$,

according to which there is something outside my present experience, cannot be thought truly, a view that might lead one to embrace the extreme solipsistic view

(Solip) $\sim(\exists y)$(I am not now acquainted with y) $[\sim(\exists y)(\sim Ay)]$,

according to which there is nothing outside of my present experience. But just as Prior argues that while (3) cannot be true, this does not establish that (4) cannot be true, and so provides no reason for holding that ~(Id) cannot be thought truly or for accepting (Id), so too Russell argues that while (3⋆) cannot be true, this does not establish that (4⋆) cannot be true, and so no provides no reason for holding that that ~(Solip) cannot be thought truly or for accepting (Solip).

2.2 Solipsism: the Theory of Knowledge *discussion*

In his 1913 manuscript *Theory of Knowledge*, Russell writes:

> How do we come to know that the group of things now experienced is not all-embracing?... This question is one of great importance, since it introduces us to the whole problem of how knowledge can transcend personal experience.... At first sight, it might seem as though the experience of each moment must be a prison for the knowledge of that moment, and as though its boundaries must be the boundaries of our present world. Every word that we now understand must have a meaning which falls within our present experience; we can never point to an object and say: '*This* lies outside my present experience.' We cannot know any particular thing unless it

[22] See e.g. Russell (1911a: 212), where he identifies the particulars with which we are acquainted with sense-data, taken 'in a large sense' so as to include the objects of both 'inner' and 'outer' awareness.

[23] See again Russell (1911a: 212) and Russell (1913: 101).

is part of present experience; hence it might be inferred that we cannot know that there are particular things which lie outside present experience. To suppose that we can know this, it might be said, is to suppose that we can know what we do not know. On this ground, we may be urged to a modest agnosticism with regard to everything that lies outside our momentary consciousness. Such a view, it is true, is not usually advocated in this extreme form; but the principles of solipsism and of the older empirical philosophy would seem, if rigorously applied, to reduce the knowledge of each moment within the narrow area of that moment's experience. (1913: 10–11)

In writing in this passage that 'we can never point to an object and say: "*This* lies outside my present experience"', Russell is, in effect, applying (PoA*) to reach the conclusion that (3*) cannot be true or that no instance of '$\sim Ay$' can be both understood and true, where 'y' is replaced by a logically proper name (here, for Russell, 'this'). And later in *Theory of Knowledge*, he reiterates this point when he writes:

[I]t is important to remember that I can never give an actual instance of a thing not now within my experience, for everything I can mention otherwise than by a description must lie within my present experience. This is involved in the very nature of experience, and is one of the most important of the obvious facts about it. (1913: 34)

indicating again that I cannot name (as opposed to describe) an item and say truly of it that it lies outside my present experience.

Moreover, Russell recognizes the tendency to infer that, since we cannot think truly (or thus know the truth of) any proposition expressed by an instance of '$\sim Ay$', we cannot think truly (or thus know) that $(\exists y)\sim Ay$ is true. For Russell, insofar as I am likely to make that inference, then 'at first sight it might seem' that, since I cannot name any item that lies outside my present experience, I can have no reason for holding that anything lies outside my present experience, in which case I am committed to regarding the 'boundaries' of my present experience as 'the boundaries of [my] present world'. Further, Russell suggests in the final sentence of the first passage I have quoted from *Theory of Knowledge* that to accept such an inference would be to accept an 'extreme form' of solipsism, according to which the solipsist accepts—or holds, at least, that we have no good reason for rejecting—(Solip). For Russell, solipsism is officially the less extreme view that no one has any good reason for believing there is anything beyond that person's 'total' experience, where one's total experience includes one's past and future experience over a lifetime in addition to one's present experience (1913: 13). However, insofar as the solipsist defends her view by inferring from the inability of anyone to state truly any instance of '$\sim Ay$' to the inability of anyone to state truly the general statement '$(\exists y)\sim Ay$', then the consistent solipsist should defend (Solip), that there is nothing beyond what she experiences at the present moment, rather than the less extreme view that there is nothing beyond what she experiences over her whole lifetime.

For Russell, however, there is a 'logical' reply to such an argument for solipsism, a reply that 'consists in pointing out a fallacy in the inference' upon which the argument depends (1913: 11). As he writes:

> [T]he logical possibility of the knowledge that there are things which we are not now experiencing... depends upon the fact that we may know propositions of the form: 'There are things having such-and-such a property', even when we do not know of any instance of such things. In the abstract mathematical world, it is very easy to find examples. For instance, we know that there is no greatest prime number. But of all the prime numbers that we shall have ever thought of, there certainly is a greatest. Hence there are prime numbers greater than any that we shall have ever thought of. But in more concrete realms, the same is true: it is perfectly possible to know that there are things which I have known, but have now forgotten, although it is obviously impossible to give an instance of such things.... To pursue this topic further would require an account of 'knowledge by description'.... For the present, I am content to have pointed out that we know that there are things outside present experience and that such knowledge raises no logical difficulty. (1913: 11–12)

For Russell, that is, the fallacy in the argument he discusses for solipsism depends upon assuming that we can think truly (or know) a proposition expressed by a sentence of the form '$(\exists x)Fx$' only if we can think truly (or know) a proposition expressed by a corresponding instance of the form 'Fx'. And in this passage, Russell produces what he takes to be obvious counter-examples—involving numbers we have never thought of or truths we have once known but now forgotten—to such a principle. So for Russell, it is wrong, as a matter of logic, for the solipsist to infer that, since no instance of '$\sim Ay$' may express a proposition that can be thought truly (or known to be true), then the proposition expressed by '$(\exists y)\sim Ay$'—that is, the denial of solipsism—likewise cannot be thought truly (or known to be true).

Later in *Theory of Knowledge*, Russell returns to this point when he writes:

> At an earlier stage, we decided that our present experience can be known not to be all-embracing. It is sometimes maintained that this cannot be known, on the ground that if a thing lies outside our experience we cannot know that there is such a thing. At the risk of repetition, it may be worth while to repeat the reasons (belonging to logic rather than to theory of knowledge) which show that this argument is fallacious. An object may be *described* by means of terms which lie within our experience, and the proposition that there is an object answering to this description is then one composed wholly of experienced constituents. It is therefore possible to know the truth of this proposition without passing outside experience. If it appears on examination that no *experienced* object answers to this description, the conclusion follows that there are objects not experienced. For example, we may know Jones and paternity and the fact that every man has a father. Then we know there is 'the father of Jones', although we may never have experienced him. (1913: 34)

Here, unlike the previous passage I have quoted, Russell gives some sense of his account of apprehending general propositions that enables him to hold that there is a fallacy in the solipsist's argument. In particular, in indicating here that any proposition that I can apprehend will be 'composed wholly of experienced constituents', Russell is alluding to his principle of acquaintance, and in this passage he suggests how, compatibly with that principle, he can hold there may be cases in which we can think truly a proposition of the form '$(\exists x)Fx$' even though we cannot think truly any of its instances.

For Russell, the proposition expressed by a sentence '*Fa*', where '*F*' designates the universal *F* and '*a*' designates the particular *a*, has two constituents—*F* and *a*—so that by the principle of acquaintance, apprehending that proposition requires being acquainted with both entities. Further, in *Theory of Knowledge*, Russell defends the view that apprehending that proposition also requires being acquainted with the 'logical form of subject-predicate complexes', an abstract entity expressed by the absolutely general sentence 'something has some predicate' or '$(\exists \varphi)(\exists x)\varphi x$'. According to Russell, by being acquainted with these three entities and relating them in the appropriate way, one will thereby know that it is 'logically possible' that *Fa*—knowledge which he holds is required for apprehending that proposition—without thereby knowing whether it is, in fact, true that *Fa*, something he holds is not (in general, at least) required in order to apprehend such a proposition (1913: 97–101, 110–17).

In contrast, for Russell, apprehending the general proposition that $(\exists x)Fx$ requires apprehending the universal *F* and the logical form of subject-predicate complexes, but does not require being acquainted with any particulars and so does not involve apprehending any instance of that general proposition. And by thus divorcing what is required to think (apprehend) the general proposition that $(\exists x)Fx$ from what is required to think (apprehend) its instances, Russell is able to maintain that we can think truly the general proposition without thinking truly—or in some cases without even being able to think truly—any of its instances. For Russell, we can think that $(\exists x)Fx$ if we are acquainted with the relevant universal and logical form and combine them in the relevant way, and we will think that general proposition truly if it has any true instances—regardless of whether we actually, or even can, think (and thus think truly) any of its true instances.

Thus, for Russell, while apprehending the general proposition expressed by 'There is one and only one father of Jones' requires being acquainted with (the universal) paternity and (the particular) Jones, it does not require being acquainted with the person who is, in fact, the father of Jones, and so does not require apprehending the instance of that general proposition in virtue of which it is true. And hence, for Russell, someone may think truly that general proposition without thinking truly any of its instances. In this case, as Russell presents the example,[24] there is no in-principle reason why we cannot think truly any of its instances; it is simply a matter of fact that one may apprehend the general proposition without actually apprehending its true instance. In contrast, with regard to the proposition expressed by $\sim$(Solip)—that is, the proposition expressed by '$(\exists y) \sim Ay$'—no one can ever think truly any of its instances. However, for Russell, the point remains that his way of applying the principle of acquaintance

[24] That is, in such a way that we may be acquainted with people, such as Jones.

enables him to hold that apprehending the general proposition that $(\exists y){\sim}Ay$ does not depend on apprehending any, let alone all, of its instances, so there is no logical difficulty in our apprehending that general proposition and its being true (and thus in our thinking truly that general proposition), even though we can never think truly any of its instances.

Having taken himself to have shown that there is no 'logical' argument establishing that $(\exists y){\sim}Ay$—that is, ~(Solip)—cannot be both thought and true, Russell recognizes that the question remains as to whether we have in fact any good reason for believing that $(\exists y){\sim}Ay$ is true. For Russell, since there are infinitely more mathematical facts than humans could ever entertain, 'it is easy to prove' that there are 'mathematical'—and thus abstract—'facts' that 'do not enter into' my present (or even my 'total') experience (1913: 13); however, he holds that there is no comparable proof that there are 'existing particulars' outside of my present (as well as 'total' experience), that is, that $(\exists y){\sim}Ay$ is true. Thus he writes that although we 'naturally suppose' that there are such particulars (including other people's bodies and minds), and 'although no reason can be alleged for believing that our supposition is mistaken, yet it would seem also that there is no *conclusive* reason for believing it is not mistaken' (ibid.). Accordingly, he sums up his defense of $(\exists y){\sim}Ay$ by writing:

> ... (a) that there is no logical reason against it, (b) that in the logical world there certainly are facts which we do not experience, (c) that the common-sense assumption that there are particulars which we do not experience has been found thoroughly successful as a working hypothesis, and that there is no argument of any sort or kind against it.... (1913: 14)

thereby taking it as reasonable to deny that either (Solip), the solipsism of my present experience, or the solipsism of my 'total' experience is true.

As should be clear, Russell's discussion of solipsism and Prior's discussion of Berkeley's master argument are structurally similar. Both agree that there are propositions of the form '${\sim}\psi y$' that no one can think truly; but both hold that their opponents wrongly infer from this that no one can think truly the corresponding general proposition of the form '$(\exists y){\sim}\psi y$'; and further both hold that, were their opponents correct in making such an inference, then they should accept more extreme versions of the idealist and solipsistic positions than those they actually accept. For both, their opponents' arguments depend on a wrong view of general propositions; and by regarding themselves as exposing a 'logical fallacy' in their opponents' arguments, both are arguing against attempts to establish idealism or solipsism by logical deduction, without thereby claiming that logical deduction can establish the falsity of these positions.

2.3 Some further Russell passages

The view—central to Russell's criticism of the purported logical argument for solipsism as well as to Prior's criticism of Berkeley's master argument—that there are cases in which we can think truly a proposition (expressed by a sentence) of the form '$(\exists x)Fx$' even when we cannot think truly any of its instances is an enduring element in

Russell's philosophy. He emphasizes the point in *The Problems of Philosophy*, both in introducing his distinction between knowledge by acquaintance and knowledge by description at the end of Chapter 4,[25] and again in Chapter 10, 'On Our Knowledge of Universals'.[26] Likewise, in his 1911 paper 'The Basis of Realism', he writes:

Sometimes . . . it is urged . . . that we cannot know that there are things we do not know; but this view rests upon a wrong analysis of general propositions. . . . When we know a general proposition that does not require that we should know all or any of the instances of it. 'All the multiplication-sums that never have been and never will be thought of by any human being deal with numbers over 1,000' is obviously a true proposition, although no instance of such a sum can ever be given. It is therefore perfectly possible to know that there are propositions we do not know, in spite of the fact that we can give no instance of such a proposition. Thus it is fallacious not merely to argue that everything must be known, but even to argue that we cannot know that there are things which we do not know. This instance, like many others, illustrates the elementary blunders which philosophers have made owing to their neglect of logic. (1911b: 130)

The point appears again in his 1936 paper 'The Limits of Empiricism' (321–3), as well as his 1940 William James Lectures published as *An Inquiry into Meaning and Truth*, where he writes:

The question of truth which transcends experience may be put as follows: suppose $a_1, a_2, \ldots a_n$ are all the names in my vocabulary, and that I have named everything I can name. Suppose $fa_1, fa_2, \ldots fa_n$ are all false, is it nevertheless possible that 'there is an x for which fx' should be true? Or, alternatively, can I infer 'fx is false whatever x may be'?. . . . (1940: 236–7)

and adds shortly thereafter:

It is clear that, unless our knowledge is very much more limited than there seems any reason to suppose, there must be basic existence-propositions [propositions of the form 'there is an x for which fx'], and that, in regard to some of these, every instance 'fa' that we can give is false. The simplest example is 'there are occurrences which I do not perceive'. I cannot in language express what makes such statements true, without introducing variables; the 'fact' which is the verifier is unmentionable. (1940: 237–8)

Again in his 1944 'Reply to Criticisms', he writes:

Let us assume that you have assigned proper names to everything that you are capable of naming; let the objects named be $a, b, c, \ldots, z$. Suppose every one of these is found by you to possess a certain property P; does this justify you in asserting: 'Everything has the property P'? Every logician would reply in the negative. . . . [I]n rejecting the above inference you are holding that there may be things which you have not experienced . . .

[25] See in particular the last five paragraphs.
[26] See in particular paragraphs 12–14.

and adds in the following paragraph:

Let our list of names consist of all those that I can assign throughout the course of my life. If, then, from the fact that '*P*(*a*),' '*P*(*b*),' ... '*P*(*z*)' are all true, I do not allow myself to infer that '*P*(*x*)' is true for all values of *x*, that is a denial of solipsism. (1944: 682)

In his 1948 book *Human Knowledge* as well as his 1950 paper 'Logical Positivism' he uses his view that 'it is possible to know existence-propositions without knowing any particular instance of their truth' (1948: 452; 1950: 379) to criticize 'some modern empiricists—in particular, the majority of logical positivists' (1948: 445; 1950: 372). And in his 1959 book *My Philosophical Development* Russell uses the point to criticize not only 'many forms of empiricism' but also idealism (1959: 131).[27]

Finally, note that in his 1946 book *History of Western Philosophy*, in his discussion of Berkeley, Russell writes:

Hylas maintains that he can conceive a house which no one perceives, and which is not in any mind. Philonous retorts that whatever Hylas conceives is in his mind, so that the supposed house is, after all, mental. Hylas should have answered: 'I do not mean that I have in mind the image of a house; when I say that I can conceive a house which no one perceives, what I really mean is that I can understand the proposition "there is a house which no one perceives", or, better still, "there is a house which no one either perceives or conceives".' This proposition is composed entirely of intelligible words, and the words are correctly put together. Whether the proposition is true or false, I do not know; but I am sure that it cannot be shown to be self-contradictory. Some closely similar propositions can be proved. For instance: the number of possible multiplications of two integers is infinite, therefore there are some that have never been thought of. Berkeley's argument, if valid, would prove that this is impossible. (1946: 627)

Thus, not only did Russell make the same sort of criticism of arguments for solipsism that Prior would later make against Berkeley's master argument for idealism, and not only was he concerned to develop an account of general propositions that supports these criticisms; he also applied that sort of criticism to the master argument itself.

3 The early Wittgenstein on solipsism

My discussion here is in five subsections. First, I argue that, like Russell, the early Wittgenstein accepts (Und★), thereby committing him to the view that no instance of '$\sim Ay$' can be both understood and true. Second, I argue that, unlike Russell, Wittgenstein accepts a view of generality that commits him to holding that, if no instance of '$\sim Ay$' can be both understood and true, then '$(\exists y)\sim Ay$'—that is $\sim$(Solip)—cannot be both understood and true. Third, I argue that Wittgenstein does not thereby hold that (Solip) is a true sentence, but rather regards both $\sim$(Solip) and (Solip) as nonsensical (*unsinnig*). Fourth, I consider in what sense, if any, Wittgenstein holds that, although

[27] As early as 1905, Russell wrote that 'empiricists (wrongly) conceive all general statements' as 'essentially ... collection[s] of particular statements' (1905a: 372).

(Solip) is not a true sentence, the 'truth' of solipsism may nevertheless be shown. Fifth, I return to the differences between Russell's and Wittgenstein's view of generality and discuss some considerations relevant to Wittgenstein's rejection of Russell's view.

3.1 Wittgenstein's Russellian view of names and the issue of solipsism

As I have suggested above, Russell's understanding of the issue of solipsism, and in particular, his acceptance of the view that no instance of '$\sim Ay$' can be both understood and true, depends upon his acceptance of views—including views of proper names and of what is involved in understanding sentences containing proper names—that Frege would reject.

Whereas Russell holds that a proper name serves only one symbolic function, that of standing for a particular, Frege holds that a proper name serves two symbolic functions, that of expressing a sense and designating an object. Whereas Russell holds that the particular designated by a name is a constituent of the proposition expressed by a sentence containing that name, Frege holds that the sense expressed by a name, not the object designated it, is a constituent of the thought, or sense, expressed by a sentence containing that name.[28] Accordingly, whereas Russell holds that a name that fails to designate a particular is altogether without significance, and a sentence containing such a name fails to express a proposition,[29] Frege holds that a name that fails to designate an object may nevertheless express a sense, in which case a sentence containing it may succeed in expressing a thought.[30]

Further, whereas Russell accepts (Und★), that understanding a sentence containing a proper name requires being acquainted with the particular designated by that name, Frege holds that, although understanding a sentence containing a proper name requires grasping the sense of that name, it need not involve being acquainted with the object designated by that name. Accordingly, whereas Russell holds that no one can understand a sentence of the form '$a = b$', where 'a' and 'b' are replaced by (genuine) proper names, and fail to know its truth-value,[31] Frege holds that we can understand such a sentence and fail to know its truth-value.[32]

Moreover, in disagreeing on these general principles as to how proper names function, Frege and Russell disagree as to which expressions are genuine proper names. Thus, for example, although Russell agrees that someone (for example, George

[28] See, for example, the 1904 Frege–Russell correspondence regarding whether or not 'Mont Blanc with all its snowfields' is a 'component part of' the proposition, or thought, expressed by 'Mont Blanc is more than 4000 metres high' (Frege 1980: 163, 169). Although Russell is there writing prior to 'On Denoting' as well as to adopting the 'multiple relation' theory of judgment, he continues to hold that the particular designated by a genuine proper name may be said to be a constituent of the proposition a sentence containing that name may be said to express; see e.g. Russell (1911a: 216–21).

[29] See e.g. Whitehead and Russell (1910: 66); Russell (1959: 168).

[30] See e.g. Frege (1892a: 162–3).

[31] See e.g. Whitehead and Russell (1910: 67) and Russell (1918: 245–6), where Russell indicates that if 'a' and 'b' are (genuine) proper names, then if '$a = b$' is true, it will be trivially true.

[32] See e.g. Frege (1892a: 176–7), where he indicates that where 'a' and 'b' are (genuine) proper names, '$a = b$' can be true without being trivially true.

IV) could understand the sentence 'Scott is the author of *Waverley*' without thereby knowing its truth-value, Russell takes this as establishing that 'the author of *Waverley*' is not a genuine proper name; and likewise, while Russell would agree with Frege that we can understand the sentence 'Hesperus is Phosphorus' without thereby knowing its truth-value, Russell would take this to show that 'Hesperus' and 'Phosphorus' are not genuine names.[33] More generally, while Frege's distinction between sense and Bedeutung enables him to regard ordinary proper names along with definite descriptions as proper names, Russell's rejection of Frege's distinction leads him to deny that definite descriptions along with most, if not all, ordinary proper names are genuine or 'logically proper' names.

On all these issues of general principle the early Wittgenstein sides with Russell, not Frege. Thus, although Wittgenstein uses Fregean terminology in remarking

> A name bedeutet an object. The object is its Bedeutung (3.203)

he also writes that 'only sentences have sense' (3.3), thus rejecting Frege's view that names have both sense and *Bedeutung*. Likewise, Wittgenstein writes that 'a sentence "*fa*" shows that the object *a* occurs in its sense' (4.1211), thereby indicating, with Russell but against Frege, that the object designated by a name is a constituent of, or 'occurs in', the propositional content (sense) expressed by a sentence containing that name. Again, for Wittgenstein, if a combination of signs 'has no sense, that can only be because we have failed to give a *Bedeutung* to some of its constituents' (5.4733); hence, since he holds that the *Bedeutung* of a name is the object which it designates, he is thereby indicating, with Russell, but as against Frege, that a sentence containing a name which fails to designate an object has no propositional content (or sense).[34]

Further, Wittgenstein remarks:

> Can we understand two names without knowing whether they signify the same thing or different things?—Can we understand a sentence in which two names occur without knowing whether they bedeuten the same or different things?
>
> If I know [*kenne*] the Bedeutung of an English word and of a synonymous [*gleichbeteudenden*] German word, then it is impossible for me not to know that they are synonymous [*gleichbedeutend*]. (4.243)

And again:

> It is impossible to *assert* the identity of Bedeutung of two expressions. For in order to be able to assert anything about their Bedeutung, I must know [*kennen*] their Bedeutung, and I cannot

[33] See e.g. Russell (1918: 245–6), where he indicates not only that 'the author of *Waverley*' is not a name, but also that in asserting 'Scott is Sir Walter' in 'the way one would mean it' (so that what is asserted is not a 'tautology'), neither 'Scott' nor 'Sir Walter' is being used as a name.

[34] Here and throughout I have appealed to both the Ogden and the Pears and McGuinness translations to arrive at the translations I have used. However, as against both translations, I have translated 'Satz' as 'sentence'.

know [*kennen*] their Bedeutung without knowing whether what they bedeuten is the same or different. (6.2322)

Thus, Wittgenstein indicates, against Frege, but with Russell, and in accord with (Und*), that since understanding a sentence containing a name requires being acquainted with (*kennen*)[35] the object designated by that name (the *Bedeutung* of that name), we could never understand an identity sentence without knowing its truth-value.[36] Again, like Russell, but not Frege, Wittgenstein distinguishes sharply between names and descriptions[37] and indicates further that which expressions are genuine names (or 'simple signs' (3.202)) is something that can only be revealed by analysis (3.201–2), so that, for Wittgenstein, as for Russell, it will turn out that ordinary proper names fail to be genuine names.

[35] Here, and throughout, I equate Wittgenstein's Tractarian uses of 'kennen' (and cognates) an object with 'to be acquainted with' (and cognates) that object. In *The Problems of Philosophy*, in distinguishing 'knowledge of *truths*'—knowledge '*that* something is the case'—from 'knowledge of *things*'—such as 'the sense in which we know sense-data'—Russell associates 'acquaintance' with 'knowledge of things', and remarks that 'the distinction involved is roughly that between *savoir* and *connaître* in French, or between *wissen* and *kennen* in German' (1912: 44). And I argue below (see especially §3.4) that in the *Tractatus* Wittgenstein likewise reserves *wissen* for knowledge that something is the case, and *kennen* for what we know in 'knowing objects'. I recognize that, in a 1922 letter to Ogden, Wittgenstein resisted translating 'ich den Gegenstand kenne' in the first sentence of 2.0123 as 'I am acquainted with an object'. However, his reason for doing so is as follows:

'To know' means both: *kennen* and *wissen* and 'to be acquainted with'—I think—hasn't exactly the meaning I want, because it seems to me to imply somehow that one knows a lot about an object, while to know here just means: I know *it* but I needn't know anything *about* it. (1973: 59)

Thus, it is clear that Wittgenstein intends, in writing of *kennen* an object, to respect Russell's distinction between knowledge of things (for which Russell reserves 'acquaintance') and knowledge of truths, and that the only reason he resists translating *kennen* as 'to be acquainted with' is because he holds that this distinction is blurred in the ordinary use of that phrase. Russell writes similarly: 'We may have knowledge of a thing by acquaintance even if we know very few propositions about it—theoretically we need not know any propositions about it' (1912: 144). However, because he identifies the particulars with which we are acquainted with sense-data, Russell holds that, as a matter of fact, in being acquainted with particulars we do thereby know some truths about them, such as, for example, what their sensory qualities are (1912: 113–14). As I argue below in this section, Wittgenstein's objects differ in certain fundamental ways from Russell's sense-data; and because of these differences he is able to maintain the distinction between 'knowledge of truths' and 'knowledge of things' more rigorously than Russell does. (For more on this point, see my 1998, especially Parts I and II.) Note also that in the letter to Ogden Wittgenstein indicates that, if 'Wenn ich den Gegenstand kenne' in the first sentence of 2.0123 is to be translated as 'If I am acquainted with an object', then in the second part of that sentence ('so kenne ich auch samtliche Moglichkeiten seines Vorkommens in Sachvalten') 'kenne' should be translated as 'know'. Here, I suggest, Wittgenstein is indicating that while *kennen* an object involves the sort of relation between a mind and entity that is involved in acquaintance, *kennen* the possible occurrences of an object in states of affairs is not likewise the relation between mind and entity, since, for Wittgenstein, possibilities are not entities. I argue below (§3.4) that, for Wittgenstein, the distinction between what we *kennen* in understanding language (including both the objects we thereby *kennen* and the possibilities we *kennen* in *kennen* those objects) and the *wissen* of truths is central to his show/say distinction.

[36] This is central to Wittgenstein's view that, rather than being sentences with sense, identity sentences are only nonsensical pseudo-sentences; see n. 48 below.

[37] See e.g. the contrast between 'descriptions of complexes' and 'names' that Wittgenstein draws in 3.24 and 3.26, a contrast that reflects Wittgenstein's way of assimilating Russell's theory of descriptions.

The central point here is that by accepting (Und*) Wittgenstein is committed, with Russell, and unlike Frege, to holding that no instance of '∼*Ay*' can be both understood and true. For Frege there is nothing incoherent in my being able to grasp the sense of a name '*a*' even though I am not acquainted with the object it designates; so, for Frege, there is nothing incoherent in '*a* is an object with which I am not acquainted' being a true sentence that I can understand. For Frege, since, as against (Und*), understanding that sentence requires grasping the sense of the name in it, not being acquainted with the object it designates, my understanding that sentence is not incompatible with its truth. In contrast, by following Russell in accepting (Und*), Wittgenstein, like Russell, is committed to holding that understanding any sentence containing a proper name requires being acquainted with the object designated by that name, in which case, understanding an instance of '∼*Ay*' precludes it from being true. On this point, which forms the starting point for Russell's discussion of solipsism, Wittgenstein is in agreement with Russell.

There is an intimation of this agreement with Russell when Wittgenstein writes, in the course of 5.61:

> [W]e cannot say in logic, 'The world has this in it, and this, but not that.'

5.61 occurs immediately after 5.6 ('*The limits of my language* mean the limits of my world') and immediately prior to 5.62, in the course of which Wittgenstein considers 'the problem, how much truth there is in solipsism' and writes 'the world is my world'. I will consider further aspects of these remarks below. Here I am only concerned to point out that, insofar as 'the world' of which we cannot say that it 'has this in it, and this, but not that' is the same as 'my world', whose 'substance' is the totality of objects that I am 'given', or that I *kennen*, then Wittgenstein's remark here reflects his agreement with Russell that no instance of '∼*Ay*' can be both understood and true. By (Und*) I cannot say truly of a named object that it is an object with which I am not acquainted; that is, I cannot say truly of a named object that it is not in 'my world'. For both Wittgenstein and Russell, determining the proper response to solipsism involves determining what does and does not follow from this point.

I hasten to add, however, that in arguing that Wittgenstein's response to solipsism should be understood in the context of his acceptance of Russellian (non-Fregean) views of names, which commit him to holding that no instance of '∼*Ay*' can be both understood and true, I am not thereby indicating that Wittgenstein accepts all of Russell's views regarding the sorts of entities designated by proper names or the nature of our acquaintance (*Kenntnis*) with those entities. In particular, I am not indicating that Wittgenstein holds, with Russell, that the only particulars with which we are acquainted—and hence the only entities designated by names (occurring in sentences) that we understand—are sense-data, or hence that being acquainted with the entities designated by proper names is the same as sensing (rather than conceiving). Rather, I am indicating only that, like Russell, and unlike Frege, Wittgenstein accepts such general principles as that the sole symbolic function of genuine names (whatever expressions these may turn out to be) is to stand for objects (whatever these may turn out to be), and that understanding a sentence containing a genuine name requires being acquainted with (*kennen*) the object designated by that name (whatever exactly that may involve).

In fact, a number of remarks in the *Tractatus* indicate that Wittgenstein's views of objects and of our acquaintance (*Kenntnis*) with them differ in a number of fundamental respects from Russell's views of sense-data and of our acquaintance with them.[38] Thus, for example, Wittgenstein remarks:

> The 'experience' that we need in order to understand logic is not that such and such is the case, but that something *is*; that, however, is *not* an experience.
>
> Logic is prior to every experience—that something is *so*.
>
> It is prior to the How, not prior to the What. (5.552)

Here, Wittgenstein is indicating both that what we experience is always an arrangement of objects ('that something is so' or 'how' things are), and not simply an object by itself ('that something is' or 'what' there is), and also that our apprehension of objects, as opposed to our apprehension of the arrangement of objects, is 'prior to experience'. Thus, whereas Russell identifies sensation with our acquaintance with particulars, so that to be given particulars is to sense them, Wittgenstein indicates that we are given, and thus *kennen*, objects prior to experience. And in that case, Wittgenstein's objects are not Russell's sense-data: they are not items our acquaintance with which consists in our sensing them, but are rather items our acquaintance with which is prior to sensory experience (of arrangements of objects).

Further, Wittgenstein writes:

> If I know [*kenne*] an object then I know [*kenne*] all its possible occurrences in states of affairs. (Every such possibility must be part of the nature of the object.)
>
> A new possibility cannot be discovered later.
>
> In order to know [*kennen*] an object, I must know [*kennen*] not its external but all its internal properties.
>
> If all objects are given, then at the same time all *possible* states of affairs are also given. (2.0123–4)

Here, in keeping with his view that we are given or *kennen* objects prior to experience, Wittgenstein indicates that to be given or *kennen* an object is to be given or *kennen* all its possible arrangements with other objects. Thus, prior to experience, we are given or *kennen* objects and thereby are given or *kennen* possible arrangements of objects, while through experience we come to know how these objects are actually arranged with each other.

Again, this view of what is involved in *kennen* objects differs dramatically from Russell's view. For, as I have mentioned above (§2.2), in *Theory of Knowledge* Russell

[38] For some discussion of how Wittgenstein is led to accept a view of the objects with which we are acquainted that is so at variance with Russell's, see my (1998, especially Parts I and II).

develops an account of our knowledge of logical possibilities according to which acquaintance with a particular is not sufficient for knowledge of the possible facts in which it may occur. On that account, in order to know, for example, that it is possible that *Fa*, we must not only have sensory acquaintance with the particular *a* and non-sensory acquaintance with the universal *F*, but must also be acquainted with the logical form 'Something has some property' and must relate these three items to each other in the appropriate way. Whereas, for Wittgenstein, our *kennen* objects is prior to experience and suffices for knowledge of possibilities, for Russell acquaintance with particulars is empirical and does not suffice for knowledge of possibilities.

Thus, although Wittgenstein accepts the Russellian, as opposed to Fregean, principles regarding proper names that I have discussed above, including (Und*), he does not apply these principles in the same way Russell does. Because Russell identifies particular objects of acquaintance with sense-data, he regards the 'logical' problem as to whether '$(\exists y){\sim}Ay$' can be both understood and true as the problem as to whether he can coherently hold that there are particulars beyond the sense-data he is currently experiencing, and so as closely linked with what he takes to be the traditional problem of solipsism. Although I deny that Wittgenstein follows Russell in making this identification, I argue that Wittgenstein does follow Russell in regarding the 'logical' problem as to whether '$(\exists y){\sim}Ay$' can be both understood and true, even though none of its instances can be, as central to arriving at the correct response to solipsism.[39] Given that his objects are not Russell's sense-data, Wittgenstein does not identify that 'logical' problem with the problem as to whether there are items beyond the sense-data he is currently experiencing; instead, given that he regards our acquaintance with objects as giving us the possible arrangements of objects that constitute the 'places' in logical space and thereby determine what is conceivable, and given that he regards that 'logical' problem as central to solipsism, the problem of solipsism—the problem as to whether there are objects beyond those with which I am acquainted—becomes for Wittgenstein the problem as to whether we can coherently conceive of there being

[39] In my view, the failure to distinguish Wittgenstein's acceptance of general Russellian principles regarding names from his applying those principles in non-Russellian ways affects, in different ways, the interpretations of the Hintikkas, on the one side, and of David Pears, on the other. Recognizing some general similarities between Russell's views of acquaintance with particulars designated by names and Wittgenstein's views of *kennen* of objects designated by names, Merrill and Jaakko Hintikka (1986: ch. 3) argue that Russell's objects of acquaintance (by which they mean sense-data) should be identified with Wittgenstein's objects. Conversely, recognizing that Wittgenstein's objects differ fundamentally, in certain respects, from Russell's sense-data, Pears assumes that Wittgenstein's treatment of solipsism in the *Tractatus*, unlike Russell's in *Theory of Knowledge*, cannot be concerned primarily with issues concerning the objects of awareness, but is rather concerned primarily with issues concerning the subject of awareness (Pears 1987: 34–5, and ch. 7). However, if, as I have argued, Wittgenstein accepts general Russellian principles regarding names without applying those principles as Russell does, then Wittgenstein may regard himself as addressing the same 'logical' issues that Russell presents in his discussion of solipsism, without thereby having to identify his objects with Russell's sense-data. (This is not to say that a full discussion of Wittgenstein's remarks on solipsism will not involve considering his views on 'the subject'; only that his views on the 'subject' should be considered in connection with his response to the 'logical' issue regarding the status of '$(\exists y){\sim}Ay$'. For some discussion in that direction, see §3.4 below.)

logical possibilities beyond those that we can conceive or of there being logical spaces alternative to our own.[40]

3.2 Wittgenstein's anti-Russellian view of generality and the issue of solipsism

While following Russell in accepting (Und⋆), which commits him to the view that no instance of '$\sim Ay$' can be both understood and true, Wittgenstein departs from Russell in accepting a view of generality which, together with his acceptance of (Und⋆), commits him to the view that '$(\exists y){\sim}Ay$' cannot be both understood and true either.

As I have emphasized above, Russell is able to avoid this conclusion because he adheres to a view of generality according to which apprehending a general proposition does not depend upon apprehending its instances. In particular, in *Theory of Knowledge*, Russell holds that apprehending a general proposition is logically simpler than apprehending any of its instances, in that apprehending the general proposition requires being acquainted with only some of the entities one must be acquainted with in order to apprehend the elementary proposition (1913: 129–31). Thus he holds that, while apprehending '$(\exists x)Fx$' requires being acquainted with the property F along with the logical form of subject-predicate complexes, apprehending an elementary proposition 'Fa' requires being acquainted not only with that property and that logical form but also with the particular a. In terms of 'logical order', we need something further besides apprehending a general proposition—namely, acquaintance with particulars—if we are to apprehend its instances. Russell is thus able to hold that there are cases in which we can think truly a general proposition '$(\exists x)Fx$' even though we cannot think truly any of its instances.

In contrast, in the *Tractatus*, Wittgenstein writes:

> It immediately strikes one as probable that the introduction of elementary sentences is fundamental for understanding the other kinds of sentences. Indeed the understanding of general sentences *palpably* depends on the understanding of elementary sentences. . . . (4.411)

and also:

> Suppose I am given *all* elementary sentences: then I can simply ask what sentences I can construct out of them. And there I have *all* propositions, and *that* fixes their limits. (4.51)

Whereas Russell holds that we can apprehend general propositions without apprehending elementary propositions, Wittgenstein indicates that understanding general sentences depends upon understanding elementary sentences. Whereas Russell holds that we could apprehend general propositions first and then would need further

[40] In contrast, in order for Russell to address fully the issue as to the sense in which we can conceive of there being propositional contents beyond those we apprehend, he will have to consider not only the coherence of thinking there are particulars with which we are not acquainted but also of thinking there are universals with which we are not acquainted and thus particulars we are in no position to describe. In my view, it is through his ramified theory of types that Russell addresses these issues. See in this connection the 'Introduction' to *Principia Mathematica*, 63–4.

experience—acquaintance with particulars—in order to apprehend elementary propositions, Wittgenstein indicates that once we are 'given' elementary sentences, we have all we need in order to 'construct' all other sentences—including general sentences—out of those elementary sentences.

More specifically, Wittgenstein introduces the notion of an 'operation' in connection with his view that all sentences are 'constructed' or 'generated' out of elementary sentences. For Wittgenstein, the 'result' of an operation on certain sentences that are the 'bases' of the operation is a sentence that is thereby 'constructed' out of the 'bases' of that operation. And while Wittgenstein writes in remark 5 that

A proposition is a truth-function of elementary sentences,

he adds further in 5.3:

Every proposition is the result of truth-operations on elementary sentences,

where 'truth-operations' are operations that result in sentences that are truth-functions of the sentences that are their 'bases'. Thus, for Wittgenstein, every sentence, including every general sentence, may be regarded as 'constructed' out of elementary sentences, and every sentence is a truth-function of the elementary sentences that it may be regarded as 'constructed' out of.

In particular, Wittgenstein indicates that every sentence may be generated out of successive applications of his *N*-operation, where the result of applying that operation to the sentences S_1, S_2, S_3, . . . , S_n is the sentence '$\sim S_1$ & $\sim S_2$ & $\sim S_3$ & . . . $\sim S_n$'. And in 5.52, he writes:

If ξ has as its values all the values of a function fx for all values of x, then $N(\bar{\xi}) = \sim(\exists x).fx$,

thereby indicating that he regards the sentence '$\sim(\exists x).fx$' as the truth-function '$\sim fa$ & $\sim fb$ & $\sim fc$ & . . . ', that is, as what Wittgenstein calls 'the negation' (5.5, 5.502) of 'all values of the function fx for all values of x'. Further, for Wittgenstein, applying the *N*-operator to '$\sim(\exists x).fx$' results in the sentence '$(\exists x).fx$', which he thereby regards as the truth-function '$fa \vee fb \vee fc \vee \ldots$'. And similarly Wittgenstein regards '$(\forall x).fx$' as the truth-function 'fa & fb & fc & . . . '.

This account of generality is fundamentally opposed to Russell's. For Russell would not only deny, for example, that '$(\exists x).fx$' means the same as '$fa \vee fb \vee fc \vee \ldots$' or that '$(\forall x).fx$' means the same as '$fa$ & fb & fc & . . . '; he would also deny that these, or any, general sentences are truth-functions of their instances. First of all, on Russell's account of generality, we can understand a general sentence without having to be acquainted with every, or indeed any, particular over which its variable ranges. But if he held with Wittgenstein that '$(\forall x).fx$' means the same as 'fa & fb & fc & . . . ', then, given (Und*), he would thereby be forced to hold that understanding '$(\forall x).fx$' requires being acquainted with the objects *a*, *b*, *c*, . . . And whereas Wittgenstein is willing to accept this consequence, Russell is not.

Thus, for example, in writing both 'that *fa* follows from $(\forall x).fx$' (6.1201) and that 'if *p* follows from *q*, the sense of "*p*" is contained in the sense of "*q*"' (5.122), Wittgenstein is indicating that the sense of '*fa*' is 'contained in' the sense of '$(\forall x).fx$', so that understanding '$(\forall x).fx$' (knowing its sense) requires understanding '*fa*' (knowing its sense); and given his view that the object *a* occurs in the sense of '*fa*' and his acceptance of (Und★), he thereby holds that the object *a* (along with every other object being quantified over) occurs in the sense of '$(\forall x).fx$' and that understanding '$(\forall x).fx$' requires *kennen* the object *a* (along with every other object being quantified over), a view that is facilitated by his holding that we are given objects a priori.

In *The Problems of Philosophy*, when discussing the proposition 'all men are mortal', Russell by contrast writes:

> [W]e can *understand* what the proposition means as soon as we understand the universals involved, namely *man* and *mortal*. It is obviously unnecessary to have an individual acquaintance with the whole human race in order to understand what our proposition means. (1912: 106; cf. Whitehead and Russell 1910: 45)

Thus Russell takes it as obvious that understanding the general sentence does not require acquaintance with the particulars over which that sentence quantifies[41] and hence (given his acceptance of (Und★)) takes it as obvious that the sort of analysis provided by Wittgenstein cannot be correct. Moreover, given that for Russell the identity of a general proposition is independent of what, if any, particulars there are, it is appropriate that he should reject the view that 'that *fa* follows from $(\forall x).fx$', as he does in his *Introduction to Mathematical Philosophy* (1919: 203–4). For if the identity of the general proposition '$(\forall x).fx$' does not depend on the object *a* existing or thus on there being a proposition '*fa*', and the truth of '*fa*' depends upon the being of '*fa*' and hence upon the existence of *a*, then the truth of '$(\forall x).fx$' does not require the truth of '*fa*'.[42]

Not only does Russell deny that '$(\forall x).fx$' means the same as '*fa* & *fb* & *fc* & . . .'; he also denies that '$(\forall x).fx$' is a truth-function of its instances—that is, he denies that the truth-value of '$(\forall x).fx$' is determined solely by the truth-values of its instances. For, given that what the general sentence '$(\forall x).fx$' means—the general proposition it expresses—is independent of what particulars there are, then if *a, b, c*, . . . , *z* are the only particulars that there are, so that '*fa*', '*fb*', '*fc*', . . . , and '*fz*' are the only instances of '$(\forall x).fx$' there are, these are general facts that are not determined by the meaning of '$(\forall x).fx$'. Accordingly, for Russell if '*fa*', '*fb*', '*fc*', . . . , and '*fz*' are, in fact, the only instances of '$(\forall x).fx$', and if they are all true, then the truth of '$(\forall x).fx$' does not follow from the truth of its instances alone, but rather follows from the truth of its instances together with the general fact that those are its only instances, or, alternatively, the

[41] In fact, for Russell, (first-order) general sentences quantify over absolutely all particulars, so that if understanding 'all men are mortal' required being acquainted with each particular over which it quantifies, it would require being acquainted not only with each human but also each particular.

[42] See also in this connection G. E. Moore (1927: 87–8), where he criticizes Ramsey's (Wittgensteinian) view of generality.

general fact that *a*, *b*, *c*, . . . , *z* are all the particulars there are. Thus, in his 1918 lectures on logical atomism:

It is perfectly clear, I think, that when you have enumerated all the atomic facts in the world, it is a further fact about the world that those are all the atomic facts there are about the world, and that is just as much an objective fact about the world as any of them are. (1918: 236; cf. 1915: 53, 55–6)

For Russell, that is, determining the truth-value of each elementary sentence (and hence identifying each atomic fact) does not by itself determine the truth-value of every general sentence. To do that one will also need the truth of the general sentence to the effect that those are the only elementary sentences there are (or the general fact that those are the only atomic facts there are).

Given his view of generality, it is not surprising that among the 'queries' Russell posed to Wittgenstein in his August 1919 letter, written after had read through the *Tractatus* 'twice carefully', is the following regarding 5.3:

All prop[osition]s results of truth-operations on elementary prop[osition]s? How about generality? (Wittgenstein 1995: 122)

By 5.3, every sentence is both 'constructed' out of, and a truth-function of, elementary sentences, but it is fundamental to Russell's view of generality that general sentences have neither of these features; hence, it is natural that he should question how 5.3 is to apply to general sentences. Moreover, given the centrality of Russell's views of generality to his response to solipsism, it perhaps no coincidence that Wittgenstein's remarks relevant to solipsism in the 5.6's are, by the numbering system of the *Tractatus*, all ultimately comments upon 5, 'A sentence is a truth-function of elementary sentences'. For given his non-Russellian account of generality, according to which general sentences are truth-functions of elementary sentences, Wittgenstein cannot respond to solipsism as Russell does. And if Wittgenstein was aware of Russell's *Theory of Knowledge* discussion of solipsism, it is plausible to suppose that he would also be aware that his rejection of Russell's view of generality commits him to rejecting Russell's response to solipsism.

In particular, given that he regards '$(\exists x).fx$' as the logical sum of its instances, '$fa \vee fb \vee fc \vee \ldots$', and thus holds that understanding '$(\exists x).fx$' requires understanding each of its instances, and given that he also holds that '$(\exists x).fx$' is true if and only if one of its instances is true, then Wittgenstein is committed to holding that if I understand '$(\exists x).fx$' and it is true then there is a true instance of 'fx' that I understand. But, by (Und*), Wittgenstein is committed to holding that there cannot be a true instance of '$\sim Ay$' that I understand, in which case, by his view of generality, he is committed to holding that it cannot be both that I understand '$(\exists y)\sim Ay$' and that it is true. Just as Berkeley's rejection of 'abstract general ideas' arguably gives him a reason for making what Prior regards as the fallacious inference that since no instance of '$\sim \tau y$' can be both understood and true, then the general sentence '$(\exists y)\sim \tau y$'—that is, ~(Id)—cannot be both understood and true, so too Wittgenstein's view of general sentences commits him to making what Russell would regard as an 'elementary blunder' of logic—namely, to holding that since no instance of '$\sim Ay$' can be both understood and true, then '$(\exists y) \sim Ay$'—that is $\sim$(Solip)—cannot be both understood and true either.

This point surfaces in 5.61, where Wittgenstein writes:

> Logic pervades the world: the limits of the world are also its limits.
>
> So we cannot say in logic, 'The world has this in it, and this, but not that.'
>
> For that would appear to presuppose that we were excluding certain possibilities, and this cannot be the case since it would require that logic should go beyond the limits of the world; for only in that way could it view those limits from the other side as well.
>
> We cannot think what we cannot think; so what we cannot think we cannot *say* either.

It also surfaces in the Preface, where he indicates that we cannot 'draw a limit to thought' because doing so would require us to 'find both sides of the limit thinkable'—that is, would require us 'to think what cannot be thought' (3)—and likewise in 4.114, where, compatibly with the view that we cannot think 'the other side' of the limits of thought, he indicates that the only way to 'limit the unthinkable' is 'from within through the thinkable'. In all these passages, Wittgenstein makes no distinction—of the sort that Russell is concerned to highlight—between the attempt to produce examples of what lies 'on the other side' of the limits of 'the world' (consisting of the totality of objects which I *kennen*) or of the limits of thought (consisting of the totality of senses I can express) and the attempt to conceive, truly or falsely, simply of there being something on 'the other side' of those limits. Instead, Wittgenstein indicates that in order make clear that it is incoherent even to conceive of the world or thought as limited, and thus even to wonder whether there is anything on 'the other side' of those limits, it is sufficient to point out that it is incoherent to attempt to produce examples from 'beyond' those limits—thus, to attempt to name objects with which we are not acquainted, or (what is closely related, given his views as to what is involved in *kennen* objects) to attempt to express possibilities that lie outside our logical space. That is, he indicates that in order to establish that I cannot coherently conceive that $(\exists y) \sim Ay$—or that '$(\exists y) \sim Ay$' cannot be both understood and true—it is sufficient to establish that I cannot coherently conceive any instance of '$\sim Ay$'—or that no instance of '$\sim Ay$' can be both understood and true; and the reason he is in a position to indicate this is his view of generality.

No less than Wittgenstein (and for reasons that influenced Wittgenstein), Russell agrees that he cannot name an object and say truly of it that it lies outside the world of entities with which he is presently acquainted; no less than Wittgenstein, Russell agrees he cannot view the 'limits of the world' from 'the other side', if this requires us to name objects with which we are not acquainted. However, by his account of generality, Russell can conceive coherently that the 'world' of particulars he is currently experiencing may be limited and thus may not be the same as 'the world' of all the particulars there are. He can coherently conceive both the truth of solipsism—according to which there are no particulars on 'the other side' of the totality of those with which he is currently acquainted—as well as the truth of its negation—according to which there are particulars 'on the other side' of that totality; that is, he can understand (Solip) and ~(Solip) without precluding the truth or falsity of either. And so, for Russell, addressing the issue of solipsism requires us to determine which of these coherently conceivable alternatives is correct, or, at least, more reasonable to believe.

In contrast, by holding that every sentence is a truth-function of elementary sentences, so that the expressive power of a language is determined or 'limited' by the elementary sentences that are 'given' (and hence, in virtue of (Und*), by the objects that are 'given'), Wittgenstein cannot understand ~(Solip) without precluding its truth and cannot understand (Solip) without precluding its falsity. And thus, unlike Russell, he can recognize no genuine question—no question the understanding of which precludes neither an affirmative nor negative answer—as to whether 'my world', consisting of the totality of objects with which I am acquainted, is the same as 'the world', consisting of the totality of absolutely all there objects there are, or as to whether 'my world' is limited. Given his view of generality, it is just as much an incoherent attempt to view the limits of 'my world' from 'the other side' to attempt to ask whether there are objects besides those with which I am acquainted as it is to attempt to provide named examples of such objects.[43]

Moreover, given his view that in *kennen* objects I thereby *kenne* the possible combinations of objects that constitute my logical space, for Wittgenstein, the incoherence of attempting to ask whether there are objects besides those with which I am acquainted is the same as the incoherence of attempting to ask whether there are possible combinations of objects besides those which I can conceive. Thus, in recognizing no genuine question as to whether all the objects with which I am acquainted are absolutely all the objects there are, he recognizes also no genuine question as to whether all the possible arrangements of objects I can conceive are absolutely all the possibilities there are, that is, the extent to which 'my logical space' does or does not match up to 'absolute logical space', consisting of absolutely all the possibilities there are.

3.3 Is solipsism an expressible truth?

Given that Wittgenstein is committed to holding, as against Russell, that ~(Solip) cannot be both understood and true, the question remains as to whether, or in what sense, he thereby holds there is 'truth in solipsism'. In this subsection, I argue (what I take to be relatively uncontroversial) that, whatever else he may hold, Wittgenstein

[43] Diamond (2000) relates—correctly I believe—differences between Wittgenstein's and Russell's views of generality to their different views regarding the coherence of holding that different subjects are acquainted with different 'private objects' and argues on that basis that there is a kind of 'private language argument' in the early Wittgenstein. As I am presenting it, the differences between Wittgenstein and Russell on generality (taken against the background of their shared acceptance of (Und*), which Diamond does not mention) lead more generally to their different views of the coherence of my holding that there are objects with which I am not acquainted, regardless of whether anyone else is acquainted with those objects and regardless of whether they are 'private' to anyone who is acquainted with them. (See n. 54 below.)

does not regard (Solip) as a true sentence that can be understood: in being unable to understand ∼(Solip) and regard it as true, he does not take himself to understand ∼(Solip) and regard it as expressing a falsehood, or thereby take himself to understand (Solip) and regard it as expressing a truth. Instead he holds that both ∼(Solip) and (Solip) are nonsensical, so that neither can be understood at all. In the following subsection, I address the question in what sense, if any, he regards solipsism as an inexpressible truth, one that may be only shown but not stated.

For Wittgenstein, there are only two kinds of genuine sentences, sentences that are 'part of the symbolism' (4.4611): sentences with sense, and sentences that are senseless, which are in turn either tautologies or contradictions. All other apparent sentences are nonsensical pseudo-sentences, expressions that look like sentences but are not 'part of the symbolism', that is, are not genuine sentences at all. While a sentence with sense, being a picture of reality, is such that understanding it does not suffice for determining its truth-value,[44] a tautology is so constructed that it cannot be false and that understanding it suffices for knowing that it is true, while a contradiction is so constructed that it cannot be true and that understanding it suffices for knowing that it is false.

Since, for Wittgenstein, understanding ∼(Solip) precludes its truth, he is thereby committed to denying that ∼(Solip) is a sentence with sense and to holding that, if ∼(Solip) is a sentence at all, it is a contradiction. Likewise, he is committed to holding that, if (Solip) is a sentence at all, it is a tautology. In contrast, given his view of generality, Russell is able to hold that neither ∼(Solip) nor (Solip) is such that understanding it precludes either its truth or its falsity, thereby meeting the Wittgensteinian standard for being a sentence with sense.

The crucial difference here is that Russell, because he does not regard general sentences as 'constructed' out of elementary sentences that we understand, is able to regard the (individual) variables in general sentences as ranging over absolutely all particulars, so that even if we are, in fact, acquainted with all particulars, this is not 'written into' the variable, in which case it remains an open question, not determined merely by understanding a general sentence, whether we are acquainted with all particulars. Because Wittgenstein regards general sentences as 'constructed' out of elementary sentences that we understand, he is committed to holding that the variables in general sentences range only over objects with which we are acquainted. Hence, for Wittgenstein, unlike Russell, the use of variables does not enable me to ask, as a genuinely open question, whether the objects with which I am acquainted are all the objects there are—to ask whether there is anything 'on the other side of' the totality of objects with which I am acquainted.[45]

[44] See e.g. 2.223–5, where Wittgenstein indicates that we cannot determine the truth-value of a picture from that picture alone.

[45] In discussing the 'novel features' of Wittgenstein's treatment of solipsism, Pears writes: 'Unlike Russell, he refuses to treat solipsism as a theory which might be true, but which the evidence showed more likely to be false' (1987: 34; cf. 1972: 61–2). As I am presenting it, Wittgenstein's view of generality *precludes* him from so regarding solipsism.

Given Wittgenstein's view of variables, it might seem that he would have a good reason for regarding (Solip) as a tautology and ~(Solip) as a contradiction. For, given his understanding of generality, it might seem that for Wittgenstein (Solip)—that is, '$\sim(\exists y)\sim Ay$' or '$(\forall y)Ay$'—is not to be read as 'For absolutely any object y, I am acquainted with y', but rather as 'For any object y with which I am acquainted, I am acquainted with y', which might seem as good a candidate as any for being a tautology. Likewise, it might seem that, for Wittgenstein, ~(Solip) should be read as 'For some object y with which I am acquainted, I am not acquainted with y', which might seem as good a candidate as any for being a contradiction. But even given Wittgenstein's view of generality it is clear that he cannot regard (Solip) as a (senseless) tautology or (~Solip) as a (senseless) contradiction, but is rather committed to regarding both as nonsensical pseudo-sentences.

One reason, for Wittgenstein, that tautologies and contradictions are 'part of the symbolism' is that they are among the sentences that can be 'constructed out of' the elementary sentences. In particular, they can be generated by successive applications of his *N*-operation to elementary sentences. For example, if p is an elementary sentence, $N(Np, p) = p \,\&\, \sim p$, while $N(N(Np, p)) = p \vee \sim p$. Thus, for Wittgenstein, even though tautologies and contradictions are not themselves sentences with sense, they are constructed out of elementary sentences—and hence out of sentences with sense—by the same general procedure that enables one to construct every sentence with sense out of elementary sentences. For Wittgenstein, tautologies and contradictions are composed of meaningful signs combined in ways permitted by the symbolism; however, in these 'limiting cases' (4.466), what results from the permissible combination of meaningful signs represents no possible situation—that is, no arrangement of objects that may either obtain or fail to obtain—and so, while 'part of the symbolism', tautologies and contradictions are not pictures of reality or, hence, sentences with sense.

The question here is whether Wittgenstein may regard ~(Solip) and (Solip) as likewise constructed out of elementary sentences. For Wittgenstein, while ~(Solip) is '$\sim Aa \vee \sim Ab \vee \sim Ac \vee \ldots$' and (Solip) is '$Aa \,\&\, Ab \,\&\, Ac \,\&\ldots$', '$Aa$', '$Ab$', '$Ac$', etc. cannot themselves be regarded as sentences with sense or hence as elementary sentences. For, by accepting (Und*), Wittgenstein is committed to holding that if an instance of 'Ax' is understood then it is true; hence, he is committed to holding that if an instance of 'Ax' is a genuine sentence, it is a tautology, and so not an elementary sentence. Thus, if Wittgenstein is to regard ~(Solip) as a contradiction and (Solip) as a tautology, it is not enough that he regard them as truth-functions of instances of 'Ax'; rather, he will have to pursue the analysis further and find some way of analyzing instances of 'Ax' so that they turn out to be tautologies. However, there would seem to be no prospect, for Wittgenstein, of such an analysis.

In order to deny that instances of 'Ax' are elementary sentences, Wittgenstein would have to find a way to analyze either 'I' or the relational expression 'am acquainted with' (since the expressions replacing 'x' are assumed to be simple names). For Russell, acquaintance is a primitive two-term relation, the relation in terms of which all other cognitive relations are to be analyzed;[46] thus, insofar as Wittgenstein accepts general Russellian principles regarding acquaintance (without necessarily applying those principles in the way Russell does), he too would be in no position to provide some analysis of acquaintance into simpler terms. Hence, in order to deny that instances of '*Ax*' are elementary sentences, Wittgenstein would, it seems, need to find a way to deny that 'I' is the name of an object with which he is acquainted. Russell adopts such a position in *Theory of Knowledge*, where he holds that 'I', on any occasion of its use, is not a genuine proper name, but is rather a disguised description of an entity with which the speaker is not acquainted, so that we can understand instances of 'Something is acquainted with *y*' despite our 'incapacity for acquaintance with the subject' (1913: 37). However, such an option is not available for Wittgenstein; for, as I have argued, given his view of generality (together with his acceptance of (Und⋆)), he cannot hold, with Russell, that the variable '*x*' in '(∃*x*)(*x* is acquainted with *y*)' includes in its range 'subjects' with which no one is acquainted. Hence, it would seem, Wittgenstein is in no position to regard instances of '*Ax*' either as elementary sentences or as analyzable: he cannot regard them as elementary, since understanding them would suffice for knowing that they are true; but there would seem to be no prospect of him finding a way to analyze them as truth-functions of elementary sentences. Rather, he has to regard them as nonsensical, and thus has to so regard both ~(Solip) and (Solip).

For Wittgenstein, to regard an apparent sentence as nonsensical is to hold that at least one sign in it has been given no meaning (5.4733); and a sign is meaningless if it is 'unnecessary' or 'useless', so that it never contributes to the formation of sentences with sense (see 3.328, 5.47321). Thus, if he is to hold that no instance of 'I am acquainted with *y*' is a genuine sentence, he has to hold that at least one sign in that expression is not needed in order to construct a language 'capable of expressing every sense' (see 4.002);[47] and since it is assumed that only meaningful names are to replace '*y*', then he must hold that either 'I' or 'am acquainted with' is meaningless, that is, not necessary for the expression of any sentence with sense. And it seems clear that Wittgenstein holds that both signs are thereby meaningless. For in 5.631, he indicates that no possible complete description of the world will mention 'the subject', in which case there is no need for 'I' in the formation of sentences with sense; and insofar as there will be no need for any form of the predicate of 'being acquainted with' if there is no need for any sign designating 'the subject' to whom objects are given, then Wittgenstein also holds that there is no need for such a predicate in the construction of a language 'capable of expressing every sense'.[48]

[46] See e.g. the first paragraph of *Theory of Knowledge* (1913: 5).

[47] Although see §3.4 for a discussion of what 'every sense' comes to for Wittgenstein.

[48] On this account, Wittgenstein's reasons for regarding the apparent predicate 'I am acquainted with *x*' as nonsensical are similar to his reasons for regarding the apparent (two-place) identity predicate '$x = y$' as nonsensical. By (Und⋆), not only does understanding any instance of 'I am acquainted with *x*' preclude its falsity—so that no instance of 'I am acquainted with *x*' can be a sentence with sense or an elementary sentence—but also understanding any instance of '$x = y$' would suffice for knowing its truth-value—so that no instance of '$x = y$' can be a sentence with sense or an elementary sentence (see 4.243). And just as I have argued here that Wittgenstein would hold that 'I' and a (two-place) predicate '*x* is acquainted with *y*' are

This account as to how, for Wittgenstein, (Solip) is nonsensical rather than tautological illustrates, I hope, some features of 'the correct method in philosophy' that Wittgenstein remarks upon in 6.53. On that method, when 'someone wants to say something metaphysical' the 'correct method' is 'to demonstrate to him that he has failed to give a meaning to certain signs in his sentences', even though 'it would not be satisfying to the other person—he would not have the feeling that we were teaching him philosophy'. In this case, the Wittgensteinian 'demonstration' that (Solip) involves signs to which no meaning has been given does not consist in any argument to the effect that we cannot 'view the limits of the world' from 'the other side'; it does not consist in presenting matters as if there is something the metaphysician (here, for example, Russell) regards as thinkable—namely, whether there are particulars beyond those with which I am acquainted—that Wittgenstein denies is thinkable; instead, it proceeds by 'establishing' that any attempt to give meaning to the signs in (Solip) in the only way (for Wittgenstein) that the signs in our language can have meaning will fail. The result is not that (Solip) transgresses the 'limits of thought' but rather that it turns out to be 'simply nonsense'. In effect, what is thus 'demonstrated' is that (Solip) turns out to be even emptier than a tautology. Given the only way (for Wittgenstein) that the variables in a general sentence can acquire meaning, then, while (Solip) is like a tautology in that it cannot be both understood and false, it does not get so far as to be a tautology, since it contains signs that are wholly unnecessary for the formation of any sentence with sense.

By regarding as (Solip) as nonsensical, rather than tautological, Wittgenstein is, in effect, accommodating Mackie's view that 'operational self-refutation' should not be confused with 'internal self-contradiction' or with 'logical impossibility'. Just as Mackie indicates that even if Berkeley were right to hold that ~(Id) cannot be thought truly, he would be wrong to conclude from that that ~(Id) is 'logically impossible', and hence that (Id) is 'logically necessary', so too even though Wittgenstein holds, with the solipsist (and as against Russell) that ~(Solip) cannot be both understood and true, he does not take this to show that ~(Solip) is therefore 'logically impossible' (a contradiction) or hence that (Solip) is 'logically necessary' (a tautology).

unnecessary for the expression for any sentence with sense, so too Wittgenstein argues that the identity-sign is unnecessary for the expression of any sentence with sense (see the 5.53s).

By accepting (Und*), Russell, like Wittgenstein, is committed to holding that no instance of '$x = y$' (where 'x' and 'y' are to be replaced by genuine names) can be understood without knowing its truth-value, in which case no instance of '$x = y$' meets Wittgenstein's standard for being a sentence with sense. However, given his view of generality, there are, for Russell, general sentences involving the identity-sign that do meet that standard. Thus, for example, if a and b are the only particulars with which I am acquainted, I can, for Russell, formulate and understand the general sentence '$(\exists x)(\sim(x = a)\ \&\ \sim(x = b))$'—a sentence that for Russell will be true in this case if and only if ~(Solip) is true—without thereby knowing its truth-value. Whereas Wittgenstein's view of generality requires him to hold that, if the identity-sign (or the phrase 'I am acquainted with') is not necessary for the expression of any *elementary* sentence, then it is not necessary for the expression of *any* sentence with sense, Russell's view of generality enables him to hold that, while the identity-sign (or the phrase 'I am acquainted with') may not be used in any elementary sentence that meets Wittgenstein's standard for expressing a sense, it still contributes to the expression of *general* sentences that meet that standard, and hence meets Wittgenstein's criterion for an expression's being meaningful.

However, without considering Wittgenstein's show/say distinction, it remains unclear as to whether Wittgenstein, like Mackie, is genuinely opposed to deriving metaphysical consequences from 'operational self-refutation'. On the one hand, Wittgenstein goes beyond anything that Mackie claims to indicate that what is in this case 'operationally self-refuting'—namely, ~(Solip)—is not a sentence at all, but is rather nonsensical, incapable of being understood. On the other hand, if, as some commentators contend,[49] Wittgenstein holds that what are shown are truths—indeed, necessary metaphysical truths—that cannot be stated, then Wittgenstein would be taking the 'operational self-refutation' of ~(Solip) to establish what (Solip) 'intends to say' as a metaphysical truth, albeit one that can only be shown but not said. Thus, determining what, if any, metaphysical consequences Wittgenstein takes the 'operational self-refutation' of ~(Solip) to have will require interpreting his show/say distinction and considering how it applies to his discussion of solipsism.

3.4 Solipsism and showing

In 5.62, which by the numbering system of the *Tractatus* is a comment on 5.6 ('*The limits of my language* mean [*bedeutet*] the limits of my world'), Wittgenstein relates solipsism to showing by writing:

> This remark provides the key to the question, to what extent is solipsism a truth.
>
> [W]hat solipsism *means* [*meint*] is quite correct; only it cannot be *said*, but shows itself.
>
> That the world is *my* world shows itself in that the limits of language (the only language which I understand) mean [*bedeuten*] the limits of *my* world.

By indicating that what solipsism means or intends is 'quite correct' but cannot be said, Wittgenstein seems to be indicating that solipsism as a doctrine involves misconstruing something that has been genuinely recognized. Thus, on interpretations according to which what are shown are truths that cannot be said, Wittgenstein is indicating that while the solipsist has recognized a truth—namely, the truth that 'the world is my world', however this is to be construed—the solipsist errs in attempting to state this truth, rather than allowing it to 'show itself'.

In this subsection I argue against this sort of interpretation. In particular, while I present Wittgenstein as holding that any attempt to state solipsism as a thesis involves misconstruing something that has been genuinely recognized, I do not present him as holding that what the solipsist recognizes is a truth that cannot be said. On the reading I recommend, what 'shows itself' are not truths at all, but rather what we *kennen* in understanding sentences—including the objects we thereby *kennen* and the possible arrangements of objects we *kennen* in *kennen* those objects; and since what we thus *kennen* is not properly expressed in the form 'that *p*', while what we say is always of the form 'that *p*', then any attempt to say what is thereby shown will distort, or misconstrue, the recognition of what is shown. I argue that, on this reading, Wittgenstein may

[49] See e.g. Anscombe (1959: ch. 13); Mounce (1997, 2001).

be regarded as holding that, while what is shown are the limits of the world and of thought, nothing can be either said or shown regarding what lies 'on the other side' of those limits, and that in any attempt to say what is shown one presents oneself, confusedly, as able to breach those limits.

To introduce this reading I begin by considering how Wittgenstein should be interpreted when, in the final sentence of 5.62, he writes:

(*) That the world is my world shows itself. . . .

Insofar as 'my world' is constituted by, or has as its 'substance' (2.021), the totality of particulars with which I am currently acquainted, then Russell would understand 'The world is my world' as he would understand (Solip)—that is, as the claim that there is nothing besides the particulars with which I am currently acquainted, a claim he regards himself as understanding without precluding either its truth or falsity, but which he also holds he has good reasons for regarding as false. Hence, if Wittgenstein intends 'the world is my world' in that way, then he would seem to be indicating by (*) that what 'shows itself' is the falsity of the anti-solipsistic view that Russell is seeking to defend. He would be indicating that while neither solipsism nor its denial, as Russell understands those positions, is capable of being expressed by a genuine sentence, nevertheless genuine sentences show that there is nothing on 'the other side' of the objects with which I am acquainted, so that the position Russell understands as solipsism is true but unstateable, while the position Russell understands as the denial of solipsism (and defends) is false but unstateable. On this interpretation, Wittgenstein would, in effect, be accepting the Berkeleyan move of taking our inability to coherently conceive ~(Solip) as true as establishing that solipsism is a metaphysical truth, albeit a truth that can only be shown but not said.

This is not, I take it, how Wittgenstein is to be interpreted. That is, I take it that just as Wittgenstein rules out Russell's account of generality as a means by which we can 'reach' in thought beyond the limits of the objects with which we are acquainted, enabling us to ask coherently whether or not there are any objects beyond those limits, so too he does not regard 'showing' as a means by which we can 'reach' beyond those limits.[50] Given that

[50] Conant (1991: 159–60; 2000: 196–8; 2002: 421–4) presents an interpretation according to which in the 'body of the text' of the *Tractatus* a 'mock doctrine' (1991: 160) is presented of the distinction between showing and saying, according to which what is 'shown' is 'ineffable truth' regarding what lies 'beyond' the limits of language and of thought (2000: 198), a 'mock doctrine' that is to be 'thrown away' once Wittgenstein is fully understood in 6.54 (see similarly Diamond 1988: 181–2). So, for Conant, the *Tractatus* as a whole presents a kind of *reductio* argument against this 'mock doctrine' (see also in this connection Kremer 2001: 42–3, as well as Sullivan 2004: 38, on the sort of *reductio* here involved). On this account, the mistake of 'proponents of the standard interpretation' is to take that 'mock doctrine' as Wittgenstein's genuine view, rather than 'throwing it away' (1991: 160). On the reading I present, and as against the readings developed by both Conant and 'proponents of the standard interpretation' he opposes, 'showing'—as it is presented in the 'body of the text'—is never such as to enable us to reach 'beyond' the limits of the world, or thought, or language and recognize 'ineffable truth'. Hence, while I agree with Conant that Wittgenstein does not hold that we are able to recognize 'ineffable truths' regarding what lies 'on the other side' of the limits of the world or of thought, I hold that 'showing' is never presented as enabling us to do so, so that there is no 'mock

Wittgenstein has just indicated in 5.61 that we cannot 'go beyond the limits of the world' and 'view those limits from the other side', it would seem that unlikely that he intends to indicate in 5.62 that while we cannot *say* that nothing lies on 'the other side' of the totality of objects with which I am acquainted, nevertheless that nothing lies 'on the other side' of that totality *shows itself*. For on such an interpretation showing, however different it may be from saying, would be taken to provide us a with way of 'viewing' what Wittgenstein has just indicated that we cannot view—namely, 'the other side' of the totality of the objects which have been given, enabling us to recognize that there is nothing there.

Nor do I take it that Wittgenstein intends (*) in such a way that what 'shows itself' is that there may be different 'worlds' for different subjects, so that while there is 'the world' that is 'my world', there also may be other worlds that are worlds for other subjects. While this interpretation may be suggested by remarks such as 6.43 ('The world of the happy man is a different one from that of the unhappy man') and 6.431 ('So too at death the world does not alter, but comes to an end', which may suggest that 'the world' of a given subject comes to an end with the death of that subject), it would, again, require Wittgenstein to hold that showing enables me to view 'the other side' of the totality of objects that I have been given. Only on this view, what is thereby revealed is not that there is nothing 'beyond' that totality, but rather that there are, or at least may be, different totalities of objects—different 'worlds'—given to different subjects.

Rather, I take it that in the final sentence of 5.62, Wittgenstein is not making any positive identification of 'the world' and 'my world' that depends upon recognizing what does or does not lie 'on the other side' of the limits of 'the' or 'my world', but is instead making a negative point that depends upon the recognition of limits. Given my inability to raise any genuine question as to whether the objects I have been given do or do not constitute all the objects there are absolutely, and, what for Wittgenstein is the same, given my inability to raise any genuine question as to whether the possibilities I can express in language are all the possibilities that can be expressed in any language, then just as I am unable to regard as language anything other than what is capable of expressing the possibilities I can express, so too there is nothing I can regard as the world other than the world of objects that has been given to me, the world to which anything I can regard as language is answerable, the world that would be completely described if all true elementary sentences (of the only language I understand) are given (see 4.26). Wittgenstein's purpose here is not to suggest that because I *cannot regard* anything else as the world, then there *cannot be* anything beyond the objects I have been given; nor is it to suggest that even though *I* cannot regard anything other than the objects I have been given as constituting the world, there may nevertheless be other

doctrine' of our being shown ineffable truths that is to be 'thrown away' by the end of the book. (In some writings, Conant distinguishes between an 'idea' of showing that is to be 'thrown away' from one that may be retained even after the 'ladder has been kicked away'; see, for example, Conant and Diamond (2004: 65–6), as well as Kremer (2001, 2004), whose views I discuss in n. 64 below. On the interpretation I am presenting there is no need for regarding the text as incorporating two such 'ideas' of showing.)

worlds for other subjects; rather, the point is to recognize the limits beyond which there are no questions to be asked and to go no further.[51] And as will be more clear below, I do not regard this negative point that Wittgenstein is making in 5.62 in identifying 'the world' and 'my world' as reflecting any commitment to truths that can only be 'shown' but not 'said'.

In remarks he makes shortly after returning to philosophy in the late 1920s and early 30s, Wittgenstein makes the same point by indicating that neither the world nor language nor thought can be 'set in relief'. Thus, for example, in his 1930 *Philosophical Remarks*, after criticizing those who 'ascribe reality' to what goes beyond 'the given' and who thus hold that 'what we neither can nor want to go beyond would not be the world', Wittgenstein writes:

> Time and again the attempt is made to use language to limit the world and set it in relief—but it can't be done. The self-evidence of the world expresses itself in the very fact that language can and does only refer to it.
>
> For since language only derives the way in which it means from its meaning, from the world, no language is conceivable which does not represent this world. (1975: 80)

Here, as in 5.62, Wittgenstein identifies 'the world' with that which 'language' represents, where 'language' is, in turn, identified with any 'conceivable' language; and here it is clear that, in making these identifications, Wittgenstein is not intending to indicate anything as to what is, or is not, beyond the limits of the world, or of thought, or of language. For it is only in the context of emphasizing that we cannot set the world 'in relief' that he identifies 'the world' with that which any conceivable language represents.[52]

[51] See in this connection, a remark Wittgenstein is quoted as making in a lecture in 1930:

> [Language] cannot express what cannot be otherwise. We never arrive at fundamental propositions in the course of our examinations; we get to the boundary of language which stops us from asking further questions. We don't get to the bottom of things, but reach a point where we can go no further, where we cannot ask further questions. (1980a: 34)

[52] The distinction suggested here between 'the world' construed as that to which anything I can regard as language is answerable and 'the world' construed 'absolutely', the conception of which would require us to 'go beyond' that which 'we neither can nor want to go beyond', has obvious Kantian resonances. However, whereas the Kantian distinction between 'things as they appear to us' and 'things as they are in themselves' is concerned with a limit of knowledge, both sides of which are 'thinkable', Wittgenstein is concerned with a limit 'the other side' of which is not thinkable. And so whereas Kant is putting the world 'as it appears to us' in relief and considering the extent to which it matches up with the world 'as it is in itself', the crucial point for Wittgenstein is that there is no way to put 'the world' to which to 'the only language I understand' is answerable in relief and consider the extent to which that world matches up with 'absolute' reality (see, in this connection, Sullivan 2003: 216ff.). In commenting on the passage just quoted from the *Philosophical Remarks*, Pears (1987: 95) remarks that it 'expresses a rejection of things-in-themselves beyond the phenomena'. However, to say that Wittgenstein, 'rejects' 'things-in-themselves beyond the phenomena' suggests that he denies—that is to say, that he regards it as false—that there is anything 'beyond' the world to which 'my language' is answerable and thereby identifies this world with the world absolutely. (And such an interpretation would support the view that Wittgenstein endorses the sort of 'basic realism' that Pears attributes to Wittgenstein; see chapter 3 of his 1987.) As I read the passage from the *Philosophical Remarks*, as well as the *Tractatus*, Wittgenstein is not 'rejecting'—that is, is not treating as false—the view that there is anything beyond the limits of 'my world', but rather regards it as nonsensical to attempt to ask whether there is

Again, in *Philosophical Remarks*, he writes that what 'is itself a space' is

> ... not something bordering on something else (from which it could therefore be limited off). And so, something language cannot legitimately set in relief. (1975: 85)

And similarly in notes taken in 1930–1, he is quoted as saying:

> Language is space; statements divide space. Language is not *contiguous* to anything else. We cannot speak of the use of language as opposed to anything else. (1980a: 112)

For Wittgenstein, just as we cannot set 'the world' 'in relief', neither can we set logical space—the space consisting of all the possibilities we can conceive—or the language capable of expressing what is in logical space 'in relief'. Indeed, in the *Tractatus*, to be given objects, or 'the [substance of the] world', is to be given all possible states of affairs, and thus the whole of logical space, and thus all that is expressible by anything I can regard as language; hence, the inability to set the world 'in relief' just is the inability to set logical space or language 'in relief' as well. And insofar as the inability to set 'the world' in relief leads Wittgenstein to identify 'the world' and 'my world', the same inability leads to an identification of 'the language' and 'my language' and of 'the logical space' with 'my logical space'.

Given that this identification of 'the' and 'my' world, logical space, and language follows not by establishing what there is or is not beyond the objects and thus the possibilities I *kennen*, but rather from my inability to set those objects and possibilities 'in relief', then for Wittgenstein, the phrases 'the/my world', or 'the/my language', or 'the/my logical space' will have to drop out. For, as I have discussed, on Wittgenstein's view, a sign is meaningful only where it contributes to the formation of sentences with sense, where a sentence with sense is capable of being understood without precluding either its truth or its falsity; and, for Wittgenstein, to deny that 'the/my world', or 'the/my language', or 'the/my logical space' can be 'set in relief' just is to deny that there can be sentences with sense, asserting or denying that there are other worlds besides 'my world', other logical spaces besides 'my logical space', other languages besides 'my language' and is thus to deny, more generally, that any of the signs 'the/my world' or 'the/my language' or 'the/my logical space' is capable of issuing in a sentence with sense.[53]

That the signs 'the world' and 'my world' are thereby meaningless and so drop out from language is, I take it, involved in 5.64, where Wittgenstein writes:

anything beyond the limits of 'my world', thereby forgoing the attempt to set 'my world' in relief, and thereby forgoing any metaphysical position, be it a 'basic realism' or any version of idealism.

[53] In a conversation with the Vienna Circle early in 1930, in considering a question of Schlick regarding 'a person [who] was locked in a red room for his whole life and could not see any colour but red' (1979a: 65–6), Wittgenstein is reported as saying:

> About that the following should be said: if everything were red and I could describe it, then I should also be able to construct the proposition that it was not red. This presupposes the possibility of other colours. Or else red is something that I cannot describe—then I have no proposition either and there is nothing I can deny. Nor would there be statements of the form 'Everything is red,' or, 'Everything I see is red,' in a world where red played the same role, as it were, as time does in our world.

Here it can be seen that solipsism, when its implications are followed out strictly, coincides with pure realism. The I of solipsism shrinks to a point without extension, and there remains the reality co-ordinated with it.

For Wittgenstein, that is, given that the objects with which I am acquainted constitute a space that cannot be set in relief, it is incoherent to attempt to speak of the objects with which I am acquainted as opposed to those with which someone else is acquainted. Thus, the 'I' drops out, and all that is left are the objects, objects which themselves cannot be 'set in relief', so that it is nonsensical to attempt to ask whether those are objects for me alone or are objects to which everyone is answerable, and so that there becomes no distinction between solipsism and 'pure realism'.[54]

My suggestion, then, is that central to Wittgenstein's distinction between what can be said and what cannot be said but only shown is his distinction between that which can and that which cannot be 'set in relief', or, to use the terminology that Wittgenstein uses in 2.173–4 and again in 4.12, the distinction between that with regard to which we can, and that with regard to which we cannot, occupy an 'outside position'. For Wittgenstein, that which we can 'set in relief', or that with regard to which we can occupy an 'outside position',[55] is that which can be said—that which can be stated by a sentence with sense—and so that which we can take a stand on correctly or incorrectly. What can be said is thus 'set in relief' against logical space, some possibilities within

Therefore, if there is a state of affairs, then it can be described and the colour red presupposes a system of colours. Or red means something entirely different and then no sense attaches to calling it a colour. In that case you cannot even speak of it. (1979a: 88–9)

Here Wittgenstein is contrasting two different ways in which a person 'could not see any colour but red': the first, in which that person may not see any other color but may nevertheless describe and conceive of other colors; the second, in which 'red plays the same role' for that person 'as time does' for us. In the first case, the person can, in effect, set red 'in relief' against other colors, can conceive of things being colors other than red, even though red is, in fact, the only color that person ever experiences; here, red is one color within 'a system of colours' or in 'colour space'. In the second case, red is not what is represented, but is rather a form of representation; it is not anything 'within a space', but is rather 'itself a space', and so is not anything that that person can 'set in relief' or occupy a position outside of. Accordingly, for Wittgenstein, in this second case, the sign 'red' becomes meaningless and drops out altogether. In a passage from a 1920 paper (in which no mention is made of Wittgenstein), Russell writes:

If my personality colours all that I observe—a view which I neither assert nor deny—then, clearly, I cannot know anything of the way in which it colours my objects. A subjectivity which can be put into words is a half-hearted subjectivity; taken seriously it defeats itself. It becomes ineffable and inexpressible.... (1920: 91)

thereby alluding to the sort of position I take to be Wittgenstein's.

[54] Hence, as I am presenting it, there is for Wittgenstein no genuine question I can raise as to whether the objects with which I am acquainted are objects private to me or whether they are objects available to other thinkers, or hence as to whether 'the only language I understand' is private to me or a public language accessible to others; to regard either as a genuine question is to attempt to set 'in relief' the objects which I *kennen* or 'the only language I understand'.

[55] In 2.173 Wittgenstein writes that 'a picture represents its object [*Objekt*] from a position outside it', adding in 2.174 that 'a picture cannot, however, place itself outside its representational form'. The German *Objekt* appears only, so far as I am aware, in 2.173, and the suggestion here (as in 4.12) is of a contrast between the sort of *Objekt*—namely, that which is stated by a sentence with sense—with regard to which we can occupy an 'outside position' and the sort of *Objekt*—including representational form, the world, logical space—with regard to which we cannot occupy an 'outside position'.

which would render what is said true, while others would render it false. In contrast, that which we cannot 'set in relief', that with regard to which we cannot occupy an 'outside position'—including the objects that I *kenne* and the possibilities that I *kenne* in *kennen* those objects—is what is shown. Here, what is shown does not divide logical space, but rather concerns the constitution of logical space itself, so that there is nothing against which we can set it in relief.[56]

More specifically, on this reading, the distinction between saying and showing corresponds to different linguistic functions. What is said is what is asserted in uttering sentences with sense, so that what is said is always that such-and-such is the case, and is thus of the form 'that *p*'; in contrast, while what is shown is symbolized by language, what is shown is never properly represented as being of the form 'that *p*'. Thus, for example, the objects that are shown are symbolized by names, but a name is not of the form 'that *p*', and it makes no sense to assert an object (see 3.221). Similarly, for Wittgenstein, while a sentence shows its sense (4.022)—that is, shows or displays a possible situation in logical space (see 2.201, 2.221, 4.021)—what it thereby shows is not of the form 'that *p*'. Thus, for example, the sentence '*a* is above *b*' shows, via the arrangement of the names '*a*' and '*b*', the possibility of *a* being above *b*—the possibility that would have to obtain for that sentence to be true. But just as the object *a* itself is not a candidate for being said, neither is 'the possibility of *a* being above *b*', and what the sentence '*a* is above *b*' says is not the possibility of *a* being above *b* that it shows, but rather that *a* and *b* are arranged in that way.[57]

Hence, on this interpretation, the distinction between saying and showing corresponds closely to the distinction between *wissen* and *kennen*. For Wittgenstein, what we *wissen* is only of the form 'that *p*', and only where '*p*' is a sentence with sense, does it make sense to say that we know that *p* (see 5.1362), so that what we *wissen* are truths. In contrast, what we *kennen* is not of the form 'that *p*', so that what we *kennen* are not truths. And, for Wittgenstein, what we *kennen* in understanding a sentence—including the objects designated by the names in that sentence, the possible situation represented by that sentence, and the logical space to which that possible situation belongs—is what

[56] Tautologies and contradictions do not divide logical space either, but neither do they attempt to set it in relief against anything else. For a tautology simply displays the whole of logical space as rendering that sentence true, while a contradiction displays the whole of logical space as rendering it false.

[57] In conversation with me a number of years ago, Tom Ricketts suggested the view that objects, rather than any truths about them, are shown; that sort of view has come to play a central role in the interpretation I present here, although I am not sure whether he would agree with the way in which I develop that view here. Some passages in his 1996 suggest that sort of interpretation; for example, '[T]he sentence, "Desdemona loves Othello," in saying what it does, in making sense, displays the possibility of Desdemona's loving Othello' (1996: 91). However, he also writes there that 'this talk of what is said and what is shown . . . suggests that there are two kinds of fact: the garden variety facts set forth in true sentences and extraordinary facts about the constitution of any possible world shown by sentences' (1996: 93). In doing so, he seems to be suggesting that, as presented in the text, the show/say distinction is not, as I have argued, Wittgenstein's way of genuinely refusing to recognize any questions of ultimate metaphysics, but is rather the sort of 'mock' doctrine, according to which what are shown are ultimate metaphysical truths, that Conant has argued has to be kicked away (see n. 50 above).

that sentence shows. Thus, in 4.021, Wittgenstein remarks that 'if I understand a sentence, I know [*kenne*] the situation it represents.' Given that the situation the sentence represents is its sense (see 2.202, 2.221), Wittgenstein is thereby indicating that in understanding a sentence, we *kennen* its sense; and in 4.022, Wittgenstein goes on to remark that 'a sentence *shows* its sense', thereby indicating that a sentence shows what we *kennen* in understanding that sentence.[58]

If correct, this reading makes clear that what is shown does not include any metaphysical truths, and does not involve 'setting in relief' that which Wittgenstein holds we cannot 'set in relief'. For if what is shown by a sentence is what we *kennen* in understanding that sentence, then since what we thus *kennen* are not truths at all, what the sentence shows are not any metaphysical truths. Moreover, in showing the objects it is about, and the possible situation it represents, and the logical space within which that possible situation belongs, the sentence is not setting those objects, or that possible situation, or that logical space 'in relief'. It is indicating nothing as to whether, for example, those objects are only objects 'for me' or whether they are objects 'absolutely'; it is indicating nothing as to whether or not that logical space encompasses all the possibilities that could be expressed in absolutely any language.[59] Instead, it simply displays those objects, that possible situation, and that logical space, without indicating anything as to their ultimate status.[60]

Thus, on this reading, the attempt to say what can only be shown will always involve presenting a certain sort of 'content'[61] in the wrong sort of linguistic form. For, on this reading, the 'contents' that are shown are non-propositional items (objects, possibilities) that we cannot 'set in relief', while the form of saying something is appropriate only where the 'content' of what is said (that such-and-such is the case) is something we can 'set in relief'. Thus, to attempt to say what can only be shown is to attempt to set in relief that which cannot be set in relief, and the result is not the attempt to state truths that can only be shown, but rather 'simply nonsense'. That is, the contrast between saying and showing is not that between truths we can state in language and truths we cannot state but can nevertheless recognize, but is rather between propositional 'contents', with regard to which we can take an outside position and thereby judge correctly or incorrectly, and non-propositional 'contents', with regard to which

[58] In 4.024 Wittgenstein does remark, 'To understand a sentence means to know [*wissen*] what is the case if it is true'; however, in doing so, he is only indicating that if I understand sentence *p* and *q* follows from *p*, then I *wissen* the tautology 'If *p*, then *q*' (see 6.1221), so that what I here *wissen* involves no knowledge of truths, where that requires knowledge of what is the case, or of facts. See further, in this connection, n. 75 below.

[59] Compare Conant and Diamond (2004: 83), who take Wittgenstein to be genuinely committed to the view that 'there is only *one* logical space and everything that can be said or thought forms part of that one space'.

[60] On the account I am presenting, it is relatively straightforward as to how a sentence shows what it shows—namely, it shows what it shows by showing its sense, by showing what anyone must *kennen* in order to understand that sentence. It would obviously be harder to explain how a sentence shows what it shows if it shows metaphysical truths concerning what does or does not lie 'on the other side' of the limits of the world and of thought.

[61] Or *Objekt*; see n. 55 above.

we cannot take an outside position, but the *kennen* of which is necessary in order for us to make any judgments at all, correctly or incorrectly.[62]

Further, on this reading, what is shown, at least after the ladder has been kicked away, is not that certain items—including, for example, the totality of objects which I *kenne* or the possibilities I *kenne* in *kennen* those objects—cannot be set in relief, but is rather simply those items themselves, without setting them in relief. And for Wittgenstein, since the objects which I *kenne* and the possibilities I *kenne* in *kennen* those objects constitute the limits of 'the/my world' and 'the/my logical space', limits that are shown by the sentences I understand (see 5.5561), then, for Wittgenstein, what is shown are the limits of 'the/my world', without setting those limits in relief, so that what is shown includes no metaphysical truths concerning what does, or does not, lie on the other side of those limits. Indeed, what is shown is not even that those objects or possibilities constitute limits, for to regard that as being shown is to invite the pseudo-question as to what does, or does not, lie on 'the other side' of those limits.[63] To kick away the ladder is not to hold on to the view that the world cannot be set in relief, but is rather to relinquish the attempt to set it in relief. It is not to have a view as to what cannot be done; it is to cease engaging in a certain sort of activity. And this is achieved when we simply allow what is shown to be shown without attempting to articulate what is shown in the form of sentences.[64]

[62] Likewise, on this reading, for other cases in which Wittgenstein is often read as holding that what are shown are metaphysical truths that cannot be stated. Thus in remarking in 5.535 that 'what the axiom of infinity is intended to say would express itself in language through there being infinitely many names with different Bedeutungen', Wittgenstein is not indicating that the metaphysical truth that there are infinitely many objects among the 'ultimate constituents' of the universe would be shown if there were infinitely many meaningful names in my language; rather, he is indicating that one who is inclined to take the axiom of infinity as an ultimate metaphysical truth has misconstrued what would be shown by those meaningful names, since they would show nothing regarding the ultimate metaphysical status of those objects. Similarly, on this reading, passages (e.g. 5.525) in which Wittgenstein suggests that a sentence shows the metaphysical status of what it expresses should not be read so that he holds, for example, that a tautology shows that what it expresses is absolutely necessary; rather, they should be read so that while he holds that a tautology shows itself as being true in all the possibilities in 'my/the' logical space, it shows nothing as to whether those are all the possibilities there are absolutely. I cannot say that it is necessary that $p \vee \sim p$, not because that $p \vee \sim p$ is necessary is a metaphysical truth that I can only show, but rather because I cannot set 'my/the' logical space in relief. (Hence, on this reading, Wittgenstein does not even take the trivial cases of 'operational self-refutation', in which what cannot be both understood and true is (unlike ~(Solip)) a contradiction, as establishing that the negation of that contradiction, a tautology, is a metaphysically necessary truth.) Likewise, where '$p \supset q$' is a tautology, I cannot say that q follows from p (see 6.1201, 6.1221), for to do so I would be representing myself as taking a position outside my logical space and declaring there is absolutely no possibility in which 'p' is true and 'q' is false. In each of these cases, the attempt to say what can only be shown—the attempt to do ultimate metaphysics—confusedly attempts to set in relief that which is not set in relief in what is shown, where no stand is taken regarding the ultimate metaphysical status of 'my/the' logical space.

[63] That is, to show objects that are limits of my world is not thereby to regard those objects *as* limits of my world.

[64] Without attempting to address anything like all the relevant interpretive issues, I make a few comments here as to how the interpretation I have outlined relates to so-called 'new' or 'resolute' readings of the *Tractatus*. First, I agree with what I take to be a fundamental tenet of 'new' readings, that Wittgenstein does not hold that what are shown are metaphysical truths that cannot be said. I regard it as a salutary feature of 'new' readings to have questioned the coherence of attributing such a view to Wittgenstein; and insofar as denying that Wittgenstein accepts such a view is constitutive of 'resolute' readings, then what I am presenting

To return, then, to Wittgenstein's view as to 'how much truth there is in solipsism', I do not take him to hold that there is an ultimate metaphysical truth of solipsism that can be shown but not said; for I do not take him to hold that any truths, let alone any metaphysical truths, are shown. Rather, I take him to hold that what the solipsist has recognized is different from what she has attempted to state, that the solipsist has misconstrued what, when taken properly, would lead to the renunciation of metaphysics altogether. In particular, for Wittgenstein, in being unable to interpret

here is one version of a 'resolute' reading. However, at least some 'new' readers seem to have agreed with some 'traditionalists' in holding that, if Wittgenstein is taken to have accepted the show/say distinction that seems to be presented in the *Tractatus*, then he would have to hold that what is shown but cannot be said are metaphysical truths; and rather than attributing what they take to be the incoherent consequent to Wittgenstein, they hold that once 'the ladder has been kicked away', the show/say distinction that seems to be presented in the *Tractatus* will be jettisoned altogether. (See the passages from Conant cited in n. 50; and see also, for example, Kremer 2001 and 2004: 60–3.) In contrast, as I have presented it, since the show/say distinction that is present in the text does not involve the view that truths are shown at all, Wittgenstein can be read (coherently) as retaining the show/say distinction after 'the ladder has been kicked away' without holding that metaphysical truths are shown. (See similarly in this connection McGinn 1999: 498; Proops 2001: 378–9; Sullivan 2002: 49–52; Moore 2003: 180–1.) Kremer does present Wittgenstein as accepting a notion of showing—one involving 'knowledge how' as distinct from 'knowledge that'—but, for Kremer, this notion of showing that Wittgenstein accepts 'must be kept sharply distinct' (2001: 61) from 'the doctrine of showing and saying' (2001: 58) that is to be abandoned once 'the ladder is kicked away' (cf. Kremer 2004: 63). Like Kremer, I hold that a coherent view of showing will avoid assimilating it to 'knowledge that'; but, unlike Kremer, I hold that this is done by associating showing with *kennen*, rather than 'knowledge how', and I deny that there is in the text a distinction between two notions of showing.

More generally, just as some 'new' readers have presented the show/say distinction as an apparent or 'mock' doctrine that is to be abandoned altogether once its incoherence is revealed, so too they have presented other apparent doctrines of the *Tractatus*, including the view of sentences as 'pictures' and the views of objects presented early in the 2s, as likewise to be jettisoned, so that the ultimate verdict of 6.54 that the remarks of the *Tractatus* are nonsensical does not depend on the distinction of sense and nonsense that seems to be adumbrated earlier in the text, but rather results from the recognition of the futility of attempting to draw such a distinction altogether, a recognition that is facilitated by recognizing the incoherence of the apparent theory of the *Tractatus*. (See e.g. Kremer 2001: 42–3.) On the reading I have outlined, these elements of the *Tractatus* are retained even after the ladder has been kicked away, albeit not in the form in which they seem to have been introduced originally in the text. Thus, on this reading, Wittgenstein has no theory of the way all possible languages function or of the way reality ultimately is, in itself; for to do so would require him to go 'beyond the limits' of language and world. Nor does he regard himself as stating a theory as to how 'his' language functions or as to the nature of 'his' world; for he could state those as theories only if it made sense for him to consider the possibility of languages that function differently than his or of worlds constituted differently than his. However, on this reading, once the ladder has been kicked away, the way in which Wittgenstein regards the sentences he understands is in accord with the 'picture theory' he seemed to present in the text. That is, he takes the sentences he understands as displaying possible arrangements of simple objects against the background of logical space (without showing anything regarding the ultimate status of those objects or possibilities or that logical space), and regards the nonsensical remarks of the *Tractatus* as the result of misrepresenting—of attempting to set in relief—what is thereby shown to him by the sentences he understands. On this reading, it is clear how Wittgenstein could later represent his earlier work, in a passage quoted approvingly by Conant and Diamond (2004: 84), as 'a theory [that] . . . does not present itself to us as a theory'. And, like Conant and Diamond, I take it that so to represent his earlier work is to represent it as having 'unwitting' theoretical commitments—commitments I would attempt to present as deriving from such assumptions as (Und*) and the views of language I suggest in §3.5 drive him to his view of generality. However, at least some 'new' readers of the *Tractatus* have taken the writings of Conant and Diamond to support the view that Wittgenstein should be regarded as having already freed himself even from such 'unwitting' theoretical commitments by 6.54. (See, for example, Read and Deans 2003, on 'strong' versus 'weak' versions of 'resolutism', and again Read 2006: 80–1; and see, in this connection, Conant and Diamond 2004: 97 n. 82.)

∼(Solip)—namely, 'There are objects with which I am not acquainted'—so that it can be both understood and true, the solipsist has come up against the objects with which she is acquainted as limits of 'the/her' world, thought, and language, and has thus recognized limits, where someone like Russell, who holds that he can understand both ∼(Solip) and (Solip) without precluding the truth or falsity of either, has not.[65] However, the solipsist has wrongly taken this brush with the limits of the world as establishing the truth of (Solip) and the falsity of ∼(Solip), when she should have taken it as revealing objects that cannot be 'set in relief' and so can only be 'shown'. That is, having 'run up against' the limits of 'the/her' world, the solipsist should not have attempted to say anything about those limits—either that that there is absolutely nothing 'on the other side' of those limits, so that 'the world' absolutely construed coincides with her world, or even that 'the world' construed as the only world to which her language is answerable coincides with 'her world'—for the attempt to do so only issues in nonsense.[66] Instead, she should have acquiesced in uttering sentences that show the objects that are those limits, in which case she would have assumed the only form of expression appropriate to the limits that cannot be set in relief that she has recognized, a form of expression that precludes her from even regarding what she has recognized as limits or as items that cannot be set in relief.

Again, on this interpretation, Wittgenstein embraces Mackie's view that we should not draw metaphysical conclusions from what we cannot coherently think as true. In Mackie's terminology, Wittgenstein holds that while the solipsist correctly regards ∼(Solip) as 'operationally self-refuting', she wrongly concludes from this that (Solip) is a necessary metaphysical truth. Further, for Wittgenstein, regarding ∼(Solip) as 'operationally self-refuting' not only does not suffice for establishing (Solip) as a metaphysical truth; it should lead us to cease to recognize any genuine question for which either (Solip) or ∼(Solip) is the answer. For Wittgenstein, the proper outcome of recognizing ∼(Solip) as 'operationally self-refuting' is to give up the attempt to occupy an 'outside position' with regard to the objects with which I am acquainted from which I could ask a genuine question as to whether those are all the objects there are; and it is to accept showing as the form of expression appropriate to those objects—a form of expression in virtue of which I am related to them but occupy no 'outside position' with regard to them. For Wittgenstein, that is, the proper outcome of 'self-refutation' arguments in the style of Berkeley is not the acceptance of idealist, or solipsistic metaphysics, but rather the renunciation of metaphysics and the acceptance as shown, and only shown, of the limits of the world and of thought that such arguments reveal.[67]

[65] Of course, as I have argued, the reason the solipsist 'recognizes' these limits while Russell does not is that they have differing views of generality.

[66] See in this connection Wittgenstein's (1929) comments to the Vienna Circle on Heidegger and Kierkegaard (1979a: 68–9) and the final sentences of his paper 'A Lecture on Ethics' (1929: 44).

[67] In his *Foundations of Arithmetic*, Frege identifies 'the widest domain of all' with 'everything thinkable' (1884: 21), thereby suggesting that he holds that there is nothing that cannot be thought-of. Further, he writes:

3.5 *Generality and holism*

If the interpretation presented above is correct, then what is perhaps most original in Wittgenstein's treatment of solipsism is his taking such 'self-refutation' arguments to reveal, not metaphysical truths, but rather 'limits' of the world and of thought that cannot be 'set in relief';[68] however, the fundamental point of difference between Wittgenstein's treatment of solipsism and Russell's remains their competing views of generality. For it is only by rejecting Russell's view of generality that Wittgenstein is committed to holding that ~(Solip) cannot be both understood and true; and whether one takes this to establish, with the solipsist, that (Solip) is therefore true, or with Wittgenstein as I have interpreted him, that both ~(Solip) and (Solip) are nonsensical attempts to 'set in relief' that which cannot be set in relief, neither option is compelling, if, like Russell, one accepts a view of generality that enables one to regard each of ~(Solip) and (Solip) as a sentence that can be understood without precluding either its truth or falsity. Thus far I have merely argued that Wittgenstein's anti-Russellian view of generality is central to his treatment of solipsism, without considering what reasons Wittgenstein may have for rejecting Russell's view of generality. It is to this task that I now turn. In particular, I argue that Russell's account of generality is incompatible with a holistic view of language to which Wittgenstein was antecedently committed, a holism which, independently of the details of his account of generality, renders Wittgenstein unable to recognize any genuine question of solipsism.

As I have mentioned above, in *Theory of Knowledge* Russell develops an account of propositions according to which apprehending more abstract propositions is logically simpler than, and logically prior to, apprehending less abstract ones. Thus, for example, he holds that while understanding

(i) *a* is similar to *b*,

I understand objective to mean what is independent of our sensation, intuition and imagination, and of all construction of mental pictures out of memories of earlier sensations, but not what is independent of the reason,—for what are things independent of reason? To answer that would be as much to judge without judging, or to wash the fur without wetting it. (1884: 36)

Here Frege seems to use a Berkeley-style argument to establish the view that nothing objective is independent of reason, since it would be incoherent to try to provide a counter-example. If so, then it would seem that, for Wittgenstein, Frege should not have taken this argument to establish that conclusion, but should have relinquished any attempt to set 'what is objective' and 'what is independent of reason' in relief against each other.

[68] In 1931 Wittgenstein wrote: '[W]hen I was in Norway during the year 1913–14 I had some thoughts of my own, or so at least it seems to me now. I mean I have the impression that at that time I brought to life new movements in thinking . . . ' (1980b: 20). In his notes to Moore, written when he was in Norway, the show/say distinction plays a prominent role (see 1979b: 108ff.); in his pre-Norway 'Notes on Logic', in contrast, the show/say distinction is absent, and Wittgenstein writes: 'Philosophy consists of logic and metaphysics: logic is its basis' (1979b: 106). If the interpretation presented here is along the right lines, it is plausible to suppose that what Wittgenstein regarded as the 'new movements in thinking' that he 'brought to life' in Norway involved the renunciation of metaphysics that is incorporated in his show/say distinction.

requires being acquainted with the particulars *a* and *b*, the relation of *similarity*, and the logical form of a dual complex, understanding the general, but not absolutely general sentence

(ii) Something is similar to something

requires being acquainted with the relation of *similarity* and the logical form of a dual complex, but no particulars, and understanding the absolutely general sentence

(iii) Something has some relation to something,

requires being acquainted only with the logical form of a dual complex.

For Russell, then, while the ability to understand some sentences (such as (i) or (ii)) depends upon the ability to understand other, more general sentences (such as (iii)), the ability to understand absolutely general sentences such as (iii) does not depend on the ability to understand any other sentence (see 1913: 129–33). Hence, Russell's account in *Theory of Knowledge* is incompatible with forms of holism according to which no sentence in a given language has its meaning in isolation from the meanings of the other sentences in that language. For on Russell's view, at least some sentences (such as (iii)) have the meanings they have—express the propositions they express—independently of any other sentence having the meaning it has. For Russell, there is nothing incoherent in supposing that there is a being who apprehends only the proposition expressed by (iii), so that the entire language of that being consists in a single sentence expressing that proposition. Further, for Russell, since the ability to understand a general sentence '$(\forall x)Fx$' is independent of the ability to understand any of its instances, then that general sentence can belong to different languages which express different elementary sentences. Hence, that sentence can belong to my language regardless of which particulars I am acquainted with and thus (given (Und*)) regardless of which elementary sentences belong to my language. And likewise that sentence can belong to the languages of different speakers, regardless of which particulars they are acquainted with and so regardless of which elementary sentences belong to their languages.[69]

This anti-holistic view is, we have seen, incompatible with Wittgenstein's view of general sentences as truth-functions of elementary sentences. Further, Wittgenstein's anti-Russellian account of general sentences is a reflection of a more general commitment to an extreme form of holism. In particular, in 3.42, Wittgenstein writes:

> A sentence can determine only one place in logical space: nevertheless the whole of logical space must already be given by it.

[69] Hence Russell can hold that, while only I can understand the elementary sentences in my language, since they are about sense-data private to me (1918: 201), and only you can understand the elementary sentences in your language, we can nevertheless understand the same general sentences.

For Wittgenstein, the 'whole of logical space' is the whole of what I can conceive—that is, all that can be expressed by sentences with sense (in the only language I understand). Thus, to indicate that to be given any one point in logical space is to be given the whole of logical space is, for Wittgenstein, to indicate that a sentence can express the sense it has only by being part of an entire language that is capable of expressing the entirety of the logical space it expresses. And from this it follows that, if a given sense is expressed by sentences in two different languages, then those languages must have exactly the same expressive powers. For, on Wittgenstein's view, if those sentences express the same sense—the same 'point' in logical space—it can only be by being parts of languages that express the same logical space. Thus, it is not merely that by regarding general sentences as truth-functions of elementary sentences, Wittgenstein rejects the Russellian view that speakers could be able to understand the same general sentences without understanding the same elementary sentences; it is rather that Wittgenstein holds more generally that speakers capable of understanding any one sentence in common will thereby be capable of understanding exactly the same sentences.

The central point here is that given this holism, and aside from the details of his account of generality, Wittgenstein cannot recognize any genuine question of solipsism in the way that Russell does; that is, he cannot recognize

(Solip?) Are the objects with which I am acquainted all the objects that there are?

as a question the understanding of which precludes neither its truth nor its falsity. If (Solip?) is thus to be a genuine question for me, I will have to possess a concept of what it is to be an object that is independent of my acquaintance with specific objects; for only if I possess such a concept will there be a conceivable gap between 'all the objects there are' and all the objects with which I am acquainted that will render (Solip?) a genuine question for me. However, if I possess such a concept of what it is to be an object, I will do so non-holistically: my possession of that concept will not depend on what objects I am acquainted with or hence (given (Und*)) on what elementary sentences I can understand. And if I can possess the concept of what it is to be an object that is thus independent of what objects I am acquainted with, I could use that concept in formulating the concept of 'objects with which I am acquainted' that is likewise independent of what objects I am actually acquainted with. But in that case, my understanding of (Solip?) is independent of what elementary sentences I understand: as against Wittgenstein's holism, it will be the same question for me regardless of which elementary sentences I can understand and hence which elementary senses are expressible in my language.

For Russell, we are able to possess a general concept of what it is to be a particular—in his terminology, we can become acquainted with the (logical) universal of being a particular—independently of what actual particulars we are acquainted with. Hence he indicates (1913: 97) that anyone who understands the word 'particular' has acquaintance with a 'logical object'. And in virtue of being acquainted with that 'logical object', we may thus understand such sentences as 'There are particulars'[70] or 'There are exactly five particulars' or 'There are particulars with which I am not acquainted' independently of which elementary sentences we understand. For Russell, it is this non-holistic acquaintance with the universal of being a particular that enables me to occupy an 'outside position' with regard to the particulars with which I am acquainted and to ask whether they are all the particulars that there are absolutely.

In contrast, and consistently with his holism, Wittgenstein denies that the word 'object' can ever function as a genuine concept-word (4.1272), denies that we can be given the 'formal concept' separately from being given an object (4.12721), and denies that apparent sentences containing 'object' as an apparent concept-word are genuine sentences. Given his holism (together with (Und*)), Wittgenstein cannot recognize us as possessing a concept of an object that is independent of the objects we have been given; but the metaphysical aspiration requires us to possess just such a concept of an object, a concept that enables us to take an 'outside position' with regard to the objects we have been given and ask whether those are the only objects there are absolutely; hence, the metaphysical aspiration is incompatible with Wittgenstein's holism.

I conclude this section by suggesting that Wittgenstein's holism should itself be regarded as reflecting his views—arguably Fregean in origin—regarding the independence of sense from facts (see 4.061).[71] For Frege, thoughts, the senses of sentences, reside in a 'third realm' distinct both from the minds of individual people and from the external world that is knowable only through sensory experience.[72] By separating thoughts from the minds of individual speakers, Frege makes thoughts equally accessible to all speakers;[73] and by separating thoughts from the external world, Frege suggests that our grasp of thoughts does not depend on making any empirical discoveries.[74] And although Wittgenstein does not reify senses in the manner of Frege as belonging to a 'third realm', he distinguishes sharply between what is involved in *kennen* senses and what is involved in *wissen* facts. In particular, for Wittgenstein, in understanding a sentence, and hence in *kennen* its sense, we do not thereby have any

[70] Here, as in the examples mentioned in n. 48, Russell holds that I can understand the general sentence without precluding either its truth or falsity, even though (given his theory of types) no instance of it can be both meaningful and false (see 1918: 268).

[71] For a more developed presentation of some characteristic features of Wittgenstein's position in the *Tractatus* as the result of a combination of Russellian views of the sort I mentioned in §3.1 together with Fregean views I sketch in this paragraph, see my (1998, Part II).

[72] See, for example, 'Thoughts' (1918: 360–3), but the same sort of view is present elsewhere, for example, 'On Sense and Bedeutung' (1892a: 159–61), where Frege distinguishes senses, both from objects designated, which are in general 'perceivable by the senses', as well as from 'subjective ideas'. I recognize that this picture of the realm of sense as sharply separated both from the realm of ideas as well as from the realm of the physical world does not sit easily with Frege's brief discussions of indexical expressions, according to which grasping the sense of such an expression requires knowing its *Bedeutung* (1918: 358).

[73] See e.g. Frege's (1892a: 160), where he writes of a 'common store of thoughts which is transmitted from one generation to another', and similarly his (1892b: 185 n.7).

[74] See e.g. 'Thoughts' (1918: 363), where Frege writes that what belongs to the 'third realm' 'cannot be perceived by the senses'.

knowledge of facts.[75] Our knowledge of facts—of how things are—is empirical; our knowledge of sense is a priori. And since, for Wittgenstein, to make a discovery is to gain new empirical knowledge, there are no discoveries to be made in the realm of sense. Hence, we cannot discover new points in logical space, so that all of logical space is given all at once.[76]

Hence, for Wittgenstein, since it is by being given objects that we are given logical space, we are given all objects all at once (5.524). On Russell's atomistic view, apprehending a given proposition requires being acquainted with certain entities—particulars, universals, and forms—that we need not be acquainted with in order to apprehend other propositions, and hence he holds that new acts of acquaintance will enable us to apprehend propositions we were not previously able to apprehend. On Wittgenstein's holism, we cannot 'discover' different regions of logical space at different times, and so cannot be acquainted with different objects at different times.[77]

Thus, not only can (Solip?) be a genuine question only if it has its meaning independently of the elementary sentences one can understand and so only if Wittgenstein's holism is false; to regard (Solip?) as a genuine question is to contemplate, as against Wittgenstein's holism, the possibility of there being unexplored regions of

[75] See for example the *Notebooks* remark: 'What do I really know [*weiss*] when I understand the sense of 'ϕa' but do not know [*weiss*] whether it is true or false? In that case I surely know [*weiss*] no more than $\phi a \vee \sim\phi a$; and that means I *know* [*weiss*] nothing' (1979b: 31). Less than one month earlier, Wittgenstein wrote: 'The sentence *expresses* what I do not know [*weiss*]: but what I must know [*wissen*] in order to be able to say it at all, *I show in it*' (1979b: 18). On the interpretation I am presenting, Wittgenstein resolves the tension in these two passages, by coming to hold in the *Tractatus* that while we know (*wissen*) nothing in merely understanding a sentence, the sentence shows what we *kennen* in understanding it.

[76] See in this connection 2.0123–2.014, as well as the distinction Wittgenstein draws in 1930 between 'empirical totalities', which correspond to 'classes of true propositions' and are determined by experience, and 'systems', which correspond to 'classes of sentences with sense' and are determined prior to experience (1979a: 214–16). In that discussion, Wittgenstein remarks that 'we cannot add a point in space, nor can we discover one' (1979a: 214) and also:

> The logical particles, the numbers, the points in space and time are systems. It is inconceivable to discover a new logical particle, a new number, a new point in space. Here we have the feeling that everything springs from *one* root. If we know the principle on which a system is based, then we know the *entire* system. (1979a: 216)

[77] While the account I present here thus incorporates a version of the 'object-centred' view Sullivan criticizes (1996: 204ff.), it should be clear, I hope, that the 'object-centred' view I am attributing to Wittgenstein does not involve the sort of 'atomistic anchoring' Sullivan associates with that view (1996: 207) and which is clearly present in Russell. One objection to the sort of interpretation I am presenting depends upon Wittgenstein's remarks in the 5.55s to the effect that we cannot 'say a priori what elementary sentences there are' (5.5571; cf. Sullivan 1996: 206–7). However, I suggest that while Wittgenstein is concerned to deny in the 5.55s that what elementary propositions there are (and hence what objects there are) is something that can 'be decided by logic' (5.551), he is not thereby concerned to deny that what elementary propositions (and objects) there are is, in a broad sense, given to us prior to experience. For Wittgenstein, what 'can be decided by logic' concerns what we can 'construct' in having been 'given' the elementary sentences (and objects), so that although logic is prior to every experience—to knowing any truths regarding how things are—it is not prior to the 'what' we are given in being given objects and the possible combinations of objects that are expressible by elementary sentences, which is thus prior to both logic and experience (see 5.552 and also 6.124, in which Wittgenstein indicates what the sentences of logic 'presuppose'). On this account, while logical analysis will reveal what elementary sentences and genuine names there are, and hence what possible states of affairs and objects are represented by those expressions, what is thus revealed is what we are given and *kennen* prior to experience, even though our apprehension of what we are thereby given is only implicit (see 4.002).

logical space. For to regard (Solip?) as a genuine question is to allow for the possibility that (~Solip) is true—the possibility that there are objects, and so senses expressed by elementary sentences, and so regions of logical space, that I have not yet been given. Whereas Russell's atomism both requires him to allow for this possibility and gives him the means to express it, Wittgenstein's holism precludes him from recognizing such a possibility and, consistently with that, denies him the means to express it.

4. Davidson and Nagel

I turn now to argue that the same sort of debate that I have discussed in connection with Prior's analysis of Berkeley's master argument and with Russell and Wittgenstein on solipsism resurfaces in a more recent dispute between Donald Davidson and Thomas Nagel. First, I argue that Davidson presents what may be taken as a 'master argument' against 'the very idea of a conceptual scheme' that mirrors not only the structure of Berkeley's 'master argument' for idealism (as presented by Prior) but also the structure and outcome of what I have presented as Wittgenstein's response to solipsism. Second, I argue that Nagel's response to Davidson mirrors the structure of Prior's response to Berkeley as well as Russell's response to solipsism. Third, I argue that the acceptance or rejection of holism is as central to the dispute between Davidson and Nagel as it is to the dispute between Wittgenstein and Russell.

4.1 Davidson on the 'very idea' of a conceptual scheme

In the course of his argument against the intelligibility of 'the very idea of a conceptual scheme', Davidson indicates that if that idea is to make sense, then conceptual schemes may be identified with 'sets of intertranslatable languages', so that people may be said to 'have different conceptual schemes if they speak languages that fail of intertranslatibility' (1974a: 185). Thus, by this standard, there are conceptual schemes different from mine if there are languages containing sentences that cannot be translated into my language. Hence, if propositions are identified with sets of intertranslatable sentences, there are conceptual schemes differing to at least some extent from mine if there is at least one proposition expressible in one language that is not expressible in mine. But Davidson has an argument intended to show that it is not intelligible to suppose that this could be so.

In particular, he writes:

> We are encouraged to imagine we understand massive conceptual change or profound contrasts by legitimate examples of a familiar sort. . . . But examples like these, impressive as they occasionally are, are not so extreme but that the changes and contrasts can be explained and described using the equipment of a single language. Whorf, wanting to demonstrate that Hopi incorporates a metaphysics so alien to ours that Hopi and English cannot, as he puts it be 'calibrated', uses English to convey the contents of sample Hopi sentences. Kuhn is brilliant at saying what things were like before the revolution using—what else?—our post-revolutionary idiom. . . . The dominant metaphor of conceptual relativism, that of differing points of view, seems to betray an underlying paradox. Different points of view make sense, but only if there is a common

co-ordinate system on which to plot them; yet the existence of a common system belies the claim of dramatic incomparability. (1974a: 183)

Thus, just as Berkeley argues that no one can produce an example of anything that is unthought-of, so too Davidson argues that no one can produce an example of a propositional content that is not expressible in my language. Just as for Berkeley producing an example of an object that is unthought-of would require thinking of that which was to be unthought-of, so too for Davidson producing an example of a 'content' inexpressible in my language would require expressing in my language that which I was to be unable to express. Here Davidson's argument may be regarded as relying on the general principle

(Und★★) If S_1 is a sentence containing another sentence S_2 as part, then understanding S_1 requires understanding S_2

from which it follows that no instance of

$\sim$*Ep* The proposition that *p* is expressible in some language but not mine,

can be both understood and true. And Davidson repeats this style of argument as late as his 1997 paper 'Seeing through Language', where he writes:

Do we understand what we mean by a real alternative to our conceptual scheme? If a scheme could be decoded by us, then it would not, by this very token, be all that different from ours except, it might be, in ease of description here or there. If we could explain, or describe, in a convincing way, how an alternative scheme deviates from ours, it would again be captured in our system of concepts. . . . (1997a: 127)

indicating again that if we can express (here 'decode') propositions of a purported alternative conceptual scheme in our language, then they do not, by this 'very token', constitute a 'real alternative to our conceptual scheme'.

Further, Davidson holds not merely that no one can produce an example of a language or sentence that cannot be translated into my language, but also that it is incoherent to suppose that there are such languages or sentences. That is, he indicates that 'conceptual relativism', taken as the view there are distinct conceptual schemes capable of expressing different propositional contents, and thereby incorporating

(CR) $(\exists p)$(The proposition that *p* is expressible in some language but not mine)

cannot be coherently conceived as true.[78] Thus, in 'On the Very Idea of a Conceptual Scheme', Davidson writes:

[78] Given that Davidson denies that there are any entities that are propositions (see e.g. 1984: xiv), it may seem (as Warren Goldfarb mentioned to me) that the only way that he could interpret (CR) is substitutionally, in which case in denying that any instance of '$\sim$*Ep*' may be understood and true, he would be forced to deny that (CR) can be understood and true. Here, however, I take no stand on the issue as to whether denying there are entities that are propositions commits one to holding that, if no instance of '$\sim$*Ep*' can be both understood and true, then (CR) cannot be both understood and true. (Prior would seem to be one who wants to deny there are any propositions without being so committed; see n. 6.)

> [N]othing, it may be said, could count as evidence that some form of activity could not be interpreted in our language that was not at the same time evidence that that form of activity was not speech behaviour. If this were right, we probably ought to hold that a form of activity that cannot be interpreted as language in our language is not speech behaviour. Putting matters this way . . . comes to little more than making translatability into a familiar tongue a criterion of languagehood. As fiat, the thesis lacks the appeal of self-evidence; if it is a truth, as I think it is, it should emerge as the conclusion of an argument. (1974a: 185–6)[79]

Thus, without considering what further argument Davidson presents for his view, the relevant point here is that he indicates not only that we cannot regard any form of activity as language unless we are capable of translating it into our language, but also accepts a 'criterion of languagehood' according to which what it is for something to be a language is for it to be translatable into ours, and hence according to which no one who understands what it is to be a language—and hence, presumably, what it is to be a sentence—may hold that there are languages, and hence sentences, that cannot be translated into one's own language.

Accordingly, in his (1988) paper 'The Myth of the Subjective', Davidson writes:

> Trouble comes when we try to embrace the idea of more comprehensive differences [between 'systems of thought' than those 'we can explain and understand'], for this seems (absurdly) to ask us to take up a stance outside our own ways of thought.
>
> We do not understand the idea of such a really foreign scheme. We know what states of mind are like, and how they are correctly identified; they are just those states whose contents can be discovered in well-known ways. If other people or creatures are in states not discoverable by these methods, it cannot be because our methods fail us, but because those states are not correctly called states of mind—they are not beliefs, desires, wishes, or intentions. The meaninglessness of the idea of a conceptual scheme forever beyond our grasp is due not to our inability to understand such a scheme, nor to our other human limitations; it is due simply to what we mean by a system of concepts. (1988a: 40)

Thus, just as Davidson indicates in 'On the Very Idea of a Conceptual Scheme' that, given what we mean by 'language', it is incoherent to suppose that there may be languages that we cannot translate into our own, so too he indicates here that, given

[79] I recognize that, in this passage, Davidson is not arguing narrowly against the coherence of supposing that the content of a single sentence of a foreign language may not be translatable into our language, but is rather arguing against 'purported cases of complete failure' of translation (1974a: 185)—cases in which no sentence of an alien language is translatable into our language. However, later in the paper, Davidson argues 'that the attempt to give a solid meaning to the idea of conceptual relativism, and hence to the idea of a conceptual scheme, fares no better when based on partial failure of translation than when based on total failure' (1974a: 197), thereby indicating that just as he holds that it is 'meaningless' or 'unintelligible' to suppose that there may be languages wholly untranslatable by us, it is likewise 'meaningless' or 'unintelligible' to suppose that there may be languages that are partly but not wholly translatable by us. Moreover, the sorts of considerations he presents in the passage preceding this note would, if valid, seem to apply to any 'form of activity [that] could not be interpreted in our language'—whether or not it is performed by beings who also engage in forms of activity that are interpretable in our language. In §4.3 below, I argue that, at least in 'Truth and Meaning' (1967), Davidson endorses the same form of holism embraced by the early Wittgenstein and that the unintelligibility of (CR) follows from that form of holism.

'what we mean by a system of concepts', it is 'meaningless' to suppose that (CR) could be true, that is, that 'we do not understand the idea of' a language or sentence that we cannot translate into our own. That is, in this passage, he indicates not merely that (CR) cannot be both understood and true, but further that (CR) cannot be understood, that it is meaningless. Likewise in 'Seeing through Language', he writes:

> Perhaps we think we can imagine a culture where creatures communicate in ways we are permanently disabled from penetrating. . . . Of course it happens in exotic settings that we recognize that we are witness to intelligible conversation, though grasping nothing that is said. . . . What these folk say has a translation we could understand, but we can't discover it for ourselves. But in such situations we have good grounds for believing that what we do not grasp we could learn to decipher: the parties look to be people like ourselves, and we are justified in thinking we descry known purposes and activities. These cases pose no problems for the anti-conceptual relativist: there is a serious problem only where no translation is possible. It is this supposed case of genuinely incommensurable languages where I sense unintelligibility in the supposition. . . . (1997a: 128–9)

indicating again that he regards 'the supposition' of there being a language that we could not translate into ours as unintelligible.[80]

Thus, just as Berkeley indicates not only that no instance of '$\sim\tau y$' can be understood and true but also that '$(\exists y)\sim\tau y$'—that is, ~(Id)—cannot be understood and true; and just as Wittgenstein indicates no only that no instance of '$\sim Ay$' can be both understood and true but also that '$(\exists y)\sim Ay$'—that is, ~(Solip)—cannot be understood and true; so too Davidson indicates not only that no instance of '$\sim Ep$' can be understood and true but also that '$(\exists p)\sim Ep$'—that is, (CR)—cannot be understood and true (although I recognize that he has not indicated that this latter follows from the former simply in virtue of a view of generality which he accepts). Furthermore, just as Wittgenstein indicates that rather than being false, in particular contradictory, ~(Solip) is nonsensical, so too Davidson regards (CR) as 'meaningless' or 'unintelligible'. And here the similarity between Davidson's position and Wittgenstein's is not merely structural. For given his view that we *kennen* the senses which can be expressed by sentences with sense in virtue of our *kennen* objects, Wittgenstein, like Davidson, is concerned with the possibility of there being propositional contents or senses that I am incapable of expressing in my language. That is to say, given that he identifies the *Kenntnis* of objects and the *Kenntnis* of possibilities that I can express in my language (an identification which, obviously, I have not attributed to Davidson), Wittgenstein is committed to holding that because no instance of '$\sim Ay$' can be both understood and true, then

[80] These passages in which Davidson denies the coherence of (CR) may suggest that he is relying on some form of verificationism, insofar as he seems to rely on claims regarding what we could never have 'evidence' for, or could never 'discover'; and Davidson has been so interpreted, by both defender (see Rorty 1972: 5ff.) and critic (see Fodor 1983: 124; Williamson 2007: 260). If, as I argue below, Davidson's denial of the coherence of (CR) follows from a form of holism to which he is antecedently committed, then it need not be regarded as following from a general commitment to verificationism, which Davidson has nowhere indicated that he would accept (see e.g. his 1997b: 69ff.).

neither can any instance of '$\sim Ep$', and because '$(\exists y)\sim Ay$' cannot be both understood and true, then neither can '$(\exists p)\sim Ep$'. Put in Wittgenstein's terminology, Davidson, like Wittgenstein, holds not only that it is self-refuting to attempt to express a propositional content that lies outside my 'logical space', but further that it is meaningless or unintelligible to attempt to say that there are any propositional contents that lie outside my logical space.

Moreover, consistently with his indicating that (CR) is meaningless or unintelligible rather than false, Davidson, like Wittgenstein, does not take the incoherence of regarding (CR) as true as establishing its negation,

(CM) $(\forall p)$(If p is a proposition expressible in some language, then p is expressible in my language),

which we might regard as incorporating a 'conceptual monism' according to which all languages are translatable into my language, so that my conceptual scheme is the only conceptual scheme.[81] At the outset of 'On the Very Idea of a Conceptual Scheme', he writes:

> Even those thinkers who are certain that there is only one conceptual scheme are in the sway of the scheme concept; even monotheists have religion. And when someone sets out to describe 'our conceptual scheme', his homey task assumes, if we take him literally, that there might be rival systems. . . . (1974a: 183)

and likewise concludes by writing:

> It would be wrong to summarize by saying we have shown how communication is possible between people who have different schemes . . . For we have found no intelligible basis on which it can be said that schemes are different It would be equally wrong to announce the glorious news that all mankind—all speakers of language, at least—share a common scheme and ontology. For if we cannot intelligibly say that schemes are different, neither can we intelligibly say that they are one. (1974a: 197–8)

In these passages, that is, Davidson takes the upshot of his argument against 'the very idea of a conceptual scheme' in the same way that I have argued that Wittgenstein takes the upshot of his reflections regarding solipsism.[82] For both, the unintelligibility of (CR) is a result of attempting to go beyond the limits of thought and to view those limits from the other side; as Davidson writes, (CR) 'seems (absurdly) to ask us to take

[81] Despite the passage I have quoted above in which he endorses the view that 'translatability into a familiar tongue [is] a criterion of languagehood', a view that sustains (CM).

[82] In his discussion of Davidson's argument in 'On the Very Idea of a Conceptual Scheme', Alex Byrne (2007: 446) writes that Davidson's 'main thesis' in that paper 'is a good candidate for explicating the Tractarian thesis that "*[t]he limits of my language* mean the limits of my world"'; however, he characterizes that 'main thesis' as the view that 'there are no alien conceptual schemes'—thereby attributing (CM) to Davidson—and adds (n. 9) that '[f]or an entirely unconvincing reason, Davidson would disavow this way of putting his main thesis'. As I am presenting it, it is precisely because Davidson takes his argument as establishing, not that (CM) is true, but rather that (CM), like (CR), is 'unintelligible' that his position is so similar to Wittgenstein's.

up a stance outside our own ways of thought'. For both, the sense in which my language is the language, or in which all languages may be translated into my language, is not that we are able to step outside 'our own ways of thought'—our own logical space—and 'discover' that there is absolutely nothing outside them, but is rather that there is nothing we can regard as a language that we cannot translate into ours. And for both, I should relinquish the very term for that which I cannot 'take up a stance outside'—'logical space', 'conceptual scheme'—once I recognize it as that which I cannot 'take up a stance outside' of.[83]

Davidson's Tractarian-style rejection of the very idea of a conceptual scheme is perhaps nowhere more succinctly presented than in his 'Reply to Solomon' given in 1971 and published, like 'On the Very Idea of a Conceptual Scheme', in 1974. There he writes:

> We cannot make good sense of the idea that there are seriously different total conceptual schemes or frames of reference, or that there may be radically 'incommensurate' languages (to use Whorf's word). . . . [I]f translation succeeds, we have shown that there is no need to speak of two conceptual schemes, while if translation fails there is no ground for speaking of two. If I am right, then, there never can be a situation in which we can intelligibly compare or contrast divergent schemes, and in that case we do better not to say there is one scheme, as if we understood what it would be like for there to be more. (1974b: 243)

Here, Davidson incorporates the three steps that I have indicated are involved in Wittgenstein's response to solipsism: first, he indicates that the attempt to provide an example of alternative conceptual scheme is self-refuting ('if translation succeeds, we have shown that there is no need to speak of two conceptual schemes'); second, he moves from that to the conclusion that we cannot intelligibly suppose that there are schemes we cannot translate into our language ('if translation fails there is no ground for speaking of two'[84]); and third, rather than drawing the Berkeleyan-style metaphysical

[83] By writing in the last of the passages I have quoted that 'it would be equally wrong to announce that all mankind—all speakers of language, at least—share a common scheme and ontology', Davidson suggests that the notions of 'conceptual scheme' and 'ontology' are in the same position, so that since we cannot intelligibly speak of alternative schemes or ontologies, neither can we intelligibly claim there is only one scheme or ontology. However, while Davidson urges us to drop 'the very idea of a conceptual scheme', he does not similarly urge us to drop 'the very idea' of ontology; on the contrary, he is concerned throughout his writings to explore 'what our language—any language—requires in the way of overall ontology' (1984: xix). In contrast, Wittgenstein, I have argued, is clear in placing 'logical space'—or all that is expressible in 'the only language I understand'—and the objects constituting the 'substance' of 'my/the world' on exactly the same footing. For Wittgenstein, I am unable to set either 'my/the logical space' or 'my/the' ontology of objects 'in relief', so that it is just as nonsensical to ask whether my logical space encompasses all that can be expressed by any language as to ask whether the objects that constitute 'the substance of' my world encompass all the objects there are absolutely; and for Wittgenstein, the only language I understand 'shows' both 'my/the logical space' and 'my/the' objects without setting either 'in relief'. Although Davidson regards his position as a version of neither antirealism nor realism (see e.g. his 1988b), he provides no account, such as the view of showing I have attributed to Wittgenstein, as to how his apparent acknowledgment that the 'very ideas' of conceptual scheme and ontology have the same status is to be accommodated.

[84] Again, seeming to invoke some sort of verificationism; but see n. 80.

conclusion that there is only one scheme, he drops, in the style of Wittgenstein, the whole use of the term 'conceptual scheme' ('we do better not to say that there is one scheme, as if we understood what it would be like for there to be more').[85]

4.2 Nagel's response to Davidson

Unlike many of those who have responded to Davidson's argument against 'the very idea of a conceptual scheme',[86] Nagel does not question Davidson's view that it is self-refuting to attempt to provide an example, in the style of Whorf or Kuhn, of an alternative conceptual scheme. Instead, he questions what I have characterized as the second step in Davidson's argument—the move from holding that no instance of '$\sim Ep$' can be both understood and true, to holding that '$(\exists p)\sim Ep$' cannot be both understood and true; and he questions this move, not so much in the service of defending 'conceptual relativism' in the style of Whorf or Kuhn, but rather in the service of defending 'realism' and opposing 'idealism'.

In particular, in his 1986 book *The View from Nowhere*, Nagel writes:

> The realism I am defending says the world may be inconceivable to *our* minds, and the idealism I am opposing says it could not be.... (1986: 91)

and again:

> [H]uman objectivity may fail to exhaust reality...: there may be aspects of reality beyond its reach because they are altogether beyond our capacity to form conceptions of the world. What there is and what we, in virtue of our nature, can think about are different things, and the latter may be smaller than the former.... (Ibid.)

and further:

> [W]hat there is, or what is the case, does not coincide necessarily with what is a possible object of thought for us. (1986: 92)

From this last passage, it is clear that, for Nagel, those 'aspects of reality' which we may be unable to conceive include both 'what there is' and 'what is the case'. That is, he holds both that there may be particulars we cannot conceive or describe and that there may be facts—and hence (true) propositions—which we not only cannot know to be true but also cannot conceive, that is, cannot express in our language.[87]

[85] For similar arguments against the possibility of the notion of alternative conceptual schemes, see Stroud (1968: 27–8); Rorty (1972: 5ff.), who cites both Stroud and Davidson as influencing his discussion; and Putnam (1981: 114–15), although compare his (2004: 50).

[86] Often the criticism is made that the Davidsonian argument fails to properly distinguish interpretation and translation. See e.g. Rescher (1980, 326ff.), Kuhn (1983, 37ff.), and Feyerabend (1987: ch. 10).

[87] Early in his discussion, Nagel considers parallels between the idealism he opposes and idealism as argued for in Berkeley's master argument (1986: 93–5). However, he bases his discussion on Williams's analysis of the master argument (see n. 14 above). Had he made use of Prior's analysis, he may have regarded the parallels as even closer than he does.

Davidson's discussion of 'the very idea of a conceptual scheme' is relevant to Nagel, because Davidson there presents a view of truth that is incompatible with this sort of realism. For, as Nagel notes, after arguing that nothing could be a conceptual scheme unless it was 'largely true', Davidson writes:

> [T]he criterion of a conceptual scheme different from our own now becomes: largely true but not translatable. The question whether this is a useful criterion is just the question how well we understand the notion of truth, as applied to a language, independent of the notion of translation. The answer is, I think, that we do not understand it independently at all. (1974a: 194)

For Davidson, that is, we do not 'understand' what it would be for there to be true sentences that are not translatable into ours. Or, as Nagel writes:

> Davidson [holds that] we do not possess a general concept of truth that goes beyond the truth of all possible sentences in any language that we could understand, or that could be translated into a language that we or others like us could understand. Our general idea of what is the case does not transcend the sum of what we could truly assert to be the case. (1986: 94)

And hence, for Nagel, Davidson's conception of truth or of what is the case, and hence of reality, is insufficiently realist. For, while Nagel agrees that we cannot provide any examples of truths that we cannot express in our language, he wants to hold that there can be such truths. That is, although he agrees that no instance of

$\sim E^{\star}p$ The proposition that p is true but not expressible in my language

can be both understood and true, he maintains that

$(\exists p)\sim E^{\star}p$ $(\exists p)$(The proposition that p is true but not expressible in my language)

can be understood without precluding its truth. Just as Davidson, like Wittgenstein, adheres to a notion of reality, or the world, as that which renders true or false the only sentences I can understand,[88] Nagel, like Russell, accepts a notion of reality, or the world, according to which there may be truths, and so aspects of reality, that are inexpressible in my language.

Further, like Russell, Nagel recognizes that in order to sustain his view he must accept an account of generality that enables him to hold, for example, that even though no instance of '$\sim E^{\star}p$' can be understood and true, understanding the general sentence '$(\exists p)\sim E^{\star}p$' does not preclude its truth. Thus, in discussing the argument of his 'idealist' opponent, Nagel writes:

> The idealist reply is that our completely general ideas of what exists, or is the case, or is true cannot reach any further than our more specific ideas of kinds of things that can exist, or be the case, or be true. We do not, in other words, possess a completely general concept of reality that

[88] Despite some remarks apparently to the contrary (1974a: 192–4), Davidson has no difficulty affirming 'the world itself makes the true sentences true' as 'harmless and trivial' (1997d: 668; cf. 1974a: 198).

reaches beyond any possible filling in of its content that we could in principle understand.... (1986: 93–4)

thereby indicating that, for the 'idealist', since we cannot coherently regard any instances of '$\sim E^{\star}p$' as true (since we cannot give any 'content' to the notion of a truth we cannot express), then we cannot coherently regard '$(\exists p)\sim E^{\star}p$' as true (and hence do not possess a concept of reality that enables us to hold that there are truths we cannot express). And in defending his own position, Nagel writes:

To be the value of a variable in our universal or existential quantifications it is not necessary to be the referent of a specific name or description in our language, because we already have the general concept of *everything*, which includes both the things we can name or describe and those we can't.... We can speak of 'all the things we can't describe', 'all the things we can't imagine', 'all the things humans can't conceive of', and finally, 'all the things humans are constitutionally incapable of ever conceiving'. The universal quantifier does not have a built-in limitation to what can be designated in some other way.... (1986: 98)

thereby indicating, like Russell, that it is through the use of variables that may range over more than we can determinately represent in our language that he can use general sentences to express his realism.[89] For Nagel, as for Russell, to accept a view of generality according to which understanding a general sentence such as '$(\exists p)\sim E^{\star}p$' (or '$(\exists y)\sim Ay$') expressing his realism (or anti-solipsism) does not preclude its truth is not thereby to hold that such a general sentence is true; however, it is the first step towards recognizing that it is likely to be true. Thus, immediately after the passage I have just quoted, Nagel continues:

Naturally the possibility of forming these ideas does not guarantee that anything corresponds to them. But in the nature of the case it is unlikely that we could ever have reason to believe that nothing does.... (Ibid.)

and later Nagel characterizes 'the view I am trying to defend' as 'the claim that there are probably things about the world that we (humans) cannot conceive' (1986: 104).

Moreover, in the course of defending his realism, according to which understanding '$(\exists p)\sim E^{\star}p$' does not preclude its truth, Nagel also defends a version of 'conceptual relativism' according to which understanding '$(\exists p)\sim Ep$' does not preclude its truth. In particular, observing that 'there are plenty of ordinary human beings who constitutionally lack the capacity to conceive of some of the things that others know about', he argues that we can

[89] Unlike Russell, Nagel presents no theory of generality to sustain the view that (in Prior's terminology) there are cases in which we can think truly the proposition expressed by a sentence of the form '$(\exists x)Fx$' without being able to think truly any of its instances. As he writes: 'I must say at the outset that I do not have an alternative theory of thought to offer in place of those that support idealism' (1986: 95). Nor would Prior agree with Russell that the way we can think truly such general propositions without being able to think truly any of their instances is through acquaintance with universals (see n. 6 above).

equally well imagine a species for whom these characteristics were normal, able to think and know about the world in certain respects, but not in all. Such people could have a language, and might be similar enough to us so that their language was translatable into part of ours. (1986: 95)

Nagel then asks us to consider whether a member of such species could form a general concept of reality that would enable him to conceive that there are aspects of reality which he cannot conceive:

Can he have, on the basis of the examples of reality with which he is acquainted, a general concept which applies beyond everything with which he and his like could conceivably become acquainted? If he can, then we ourselves can have the same concept which will apply to features of the universe that we are incapable of comprehending. (1986: 96)

And Nagel endorses this conclusion. Thus, for Nagel, the member of the conceptually limited species (relative to ours) who states '$(\exists p){\sim}Ep$' would be stating something that is not only significant but also true—for, as Nagel describes the case, there are propositions that are expressible in developed human languages that are not expressible in any language of that conceptually limited species. And thus, for Nagel, if I state '$(\exists p){\sim}Ep$', what I state will also be significant—will be conceivably true and conceivably false—whether or not there actually are creatures whose conceptual capacities stand to ours as ours stand to the conceptually limited species that Nagel imagines.[90]

4.3 Holism again

So far I have argued that just as Wittgenstein holds that neither any instance of '${\sim}Ay$' nor the general sentence '$(\exists y){\sim}Ay$' can be understood and true, so too Davidson holds that neither any instance of '${\sim}Ep$' nor the general sentence '$(\exists p){\sim}Ep$' can be both understood and true. As yet, I have not considered Davidson's reasons for so regarding '$(\exists p){\sim}Ep$'. In particular, I have not argued that Davidson so regards '$(\exists p){\sim}Ep$' in virtue of holding with Wittgenstein that general sentences are truth-functions of their instances. However, as I argue now, Davidson like Wittgenstein adheres to a holistic view of language from which it follows that it is unintelligible to suppose that there are propositional contents not expressible in my language.

In 'Truth and Meaning' Davidson writes:

If sentences depend for their meaning on their structure, and we understand the meaning of each item in the structure only as an abstraction from the totality of sentences in which it features, then we can give the meaning of any sentence (or word) only by giving the meaning of every sentence (and word) in the language. Frege said that only in the context of a sentence does a word have meaning; in the same vein, he might have added that only in the context of a language does a sentence (and therefore a word) have meaning. (1967: 22)

[90] More recently, McDowell (1994: Lecture II) has defended the sort of 'idealism' Nagel opposes (while denying that it is properly called 'idealism'), while Williamson (2007: 15–17) argues against McDowell in favor of the sort of 'realism' Nagel defends.

For Davidson, that is, a sentence has the meaning it has only in the context of the language in which it occurs, so that to understand a given sentence (to know the meaning of that sentence) is to understand a language (to know the totality of meanings expressible in the language in which that sentence occurs). Hence, given that we grasp propositional contents only insofar as we understand sentences expressing those contents, Davidson is thereby indicating that to be able to grasp any one propositional content is to be able to grasp the totality of propositional contents expressible by a language capable of expressing that content. In Wittgenstein's terminology, he is indicating that to be given a point in logical space is to be given the whole of logical space.[91]

Throughout his writings, Davidson recognizes that one consequence of this holism is that we cannot regard creatures who seem capable of understanding only some of our sentences as genuinely understanding those sentences. Thus, as early as 1965, he writes:

> [I]n so far as we take the 'organic' character of language seriously, we cannot accurately describe the first steps towards its conquest as learning part of the language; rather it is a matter of partly learning. . . . (1965: 7)

thereby indicating that because understanding a sentence requires understanding a whole language, creatures who apparently understand only some but not all our sentences have not genuinely learned part of our language, but have only 'partly learned' our language. Again in 1976, he writes:

> We tend to assume that we can interpret quasilinguistic behavior, whether in children or chimpanzees, by using some elementary subset of sentences drawn from our own developed language: the idea is that these simple sentences of ours give the meaning of the behavior or signs to be interpreted. But this idea, and the assumptions behind it, are not easy to justify. The general source of trouble is easy to identify: it is that even the simplest sentences of a language depend for their meaning on relations that hold between them and sentences with more complex structure,

[91] In more recent writings, Davidson has disavowed the extreme form of holism he endorsed in 'Truth and Meaning'. Thus, while he writes in 'Radical Interpretation Interpreted' that 'what one sentence means depends on the meanings of other sentences', he continues: 'I am not an unbuttoned holist in that I do not say the meaning of a sentence depends on the meanings of all sentences.' In a note to this remark, after quoting from the passage in 'Truth and Meaning' in which he endorsed just that sort of 'unbuttoned' holism, he writes: 'I have subsequently been more restrained on this topic' (1994: 124 and 128 n. 3). Similarly, in his 'Reply to Evnine', he denies that his holism commits him to the view that 'the meaning of an expression is so tightly tied to the meaning of all the rest of one's words that two speakers can mean the same thing by a word only if they control exactly the same linguistic and conceptual resources', and indicates that Nagel is wrong to suppose that he (Davidson) is forced to deny that there are regions (for Davidson peripheral 'suburbs') of one language that might be untranslatable into another language (1997c: 307–9). Instead, he indicates that his position is that there is a 'central core' that is intertranslatable among all languages, regardless of whether absolutely every sentence in a given language has a translation into every other language. Without considering how this position bears on Davidson's original arguments against 'the very idea of a conceptual scheme' or on Nagel's criticisms of those arguments, my concern here is to consider the 'unbuttoned' holism that is common to the early Wittgenstein and Davidson of 'Truth and Meaning' and its relation to their positions in the *Tractatus* and 'On the Very Idea of a Conceptual Scheme', respectively, as well as to the ways in which Russell and Nagel reject those positions.

> or sentences based on more sophisticated concepts. Because of the holistic character of the semantic aspect of language, it is wrong to suppose that a sentence (or word) belonging to a natural language would have the same meaning if its linguistic environment were radically impoverished.... Philosophy, at least since the time of Kant, has abounded with arguments designed to show that if one or another important part of the system of concepts with which we operate is lopped off, what remains cannot be made intelligible simply as the remainder; we may be quite unable to make clear sense of the truncated structure. The same point holds for language.... [U]ntil a language, or language-like structure, is found to have expressive powers very similar to the most highly developed languages, we are apparently at a loss to make clear sense of what its sentences, or utterances, mean.... (1976: 18–19)

thus making explicit the connection between his semantic holism and his denial of the view that a sentence could have the same meaning it does outside of the context of the whole language in which it occurs. More recently, in his (1999) paper 'The Emergence of Thought', Davidson writes similarly that 'it is the holism of the mental that makes its emergence so difficult to describe', and repeats his argument that because 'words, like thoughts, have a familiar meaning, a propositional content, only if they occur in a rich context', it would be wrong to hold 'that the child is talking if it makes sounds which, if made by a genuine language-user, would have a definite meaning' (1999: 127).

Given the holism that is common to Davidson in 'Truth and Meaning' and the early Wittgenstein, sentences from different languages can be translations of each other only if those languages have exactly the same expressive power. For if the meaning (sense) of sentence *S* in language *L* depends upon the meanings (senses) that are expressible in *L*, then no sentence from another language *L** can be a translation of *S* unless *L** has the same expressive power—that is, encompasses the same logical space—as *L*. Hence, the sort of example of 'conceptual relativism' that Nagel presents in order to defend his realism is incompatible with the sort of holism that both Davidson and Wittgenstein accept. For by that holism, if sentences from two different languages are capable of expressing the same propositional content, then those languages are capable of expressing exactly the same propositional contents, in which case one language cannot be capable of expressing some but not all of the propositional contents that are expressible in another language. The same sort of holism that precludes a creature from being able to understand only part of our language, also precludes a creature from speaking a language only parts of which can be translated into our language.

Thus, on the holism common to Davidson and the early Wittgenstein, the only possible sort of case of 'conceptual relativism' would be one in which there are languages capable of expressing none of the same propositional contents—that is to say, languages capable of expressing wholly different logical spaces with no points in common, so that there would be no translatability at all between those languages. However, the very holism that permits that to be the only sort of case of 'conceptual relativism' also precludes one from being able to express it. For if I am to be able to regard that sort of conceptual relativism as possible, I will have to be able to understand

(CR?) Are there languages L_1 and L_2 such that, for every proposition p, if p is expressible in L_1 it is not expressible in L_2?

and regard it as a genuine question—a question my understanding of which precludes neither an affirmative nor negative answer. But if (CR?) can be expressed in my language as a genuine question, then I possess general conceptions of what it is to be a propositional content and of what it is to be a language that are independent of the propositional contents that are expressible in my language or of any language I can understand, so that (CR?) has its meaning non-holistically. In that case, another creature possessing a language that is wholly untranslatable into mine could possess the same general conceptions of propositional contents and of language that enables that creature to express a sentence that is translatable as (CR?). But then, contrary to the holism shared by Wittgenstein and Davidson, the propositional content expressible by (CR?) would be expressible by different languages that did not have exactly the same expressive powers. That is to say, we can express (CR?) only if that holism is false.

Thus, just as Wittgenstein's holism precludes him from regarding us as possessing a concept of what it is to be an object that would be needed if we are to recognize (Solip?) as a genuine question (or hence (Solip) and (~Solip) as genuine claims), so too the holism shared by Wittgenstein and Davidson precludes them from regarding us as possessing concepts of what it is to be a language or what it is to be a propositional content that would be needed if we are to recognize (CR?) as a genuine question (or similarly (CR) and (CM) as genuine claims). And while Nagel, like Russell, regards us as possessing general conceptions that we could not possess if that holism were correct but that enable us to raise as a genuine question whether what we can express is all there is to be expressed, Davidson, like Wittgenstein, holds, consistently with their shared holism, that we do not possess the general conceptions that we would need to possess in order to recognize genuine questions that require us 'to take up a stance outside our own ways of thought'.

5 Conclusion

As I have presented their views, neither Russell nor Wittgenstein, and neither Nagel nor Davidson, take our inability to provide examples of a certain sort of entity (be it an object with which I am not acquainted or a proposition I cannot express in my language) to establish—in the style of Berkeley's master argument—the metaphysical conclusion that there can be no entity of that sort. Like Mackie, both sides in each debate agree that 'self-refutation' arguments in the style of Berkeley do not successfully provide an a priori basis for recognizing necessary metaphysical truths; rather, the disagreements in these debates concern what is wrong with such 'self-refutation' arguments and the bearing such arguments have on the prospects for metaphysics.

For Russell and Nagel, such arguments fail because we possess the conceptual resources to raise general questions ('Are there particulars with which I am not acquainted?' 'Are there propositions I cannot express in my language?') even in cases where it is incoherent to attempt to provide examples that would support affirmative answers to those questions; and for Russell and Nagel, this failure of such 'self-refutation' arguments shows that while we can legitimately raise the relevant metaphysical questions, there is no way to answer them a priori. For Wittgenstein and Davidson, in contrast, we lack the resources to raise such questions, and because we do so, the proper way to take such 'self-refutation' arguments is not as establishing necessary metaphysical truths a priori, but rather as revealing limits that cannot be set 'in relief' and as rendering unintelligible the apparent metaphysical questions.

Similarly, as I have presented their views, Russell and Nagel are no less sensitive than Wittgenstein and Davidson to avoiding what John McDowell has characterized as the confusion that we can occupy 'a standpoint independent of all . . . human activities' and can consider 'from sideways on' the 'relation of . . . our thought and language to the reality it characterizes' (1981: 44).[92] For as I have presented their views (at least with regard to the issues I have considered), Wittgenstein and Davidson, as well as Russell and Nagel, are all concerned in a fundamental way with how, or whether, we can raise questions of ultimate metaphysics from within our position as thinking creatures; and, for all of them, Berkeleyan-style 'self-refutation' arguments for metaphysical conclusions fail to take that concern seriously. For such arguments assume uncritically—what we could know, in effect, only from 'sideways on'—that where we cannot coherently provide examples of entities of a certain sort, there cannot be, as a matter of ultimate metaphysics, entities of that sort.

What divides Russell from Wittgenstein, or Nagel from Davidson, is not whether we can 'take up a stance outside our own ways of thought' and attain a 'sideways on' view of the relation between our thought or language and reality; rather, what divides them is whether we can, from within our position as thinking creatures, so much as recognize such questions as to whether our ways of thinking are the only ways of thinking or are adequate to all that there is. Hence, to hold that Wittgenstein, both early and late, regards it as incoherent to take up such a stance is not yet to determine how Wittgenstein's later writings bear on the sorts of differences I have discussed here between Russell and the early Wittgenstein, or between Davidson and Nagel. Nor, more substantively, does denying that Berkeley-style 'self-refutation' arguments succeed in establishing metaphysical conclusions thereby determine one's view toward metaphysics. For even supposing that it precludes one from recognizing an a priori route to metaphysical conclusions, it leaves open whether we cannot occupy a position from which we can raise genuine metaphysical questions at all or whether we can raise such questions but cannot answer them a priori.[93]

[92] While McDowell is here discussing Wittgenstein's later view of rule-following, both Diamond (1988: 185) and Conant (1991: 157; 2002: 422) cite this passage to clarify the stance of the early Wittgenstein.

[93] My interest in the topic of Wittgenstein on the limits of thought goes back to my first encounter with the *Tractatus* in a course taught by Warren Goldfarb. As a graduate student at Berkeley, I had many formative

References

Anscombe, G. E. M. (1959) *An Introduction to Wittgenstein's Tractatus*. London: Hutchinson University Library.

Armstrong, D. M. (1965) 'Editor's Introduction' in Berkeley (1965), 7–34.

Berkeley, George (1710/34) *A Treatise concerning the Principles of Human Knowledge*. Reprinted in Berkeley (1965), 41–128.

—— (1713) *Three Dialogues between Hylas and Philonous*. Reprinted in Berkeley (1965), 129–225.

—— (1965) *Berkeley's Philosophical Writings*, ed. D. M. Armstrong. New York: Collier Books.

Byrne, Alex (2007) 'Soames on Quine and Davidson', *Philosophical Studies* 135, 439–49.

Conant, James (1991) 'The Search for Logically Alien Thought: Descartes, Kant, Frege, and the *Tractatus*', *Philosophical Topics* 20, 115–80.

—— (2000) 'Elucidation and Nonsense in Frege and the early Wittgenstein', in Crary and Read (2000), 174–217.

—— (2002) 'The Method of the *Tractatus*', in E. Reck, ed., *From Frege to Wittgenstein*. Oxford: Oxford University Press, 374–462.

Conant, James and Diamond, Cora (2004) 'On Reading the *Tractatus* Resolutely', in M. Kölbel and B. Weiss, eds, *Wittgenstein's Lasting Significance*. London and New York: Routledge, 46–99.

Crary, Alice and Read, Rupert, eds (2000) *The New Wittgenstein*. London and New York: Routledge.

Davidson, Donald (1965) 'Theories of Meaning and Learnable Languages'. Reprinted in his (1984), 3–15.

—— (1967) 'Truth and Meaning'. Reprinted in his (1984), 17–36.

—— (1974a) 'On the Very Idea of a Conceptual Scheme'. Reprinted in his (1984), 183–98.

—— (1974b) 'Reply to Solomon'. Reprinted in his (1980), 243–4.

—— (1976) 'Introduction', in S. Harnad, H. Steklis, and J. Lancaster, eds, *Origins and Evolution of Language and Speech*, vol. 280 of *Annals of the New York Academy of Sciences*. New York: New York Academy of Sciences, 18.

—— (1980) *Essays on Actions and Events*. Oxford: Clarendon Press.

—— (1984) *Inquiries into Truth and Interpretation*. Oxford: Clarendon Press.

—— (1988a) 'The Myth of the Subjective'. Reprinted in his (2001), 39–52.

—— (1988b) 'Epistemology and Truth'. Reprinted in his (2001), 177–91.

discussions with Paul Loeb on issues central to this paper, including relations not only between Wittgenstein's position and Davidson's but also (and at Paul's suggestion) between Wittgenstein's position and Berkeley's. Earlier versions of this material were presented at colloquia at the University of California, San Diego; the University of Ulster, Coleraine; the National University of Ireland, Maynooth; University College, Cork; a 2003 Stirling workshop on the *Tractatus*; and the 2005 conference at Stirling on 'The *Tractatus* and its History'. I am grateful to members of those audiences, including Jim Conant, Warren Goldfarb, Marie McGinn, Michael Potter, Thomas Ricketts, and Steven Yalowitz for helpful discussion. One comment from the late Stephen Mills significantly influenced the overall shape of this chapter: whereas my earliest versions of the chapter made reference to Prior's views on modal logic (see n. 13), Stephen suggested I look at Prior's discussion of Berkeley's master argument. Thanks also to Dan Watts for helpful discussion and comments. I am especially indebted to Peter Sullivan for detailed comments on earlier versions of this chapter and for organizing the Stirling workshops and conference. I gratefully acknowledge my receipt of a Senior Research Fellowship from the Irish Council, which facilitated work on this chapter.

—— (1994) 'Radical Interpretation Interpreted', *Philosophical Perspectives*, vol. 8: *Logic and Language*, 121–8.

—— (1997a) 'Seeing through Language'. Reprinted in his (2005), 127–41.

—— (1997b) 'Indeterminism and Antirealism'. Reprinted in his (2001), 69–84.

—— (1997c) 'Reply to Simon J. Evnine', in Hahn (1997), 305–10.

—— (1997d) 'Reply to Stephen Neale', in Hahn (1997), 667–9.

—— (1999) 'The Emergence of Thought'. Reprinted in his (2001), 123–34.

—— (2001) *Subjective, Intersubjective, Objective*. Oxford: Clarendon Press.

—— (2005) *Truth, Language, and History*. Oxford: Clarendon Press.

Diamond, Cora (1988) 'Throwing Away the Ladder: How to Read the *Tractatus*'. Reprinted in her *The Realistic Spirit*. Cambridge, Mass.: MIT Press, 1991, 179–204.

—— (2000) 'Does Bismarck Have a Beetle in his Box?', in Crary and Read (2000), 262–92.

Feyerabend, Paul (1987) *Farewell to Reason*. London and New York: Verso.

Fodor, Jerry (1983) *The Modularity of Mind*. Cambridge, Mass.: MIT Press.

Frege, Gottlob (1884) *Die Grundlagen der Arithmetik*, trans. J. L. Austin as *The Foundations of Arithmetic*. Evanston, Ill.: Northwestern University Press, 2nd revised edition 1978.

—— (1892a) 'On Sense and Meaning [Bedeutung]'. Reprinted in Frege (1984), 157–77.

—— (1892b) 'On Concept and Object'. Reprinted in Frege (1984), 182–94.

—— (1918) 'Thoughts'. Reprinted in Frege (1984) 351–72.

—— (1980) *Philosophical and Mathematical Correspondence*, ed. G. Gabriel *et al.*, trans. H. Kaal. Chicago: University of Chicago Press.

—— (1984) *Collected Papers on Mathematics, Logic, and Philosophy*, ed. B. F. McGuinness. Oxford: Basil Blackwell.

Gallois, André (1974) 'Berkeley's Master Argument', *Philosophical Review* 83, 55–69.

Hacker, P. M. S. (2000) 'Was he Trying to Whistle it?', in Crary and Read (2000), 353–88.

Hahn, Lewis E., ed. (1997) *The Philosophy of Donald Davidson*. Chicago and LaSalle, Ill.: Open Court.

Hay, W. H. (1953) 'Berkeley's Argument from Nominalism', *Revue Internationale de Philosophie* 7 (23–4), 19–27.

Hintikka, Merrill B. and Hintikka, Jaakko (1986) *Investigating Wittgenstein*. London: Basil Blackwell.

Kremer, Michael (2001) 'The Purpose of *Tractarian* Nonsense', *Nous* 35, 39–73.

—— (2004) 'To What Extent is Solipsism a Truth?', in B. Stocker, ed., *Post-Analytic Tractatus*. Aldershot: Ashgate, 59–84.

Kuhn, Thomas (1983) 'Commensurability, Comparability, Communicability'. Reprinted in his *The Road Since Structure*, ed. J. Conant and J. Haugeland. Chicago and London: University of Chicago Press, 2000, 33–57.

Levine, James (1998) 'The What and the That: Theories of Singular Thought in Bradley, Russell, and the Early Wittgenstein', in G. Stock, ed., *Appearance versus Reality*. Oxford: Clarendon Press, 19–72.

—— (2001) 'Logical Form, General Sentences, and Russell's Path to "On Denoting"', in R. Gaskin, ed., *Grammar in Early Twentieth-Century Philosophy*. London: Routledge, 74–115.

—— (2005) 'Aboutness and the Argument of "On Denoting"', in G. Imaguire and B. Linsky, eds, *On Denoting: 1905–2005*. Munich: Philosophia Verlag, 29–97.

McDowell, John (1981) 'Non-Cognitivism and Rule-Following'. Reprinted in Crary and Read (2000), 38–52.

—— (1994) *Mind and World*. Cambridge, Mass.: Harvard University Press.

McGinn, Marie (1999) 'Between Metaphysics and Nonsense: Elucidation in Wittgenstein's *Tractatus*', *The Philosophical Quarterly* 49, 491–513.

Mackie, J. L. (1964) 'Self-Refutation—A Formal Analysis', *The Philosophical Quarterly* 14, 193–203.

—— (1976) *Problems from Locke*. Oxford: Oxford University Press.

Moore, A. W. (1990) *The Infinite*. London and New York: Routledge.

—— (2003) 'Ineffability and Nonsense', *Proceedings of the Aristotelian Society*, suppl. vol. 77, 169–93.

Moore, G. E. (1927) 'Facts and Propositions'. Reprinted in his *Philosophical Papers*. London: George Allen & Unwin, 60–88.

Mounce, H. O. (1997) 'Philosophy, Solipsism and Thought', *The Philosophical Quarterly* 47, 1–18.

—— (2001) 'Critical Notice of The New Wittgenstein', *Philosophical Investigations* 24, 185–92.

Nagel, Thomas (1986) *The View from Nowhere*. New York and Oxford: Oxford University Press.

Pears, David (1972) 'Wittgenstein's Treatment of Solipsism in the *Tractatus*', *Critica* 6, 57–84.

—— (1987) *The False Prison, Volume One*. Oxford: Clarendon Press.

Priest, Graham (2002) *Beyond the Limits of Thought*, 2nd edition. Oxford: Clarendon Press.

Prior, A. N. (1955) 'Berkeley in Logical Form'. Reprinted in his (1976), 33–8.

—— (1957) *Time and Modality*. Oxford: Clarendon Press.

—— (1971) *Objects of Thought*, ed. P. T. Geach and A. J. P. Kenny. Oxford: Clarendon Press.

—— (1976) *Papers in Logic and Ethics*, ed. P. T. Geach and A. J. P. Kenny. London: Duckworth.

Proops, Ian (2001) 'The New Wittgenstein: A Critique', *European Journal of Philosophy* 9, 375–404.

Putnam, Hilary (1981) *Reason, Truth and History*. Cambridge: Cambridge University Press.

—— (2004) *Ethics without Ontology*. Cambridge, Mass.: Harvard University Press.

Read, Rupert (2006) 'A No-Theory? Against Hutto on Wittgenstein', *Philosophical Investigations* 29, 73–81.

Read, Rupert and Deans, Rob (2003) '"Nothing is Shown": A "Resolute" Response to Mounce, Emiliani, Koethe and Vilhauer', *Philosophical Investigations* 26, 238–68.

Rescher, Nicholas (1980) 'Conceptual Schemes', *Midwest Studies in Philosophy*, vol. 5, ed. P. French, T. E. Uehling, and H. K. Wettstein. Minneapolis: University of Minnesota Press, 323–45.

Ricketts, Thomas (1996) 'Pictures, Logic, and the Limits of Sense in Wittgenstein's *Tractatus*', in H. Sluga and D. Stern, ed, *The Cambridge Companion to Wittgenstein*. Cambridge: Cambridge University Press, 59–99.

Rorty, Richard (1972) 'The World Well Lost'. Reprinted in his *Consequences of Pragmatism*. Minneapolis: University of Minnesota Press, 1982, 3–18.

Russell, Bertrand (1903) *The Principles of Mathematics*. Cambridge: Cambridge University Press, 2nd edition. London: George Allen and Unwin, 1937.

—— (1905a) 'On Fundamentals'. In Russell (1994), 359–413.

—— (1905b) 'On Denoting'. Reprinted Russell (1994), 414–27.

—— (1911a) 'Knowledge by Acquaintance and Knowledge by Description'. Reprinted in his *Mysticism and Logic*. London: Longmans, Green and Co., 1919, 209–32.

—— (1911b) 'The Basis of Realism'. Reprinted in *The Collected Papers of Bertrand Russell*, vol. 6, ed. J. Slater. London and New York: Routledge, 1992, 125–31.

—— (1912) *The Problems of Philosophy.* (Page references are to the 1959 Oxford University Press edition.)
—— (1913) *Theory of Knowledge: The 1913 Manuscript*, ed. E. Eames. London: George Allen and Unwin, 1984.
—— (1915) *Our Knowledge of the External World.* Chicago and London: Open Court.
—— (1918) 'The Philosophy of Logical Atomism'. Reprinted in Russell (1956), 177–281.
—— (1919) *Introduction to Mathematical Philosophy.* London: George Allen and Unwin.
—— (1920) 'The Meaning of "Meaning"'. Reprinted in *The Collected Papers of Bertrand Russell*, vol. 9, ed. J. Slater. London and New York: Routledge, 1988, 87–93.
—— (1936) 'The Limits of Empiricism'. Reprinted in *The Collected Papers of Bertrand Russell*, vol. 10, ed. J. Slater. London and New York: Routledge, 1996, 313–28.
—— (1940) *An Inquiry into Meaning and Truth.* London: George Allen and Unwin.
—— (1944) 'Reply to Criticisms', in P. A. Schilpp, ed., *The Philosophy of Bertrand Russell.* La Salle, Ill.: Open Court, 680–741.
—— (1946) *History of Western Philosophy.* London: George Allen and Unwin.
—— (1948) *Human Knowledge: Its Scope and Limits.* New York: Simon and Schuster.
—— (1950) 'Logical Positivism'. Reprinted in Russell (1956), 367–82.
—— (1956) *Logic and Knowledge*, ed. R. Marsh. London: George Allen and Unwin.
—— (1959) *My Philosophical Development.* New York: Simon and Schuster.
—— (1996) *The Collected Papers of Bertrand Russell*, vol. 4, ed. A. Urquehart. London: Routledge.
Stroud, Barry (1968) 'Conventionalism and the Indeterminacy of Translation'. Reprinted in his *Meaning, Understanding and Practice.* Oxford: Oxford University Press, 2000, 17–32.
Sullivan, Peter (1996) 'The "Truth" in Solipsism and Wittgenstein's Rejection of the A Priori', *The European Journal of Philosophy* 4, 195–219.
—— (2002) 'On Trying to be Resolute', *The European Journal of Philosophy* 10, 43–78.
—— (2003) 'Ineffability and Nonsense', *Proceedings of the Aristotelian Society*, suppl. vol. 77, 195–223.
—— (2004) 'What is the *Tractatus* about?', in M. Kölbel and B. Weiss, eds, *Wittgenstein's Lasting Significance.* London and New York: Routledge.
Whitehead, Alfred North and Russell, Bertrand (1910) *Principia Mathematica*, vol. 1. Cambridge: Cambridge University Press, 2nd edition 1927.
Williams, Bernard (1966) 'Imagination and the Self'. Reprinted in his *Problems of the Self.* Cambridge: Cambridge University Press, 1973.
Williamson, Timothy (2007) *The Philosophy of Philosophy.* Oxford: Blackwell Publishing.
Wittgenstein, Ludwig (1922) *Logisch-Philosophische Abhandlung/Tractatus Logico-Philosophicus*, trans. C. K. Ogden, London: Routledge and Kegan Paul Ltd, 2nd edition 1933; trans. by D. F. Pears and B. F. McGuinness, London: Routledge and Kegan Paul Ltd, 1961, 2nd edition 1974.
—— (1929) 'A Lecture on Ethics', in his *Philosophical Occasions: 1912–1951*, ed. J. Klagge and A. Nordmann. Indianapolis and Cambridge: Hackett Publishing Company, 36–44.
—— (1973) *Letters to C. K. Ogden*, ed. G. H. von Wright. Oxford: Basil Blackwell.
—— (1975) *Philosophical Remarks*, ed. R. Rhees, trans. R. Hargreaves and R. White. Chicago: University of Chicago Press.
—— (1979a) *Wittgenstein and the Vienna Circle*, conversations recorded by Friedrich Waismann, ed. B. F. McGuinness, trans. J. Shulte and B. F. McGuinness. New York: Barnes and Noble Books.

Wittgenstein, Ludwig (1979b) *Notebooks 1914–1916*, 2nd edition, ed. G. H. von Wright and G. E. M. Anscombe, trans. G. E. M. Anscombe. Chicago: University of Chicago Press.

—— (1980a) *Wittgenstein's Lectures: Cambridge, 1930–1932*, ed. D. Lee. Chicago: The University of Chicago Press.

—— (1980b) *Culture and Value*, ed. G. H. von Wright, trans. P. Winch. Chicago: The University of Chicago Press.

—— (1995) *Cambridge Letters: Correspondence with Russell, Keynes, Moore, Ramsey, and Sraffa*, ed. B. F. McGuinness and G. H. von Wright. Oxford: Blackwell.

10

Was the author of the *Tractatus* a transcendental idealist?[1]

A. W. Moore

> The limit of language is shown by its being impossible to describe the fact which corresponds to . . . a sentence, without simply repeating the sentence.
>
> (This has to do with the Kantian solution to the problem of philosophy.)
>
> Ludwig Wittgenstein

I

Imagine a dispute between two readers of the *Tractatus*[2] that would be readily settled once both parties had acknowledged that Wittgenstein holds the propositions of logic to be true. An example would be a dispute about whether he advocates a principle of bipolarity for all propositions, whereby only what could be false could be true. There is a clear sense in which any such dispute ought not to survive an appeal to the 4.46s. These numbered remarks indicate clearly that Wittgenstein does indeed hold the propositions of logic to be true. It would be captious to object that, by the very lights of the *Tractatus*, these remarks are themselves nonsensical, and that, in his most authentic mode, Wittgenstein would not have assented to 'The propositions of logic are true,' but would rather have urged that we had not yet given any meaning to certain signs in this pseudo-proposition (cf. 6.53). There *could* be an interesting exegetical dispute about *that* too. Two readers could disagree about why, or even whether, Wittgenstein would indeed have regarded this as a nonsensical pseudo-proposition. But relative to the dispute about bipolarity, even to raise this issue would be cavilling. Relative to *that* dispute, the idea that the propositions of logic are true is simply there in the text. And its being simply there in the text is a matter of our prescinding from the book's self-renunciation at the end and pretending that what

[1] A much abridged version of this chapter has already appeared as Pt I of Moore (2007). I am extremely grateful to Blackwell for permission to reproduce material from that essay.

[2] Wittgenstein (1961); all unaccompanied references to Wittgenstein are to this.

we have before us in the 4.46s are what we appear to have: straightforward affirmations. The author of the *Tractatus*, in a clear and unproblematic sense, held that the propositions of logic are true.[3]

Would it be fair to say that, in just the same sense, the author of the *Tractatus* was a transcendental idealist; that, subject to all the obvious caveats about the book's self-renunciation, transcendental idealism is 'simply there in the text'? There are all sorts of reasons why it would not. First, there is the entirely non-trivial issue of what transcendental idealism *is*. Second, even given an account of what transcendental idealism is, there will still be an issue about whether it is 'there in the text'. For there is no reasonable account of what transcendental idealism is on which it can be said to receive unambiguous expression there, of however vicarious or disingenuous a kind. And third, most significantly, even given some compelling reason for thinking that transcendental idealism *is* there in the text, there will still be, or there should still be, an issue about whether it is 'simply' there in the text, in other words—in Peter Sullivan's words—about 'what it's doing there' (2003: n. 4).

Why do I describe this third issue as the most significant? Partly, though not exclusively, for a reason that is admittedly parochial. As Sullivan indicates in the context from which I have just quoted, he and I would be in substantial agreement about the first two issues; and this chapter is largely a continuation of a dialogue between us about the third.[4]

Let me say, then, far more briefly than is really warranted, what I take to be the gist of our agreement about the first two issues. There are, in the *Tractatus*, notably in the 5.6s, remarks in which the limits of language and the limits of the world appear not

[3] This is an example only, not the main topic of my chapter. Even so, I cannot resist remarking on the surprising number of commentators who deny that the author of the *Tractatus* held that the propositions of logic are true. Thus Brian McGuiness (1990: 312) attributes the following view to Wittgenstein: 'Logic . . . [consists] of tautologies . . . which, when properly viewed, manifest . . . their own lack of truth-or-falsity, their own status as pseudo-propositions.' And Juliet Floyd (2000: 241) writes as follows: 'In the *Tractatus* . . . "logical propositions" . . . are seen to be lacking in sense, lacking the capacity for truth or falsity altogether,'—bizarrely citing 4.46 in support. (She compounds the error by going on to attribute to Wittgenstein the view that 'it is . . . *unsinnig* . . . to assert that a person knows that . . . war is war,' this time citing 5.1362 in support—presumably taking 'war is war' to be a tautology. But 5.1362 states that '"*A* knows that *p* is the case" . . . [is *sinnlos*] if *p* is a tautology,' and Wittgenstein, as Floyd herself has earlier in effect noted, carefully distinguishes between that which is *unsinnig* and that which is *sinnlos*, one cardinal difference being that the truths of logic, although they are *sinnlos*, are not *unsinnig*: see 4.461–4.4611.) To be sure, there is material in the 4.06s that may appear to contradict the idea that the propositions of logic are true. But 4.064 states that 'every proposition must *already* have a sense: it cannot be given a sense by affirmation' (Wittgenstein's emphasis) and this suggests that the 4.06s are tacitly concerned only with those propositions which, unlike the propositions of logic, do indeed have senses.

[4] This dialogue was initiated by the pair of essays that constitute the broader context from which the quotation from Sullivan is taken: Moore (2003) and Sullivan (2003). It is an incidental pleasure for me to take this opportunity to thank Sullivan both for his very generous reply to my essay and for putting me right on a number of points. For instance, on p. 212 he queries my claim that, where illusions of sense are concerned, there are always relevant intentions about which concepts are being exercised. And on pp. 197 and 215 he hints—though he is too kind to say—that it is unhelpfully simplistic to contrast 'the' traditional reading of the *Tractatus* with 'the' new reading of it in the way in which I do. He is quite right in each case.

merely as *limits*, not merely as essential features, but as *limitations*, as features that at some level exclude certain possibilities. Because these features appear to exclude certain possibilities, they also appear to admit the question why they are as they are. The answer implied in those same remarks involves the subject, who both understands language and has thoughts about the world, and who somehow sets the limits—*is* a limit—of each. And this is a version of transcendental idealism.

Granted this broad characterization of our shared reaction to the first two issues, the third, to repeat, is what such transcendental idealism is doing in the text. It is easy to overlook this third issue. It is easy to treat the *Tractatus* as a relatively uniform sequence of apparent affirmations, 'framed' by some remarks near the very beginning and near the very end which are themselves genuine affirmations concerning the status of what they 'frame'.[5] On this crude way of regarding the *Tractatus*, whether a view can be attributed to Wittgenstein, in the relatively relaxed sense that is of concern to us here, is a matter of whether there is, among the apparent affirmations that constitute the bulk of the book, an apparent affirmation of the view in question, or at least a set of apparent affirmations that apparently commit Wittgenstein to it.

I said that it is easy to treat the *Tractatus* in this way. Perhaps I should have made a more cautious autobiographical claim: I myself have slipped into treating the *Tractatus* in this way. I have in the past (e.g. 1997: 149, 206) attributed transcendental idealism to the author of the *Tractatus* as though this were no more contentious than attributing to him the belief that the propositions of logic are true, or the belief that objects are simple (2.02). But, as Sullivan has helped me to appreciate, to proceed as though these attributions are of a piece is to do scant justice to the subtleties of this extraordinary text, a text in which ideas are variously developed and suppressed, and temptations are variously indulged and dispelled, all within a context that allows for the eventual recognition that some of the words being put to critical use in these tasks are being put to use without meaning.

Sullivan has in various places developed a reading of the *Tractatus* whereby these apparent affirmations of transcendental idealism are altogether more dissembling than the apparent affirmations of various other doctrines that Wittgenstein wants us to throw away once we have used them to climb up beyond them (cf. 6.54). Sullivan draws a contrast with Wittgenstein's treatment of logical category distinctions (2003: 217ff.). For Wittgenstein, the logical category to which anything belongs is determined by the formal variable within whose range it lies. But this means that the very use of the word 'anything' in this formulation of the doctrine—'the logical category to which *anything* belongs'—cannot have its intended generality. It is impossible to generalize about things of different logical categories (as indeed this very admonishment purports to do). Wittgenstein brings us to an understanding of these matters

[5] This is an allusion to the kind of reading of the *Tractatus* that has come to be associated with Cora Diamond and James Conant; cf. Conant (1991: 159); Conant (2000: 198); and Diamond (2000: 149–51). But I am not suggesting that either Diamond or Conant treats the *Tractatus* in this overly crude way.

which involves our 'coming to share with him an appreciation that the best that can be done in the way of trying to express this understanding will be to produce formulations which the understanding itself enables [us] to recognize as nonsense' (2003: 218). Sullivan contrasts this with the case of transcendental idealism. The two cases may look analogous. We may think that, for Wittgenstein, language has certain limitations which, for that very reason, it cannot be *said* to have; hence that, here too, Wittgenstein brings us to an understanding of the matter which involves our 'coming to share with him an appreciation that the best that can be done in the way of trying to express this understanding will be to produce formulations which the understanding itself enables [us] to recognize as nonsense.' But on Sullivan's view, this is seriously to understate Wittgenstein's eventual repudiation of transcendental idealism. On Sullivan's view, although Wittgenstein both feels and indulges a sympathy for transcendental idealism, his ultimate aim is to dispel it completely, that is in such a way that it no longer serves even as a 'self-consciously nonsensical and knowingly futile attempt to express the understanding [we have] of its sources' (2003: 216). Kant, from whom of course transcendental idealism derives, remains very much in the picture for Sullivan, but only as the enemy. Once transcendental idealism has collapsed into its opposite, which it does so spectacularly in 5.64, Wittgenstein can be seen as proceeding to a case-by-case repudiation of a Kantian conception of the a priori.[6] Transcendental idealism, by the end of this exercise, can in no way be said to be 'simply' there in the text.

The contrast with Wittgenstein's treatment of logical category distinctions is one of two that Sullivan draws. The second is with Wittgenstein's later work.[7] Sullivan avers that there is a fairly direct route from some fundamental ideas in Wittgenstein's later work to transcendental idealism. That route is roughly as follows. It is because of 'how we go on' that our concepts are as they are. Hence it is because of how we go on that the necessities that hold in virtue of the interconnections of our concepts hold. But how we go on is grounded in a complex of biological and cultural contingencies to which we are subject. So the necessities in question are ultimately grounded in those same contingencies. So the limits of language and the limits of the world are indeed, at some level, *limitations*, set by us, those who understand language and who think about the world. And that is transcendental idealism.

I am suspicious of both contrasts. I think that the relation of the *Tractatus* to transcendental idealism *is* analogous to its relation to the doctrine of logical categories, or at least more analogous than Sullivan suggests. And I think that the relation of the *Tractatus* to transcendental idealism is both more straightforward and more

[6] See Sullivan (1996: 197–8); and note in particular the wonderful footnote 9 in which he makes clear how the organization of the 6s reflects topics that are of central concern to Kant. See Sullivan (1996) more generally, and Sullivan (2002, esp. §3.3), for his detailed defence of such a reading.

[7] Sullivan (2003: 221–2). It is, if I may be excused a statement of the obvious, precisely because of potential contrasts with the later work that I have cast my title as I have, although, as will become apparent, I may have a diametrically opposed view of what the principal contrast is to that of Sullivan.

straightforwardly one of commitment than that of the later work, whose avenue to transcendental idealism seems to me altogether less direct than Sullivan suggests.

I do not want to say much more about the first contrast now, beyond making the point that, in the case of the doctrine of logical categories just as in the case of transcendental idealism, 'seeing the world aright' (6.54) involves acquiescing in the lack of whatever alternatives would make limitations out of limits, for instance an alternative in which a formal variable can correspond to more than one logical category; and that, again in the former case just as in the latter case, this is a prize that we can be helped to win by watching claims whose implications are being followed out strictly, for instance the claim that things belong to different logical categories, collapse into their opposites, in this instance the claim that things do not belong to different logical categories (cf. 5.64).[8] So when Sullivan offers me the analogy with the doctrine of logical categories as one that he himself rejects, I am not inclined to be proud: I am inclined to welcome his cast-off.

As far as the contrast with Wittgenstein's later work is concerned, again I do not want to say much, partly because I have already discussed this matter elsewhere.[9] But I do want to say just enough to correct a false impression that I may have given about how I see the relation between transcendental idealism and the earlier work.[10] I suggested that, in the *Tractatus*, Wittgenstein brings us to transcendental idealism by indulging an urge that we have to transcend our limitations; and I further suggested

[8] Cf. the transition from the first sentence of 4.126 to the first sentence of 4.1274. Note that we must regard the 'opposite' of any such claim, which in the case of transcendental idealism is some kind of realism, as being no less a nonsensical pseudo-proposition than the claim itself.

[9] See esp. Moore (1997: ch. 6) and Moore (2007: Pt II).

[10] I may also have given a false impression about how I see the relation between transcendental idealism and the later work. When Sullivan traces a route from ideas in the later work to transcendental idealism, he credits me, among others, with having mapped out that route (Sullivan 2003: 221). He cites my (1997). He does not say that I take the author of the later work to have *been* a transcendental idealist. He knows that I would have demurred if he had. But other readers of my book have thought that I take the author of the later work to have been a transcendental idealist, which is what makes me think that I may have given a false impression. (See e.g. Travis 2006: 143,149. I think that Travis and I are far closer in our understanding of the later work than he suggests.) Not that I am inclined to be especially apologetic. The very first claim that I make about Wittgenstein in the main text of my book, a claim whose subsequent context makes clear that I mean the later Wittgenstein, is precisely that he was not a transcendental idealist (Moore 1997: 126). What I do think is that there is, in his later work, an inducement to embrace transcendental idealism, and that transcendental idealism accordingly has an important role to play in exegesis of his later work. But I also think that the inducement can be resisted. This is why I take the route to transcendental idealism to be less direct than Sullivan makes it out to be. Certainly there are plenty of side routes that we can take that will keep us from reaching that problematical destination.—It is worth noting in this connection *Philosophical Investigations*, Pt I, §§90ff., where Wittgenstein is engaging with his former self, and, I would contend, trying to counteract some of what made transcendental idealism so enticing to him. True, he himself makes claims in the course of this discussion which can themselves sound very transcendentally idealistic: see e.g. §§103 and 114. But I take him merely to be emphasizing the contingency of 'how we go on', and, with it, the contingency of the concepts we possess. To hear these claims as *already* transcendentally idealistic—to hear them as already making limitations out of limits—is to commit the very error that he is trying to guard against (see e.g. §108). Indeed this is an error that he tries to guard against not only here, but again and again throughout his later work: thus PI, Pt I, §§241–2, and OC, §98.

that the limitations in question are those that ground and shape our concepts (Moore 2003: 189–90). I may thus have given the impression that I see Wittgenstein as grappling with the complex of biological and cultural contingencies to which Sullivan alludes and which are undoubtedly crucial to any proper understanding of how transcendental idealism beckons in the later work, however resistibly and with whatever actual efforts to resist it on the part of the author of that work.[11] If I had seen Wittgenstein in this way, then I would of course have owed an explanation for why, in the *Tractatus*, despite an allusion to the complexity of the human organism (4.002), Wittgenstein does not seem to be the least bit interested in reckoning with any of the actual manifestations of that complexity. But I do not see Wittgenstein in this way. I had in mind something more Kantian.

There is a kind of scope distinction here. My idea was not that, given whatever such limitations we are subject to, we aspire not to be subject to them. It was rather that we aspire not to be subject to any such limitations at all. It is ultimately the very finitude of our thinking that we have an urge to transcend: the very fact that our thinking is answerable to something.[12] This is not to suggest that there is any alternative to our thinking's being thus finite. Our urge may be utterly ill-conceived. What *is* true is that for our thinking to be thus finite is for there to be an alternative to whatever *qualifies* it as thinking which is thus finite—to whatever equips it to answer to that to which it is answerable, namely reality, or the world. (By way of analogy: even if there is no alternative to a stick's having some finite length, for a stick to have some finite length is for there to be an alternative to its having whatever finite length it has.) So, for example, our thinking involves our using signs in some particular way (3.1ff.), where what this means is our using them in some way rather than some other that would have equipped us to think just as well (3.326–3.327).

One example of our using signs is, in Tractarian terms, our naming objects. Let us reflect for a while on the naming of objects. This will provide us with an indication of how these ideas connect with transcendental idealism. Our thinking does not depend on our actually naming objects (5.526; cf. also 4.002). But it does depend on our being, so to speak, at one remove from naming objects, in as much as each of our thoughts can be analysed into simple thoughts in which objects are named. Our thinking is *about* objects (3.2ff.). This is how it touches reality. If our thinking did not touch reality in this way, then it would not be able to answer to reality (2.15–2.1515 and 5.542). For our thinking to be about objects in this way, we need to know the objects it is about (6.2322). But our knowing an object does not itself consist in our thinking anything: we cannot *say* what we know when we know an object.[13] That our thinking depends

[11] See above, n. 10. For examples of how these contingencies figure in the later work see PI, Pt II, §xii, and RFM, Pt I, §§143–50.

[12] Cf. Kant (1998: B145). Henceforth all unaccompanied references to Kant are to this book.

[13] The distinction between the two German words that are both rendered as 'know', namely '*kennen*' and '*wissen*', is obviously pertinent here. Cf. also PI, Pt I, §78.

on our having such inexpressible knowledge is therefore another fundamental mark of its finitude. But it is just such fundamental marks as this that we have an urge to transcend. Thus one particularly significant way in which our urge is liable to manifest itself is through the attempt to say what we know when we know objects; to say *what* objects are, not just *how* they are; to put objects themselves into words (cf. 3.221). This connects with transcendental idealism in two ways.

The first connection is a direct one. This attempt is itself liable to issue in some version of transcendental idealism. For it is an attempt to say what grounds the very essence of reality, conceived as something to which not even logic is prior (5.552). It is an attempt to say, not how things are, nor yet how things must be, but what it is, in some deep metaphysical sense, for things to be how they are. It is an attempt to say, not how *things* are, but how their limits are. And this is liable to involve casting those limits as limitations. It is liable to involve embracing some version of transcendental idealism.[14]

The second connection with transcendental idealism is less direct, but more significant as far as the project of the *Tractatus* is concerned. Our understanding of why the attempt to express our inexpressible knowledge of objects is ill-conceived is itself inexpressible. Like our knowledge of objects, it does not itself consist in our thinking anything: it is part of our understanding of what thinking is. And the attempt to express *it* is liable to issue in some version of transcendental idealism. We are liable to say that reality consists of how objects are, not of what they are, not of the objects themselves—or, in somewhat more familiar terms, that the world is the totality of facts, not of things (1.1)—intending this as a corrective, as a way of signalling that only our knowledge of *facts* is a form of thinking, answerable to reality, and apt to be expressed. But if we do say this, then we shall be violating an admonishment that Wittgenstein himself gives in the *Tractatus*:

> We cannot say in logic, 'The world has this in it, and this, but not that.'
>
> For that would appear to presuppose that we were excluding certain possibilities, and this cannot be the case, since it would require that logic should go beyond the limits of the world; for only in that way could it view those limits from the other side as well. (5.61)

Wittgenstein's admonishment, which of course he himself violates, is an admonishment precisely against treating limits as limitations.[15] In saying that the world is the totality of facts, not of things, we do exclude certain possibilities; and we exclude them,

[14] Cf. in this connection various currents in 2.02–2.0272 and 5.55–5.5571.

[15] In saying that Wittgenstein himself violates this admonishment, I am referring back to 1.1. An obvious reply on Wittgenstein's behalf is that he uses the word 'world' differently in 1.1 from how he uses it in 5.61: whereas in the former case he uses it to refer to the realm of the actual, in the latter case he uses it to refer to the realm of the possible. I incline to the view that he uses it to refer to the realm of the actual throughout the *Tractatus*; and that what enables him to refer to the realm of the possible in 5.61 is his use of other words and phrases, notably 'limits' and 'in logic'. But even if I am wrong about that—even if Wittgenstein's use of 'world' is ambiguous in the way proposed—what he says in 1.1, with its clearly implied application to any other possible world, is still surely offensive to the spirit of what he says in 5.61.

moreover, on the grounds that they are not consonant with the nature of our thinking. In sum, we embrace a version of transcendental idealism.[16]

II

It is clear, then, that I see some sort of commitment to transcendental idealism in the *Tractatus* that Sullivan does not. To try to get a better sense of what is at issue between us, I shall now approach these matters from a slightly different angle. I begin with the error that we have just been considering, the error of misconstruing limits as limitations.

This is an extremely natural error. We easily hear the claim that aunts have to be female not as *denying* the possibility of non-female aunts but as *presupposing* that possibility—and then excluding it (rather like the injunction that dogs have to be on a lead). Indeed, there is a certain sort of limit that is quite properly construed in this way. Here I am alluding to the fact that there are different kinds of possibility, some of which are strictly subsumed by others. Thus whatever is economically possible is technologically possible, but not vice versa; whatever is technologically possible is physically possible, but not vice versa. These different kinds of possibility can be pictured as a series of concentric circles, in which larger circles include more possibilities than smaller circles. Given any two of these circles, it is entirely appropriate to construe the one as excluding some of what lies within the other. (If a politician says, 'There are some ways of improving the safety of our railways that are unaffordable,' he is alluding to what is technologically possible. He would not be vindicated by the technological impossibility of a completely failsafe automated signalling system.) But these are relative limits. There are also absolute limits, the limits of what we might call 'logical' possibility, limits *simpliciter* (cf. 6.3 and 6.37). And, however tempting it may be to construe these in the same way, it is incoherent to do so. We cannot say, except as a kind of joke, that *logical* possibility excludes possibilities of such and such another kind, as it may be the 'illogical' possibility of non-female aunts, or of grass that both is and is not green. Logical possibility is not just another circle in the space that we have been considering. It *is* the space that we have been considering.[17]

[16] Insofar as the material in this paragraph is not just a play of Tractarian ideas, but has serious exegetical intent, a notebook entry for 27.5.15 (NB 51–2) poses a threat to me. Wittgenstein there suggests that his concern with the inexpressible is a concern with that which transcends even objects; and indeed, as far as objects (or things) are concerned, he explicitly says that '*we feel no demand at all* to express them in propositions' (his emphasis). That his concern with the inexpressible is, in part, a concern with that which transcends even objects does not worry me: what I have said is not meant to exhaust the concern (see further below, §III). But the idea that we feel no demand at all to express objects in propositions does run counter to what I have said. To this I can only reply, first, that Wittgenstein's comment here is (after all) taken from his notebooks; and second, that the material from the *Tractatus* cited above seems to me to bear witness to the demand in the form in which I take Wittgenstein to think we feel it, as a demand to say *what* objects are, not just how they are.

[17] In this paragraph, and again subsequently, I have borrowed material from Moore (2006).

How then are we to characterize it? One thing that we can do is to ascend to the metalanguage. We can say that certain combinations of words do not represent possibilities at all. This is clearly connected to what Wittgenstein says in the Preface to the *Tractatus*, where he tells us that 'the aim of the book is to draw a limit to thought,' and then immediately adds, 'or rather—not to thought, but to the *expression* of thought: for in order to be able to draw a limit to thought, we should have to find both sides of the limit thinkable (i.e. we should have to be able to think what cannot be thought).'[18] Nevertheless, as Sullivan has emphasized (2003: 209ff.), if we do ascend to the metalanguage and distinguish between combinations of words that represent possibilities and combinations of words that do not, or between combinations of words that make sense and combinations of words that do not, we do not thereby provide ourselves with a way of construing absolute limits as, after all, limitations. They are not limitations. The metalanguage no more provides us with the wherewithal to describe what lies *beyond* such limits, or in other words to say what we cannot think and thus to say what we cannot say,[19] than the object language does.

Very well; given that absolute limits, or limits simpliciter, are not limitations, why is there such a powerful and pervasive temptation to construe them as such? In §I I tried to identify one source of this temptation: the urge to transcend our own finitude, in particular by saying what objects are. But I also suggested that another source is the very urge to counter the first urge. At some level we recognize the incoherence of construing limits as limitations, and, in recognizing this incoherence, we have an urge to forbid any reference to the possibilities that limits exclude, in such a way that we ourselves make reference to the possibilities that limits exclude, and hence in such a way that we ourselves construe limits as limitations. There is a general pattern here. Having seen through the appearance of sense in some piece of nonsense, we have an urge to repress the appearance by redeploying the nonsense, using some such formula as 'It does not make sense to say that . . . ' And if, furthermore, we attempt to say *why* it does not make sense to say this thing that we have just said, then we are liable to indulge in yet more nonsense of the same general sort, trying to characterize both sides of a border between that which we can characterize and that which we cannot. This is how we get to transcendental idealism: a kind of nonsense born of the urge to combat nonsense of that very kind; a disease for which it itself purports to be the cure.[20] And that is what I think we find in the *Tractatus*.[21]

[18] P. 3, my emphasis. Cf. 4.113–4.116. Cf. also some comments in his later work: e.g. in PG he insists that we should not say, 'There is no last cardinal number,' as though we were excluding some possibility, but should rather say, 'The expression "last cardinal number" makes no sense' (p. 465); and in PI, shortly after having remarked that '*essence* is expressed by grammar,' and commenting on a puzzle that he is wrestling with there, he says: 'The great difficult here is not to represent the matter as if *there were something* one *couldn't* do' (Pt I, §§371 and 374, first and third emphasis his, second emphasis mine).

[19] This is an allusion to 5.61, whose reversal Wittgenstein would surely accept.

[20] Cf. Karl Kraus's famous remark about psychoanalysis, which Bernard Williams (2006: 208) applies to Wittgenstein's later philosophy. For further discussion of this idea, see Moore (1997: 248).

[21] Cf. Moore (2003: 189–90); and cf. Sullivan's helpful gloss on this (2003: 219).

But Sullivan has convinced me that, even if I am right about this, it does not settle any dispute between us. For there is still the question of what such transcendental idealism is *doing* in the *Tractatus*.

My own view is that Wittgenstein has put it there *in propria persona*—or at least, as much *in propria persona* as is compatible with an exhortation to recognize it as itself nonsense that we must ultimately transcend (6.54). It is there as a result of an unsuccessful attempt to express certain inexpressible insights into the limits of language and into the incoherence of attempts to violate those limits, insights which, on the one hand, are supposed to be fostered by seeing that such nonsense is the result of an attempt to express them and, on the other hand, are supposed to foster seeing that such nonsense is nonsense. (Again I appeal to the analogy with the doctrine of logical categories.)

Sullivan thinks that this is to accredit Wittgenstein with more sympathy for transcendental idealism than he has. On Sullivan's view, there is a much greater critical distance between Wittgenstein and the transcendental idealism in his book. It is there as if in scare quotes. Wittgenstein does feel its allure, and he quite deliberately exploits that allure, but only so that it will eventually no longer have any hold on us. His aim is to repudiate the doctrine entirely. He wants us to acquiesce in viewing the limits of language as precisely that: limits, not limitations (Sullivan 2003: 219ff.). We are to recognize that there is no way for things to be which is not how they can be represented as being. Language is the totality of propositions; the world is the totality of facts; and the limits of these totalities coincide, that is to say any possible proposition expresses some possible fact and any possible fact is expressed by some possible proposition. But they coincide not because the limits of either are limitations somehow set by the other, nor yet because the limits of both are limitations somehow set by the subject, but precisely because they are the *limits* of two formal totalities that are to be conceived in terms of each other.[22]

In sum, then—and this is a cartoon sketch—where I take Wittgenstein to be a transcendental idealist, Sullivan takes him not to be. But of course, that *is* a cartoon sketch, and one of the reasons why settling our dispute presents such a challenge, or so it seems to me, is that it is enormously difficult to say what *exactly* is at issue between us. I have now approached this dispute from two somewhat different angles. But I have still not really pinpointed it. In the outline of Sullivan's view which I just provided (and in which I naturally hope that there was nothing to which he would take exception) I made a series of claims about Wittgenstein's vision, for instance that Wittgenstein wants us to acquiesce in viewing limits as limits and that he sees the limits of language as coincident with the limits of the world, which could just as well have appeared in an outline of my own view of his vision. And this is something that I think Sullivan himself may not have fully taken on board. At one point he says, supposedly *contra* me,

[22] See Sullivan (1996, principally §IV, and 2002: 59–60). Relevant remarks in the *Tractatus* include 1, 4.001, 4.5, and 5.4711.

that, for Wittgenstein, 'the only stable understanding of language and its limits . . . [is] one that . . . [does not misconstrue] those limits as limitations' (2003: 220). I entirely agree. My attribution of transcendental idealism to Wittgenstein is not meant to suggest that Wittgenstein recognizes a *stable* understanding of the limits of language as limitations. Precisely not.

Still, I am reasonably confident (not least because I have been persuaded by Sullivan himself) that there is something at issue between us, even though I am altogether less confident that I can state it with the precision I would like. In the final section of this chapter I shall approach it from yet a third angle, in the hope that this will go at least some way towards identifying it. I shall also try to give an indication of why I am still inclined to take the same side. This will involve a brief digression on Kant. It will also involve an attempt to take due account of what seems to me to be in many respects the most striking and most significant fact about the appearance of transcendental idealism in the *Tractatus*, a fact that has been completely absent from the discussion so far.

III

Kant too is exercised by the limits of thought and the limits of reality. That is, he is exercised by the limits of thought which, in his terms, has *content*, thought which is about objects of sensible intuition (A51/B75); and he is exercised by the limits of *empirical* reality, the reality to which such thought is answerable (A26/B42ff.).[23] He takes the limits of each of these to be limitations. He contrasts thought which has content with thought which lacks content; and he contrasts empirical reality with the reality of things in themselves (Bxxv–xxvii).[24]

The fact that Kant effects these two contrasts, and takes the limits of thought and the limits of reality to be limitations, signals that he is an unregenerate transcendental idealist. And one of the ways in which he arrives at his transcendental idealism is by seeking an explanation for why these limits coincide in the way they do. This is something that he insists demands explanation. Furthermore he sees only three possible explanations: one is that the limits of reality are determined by the limits of thought; a second is that the limits of thought are determined by the limits of reality; and a third is that the limits of each have some common determinant. But it is only the first of these

[23] Henceforth, whenever the two terms 'thought' and 'reality' are used without qualification, in connection with Kant, they are to be understood elliptically, as standing for thought which has content and for empirical reality respectively.

[24] John McDowell, in the opening section of his *Mind and World*, denies that Kant effects the first of these contrasts. Commenting on the famous passage in which Kant declares that thoughts without content are empty (A51/B75), McDowell writes, 'For a thought to be empty . . . would be for it not really to be a thought at all, and that is surely Kant's point; he is not, absurdly, drawing our attention to a special kind of thoughts, the empty ones' (1996: 3–4). But that is precisely what Kant is doing, or at least what he takes himself to be doing: see e.g. A253–254/B309. In philosophical terms, I am sympathetic to the point that McDowell is making. But that is because I am unsympathetic to Kant's transcendental idealism. In exegetical terms, I take McDowell to be precisely wrong. For amplification, see further below.

(that the limits of reality are determined by the limits of thought) which he thinks can account for our a priori knowledge of their coincidence. In neither of the other two cases, where the limits of thought would be determined by something else, could we have any way of foreseeing—to use a Wittgensteinian word (4.5 and 5.556)—that the limits of reality would always conform to them. Hence Kant's transcendental idealism.[25]

But, as Wittgenstein helps us to appreciate, if these limits really were *limits*, *simpliciter*, and not limitations, then there would be no question of any such determination and no need for any such explanation. The limits would be as they are because, in the most austere sense of the word 'could', they *could* not be otherwise. And their coincidence would still be knowable a priori. Kant arrives at his transcendental idealism by taking as a datum the existence of a priori knowledge that is not merely of limits, but of limitations; not analytic, but synthetic (Introduction, §§IV and V). Wittgenstein, in direct opposition to this, insists that 'there is no a priori order of things' (5.634; cf. 2.225, 3.04–3.05, 6.31, 6.3211). That is, there is no a priori knowledge of limitations. And this is how, for Wittgenstein, transcendental idealism is to be resisted. On this Sullivan and I are in agreement. And yet...

Why does Kant see limitations where Wittgenstein sees only limits? Well, for Kant, thought which has content is thought which is capable of becoming knowledge; thought which lacks content is thought which is not capable of becoming knowledge.[26] *Kant needs the contrast.* As he famously says in the Preface to the second edition of *Critique of Pure Reason*, 'I had to deny *knowledge* in order to make room for *faith*,' (Bxxx, his emphasis). Faith, for Kant, 'is reason's moral way of thinking... [It] is trust in the attainability of an aim whose promotion is a duty but the possibility of whose realization we cannot have any insight into.'[27] It is directed at our freedom, at our capacity to exercise our wills either by obeying the moral law or by disobeying it, at our immortality, at the existence of God, and, quite generally, at whatever is ultimately of value, all of which Kant sees as lying beyond the reach of discursive knowledge. Freedom; the good or bad exercise of the will; the moral law; immortality; God; value: these cannot but ring extremely loud bells for any student of the *Tractatus*. They also provide me with a good cue to reveal what I had in mind when I referred earlier to what seems to me to be in many respects the most striking and most significant fact about the appearance of transcendental idealism in the *Tractatus*. I had in mind the fact that it appears, not only in the 5.6s, on which we have so far been concentrating, but just as blatantly, if not more so, in the 6.4s.[28]

[25] A92–93/B124–126 and B166–168. Note that only in the second of these two passages does Kant touch on the third possible explanation, specifically in theistic terms.

[26] See Bxxvi, footnote; B146; B166, footnote; A492–493/B521–522; and A771–772/B799–800.

[27] Kant (2000: 5:471–2, translation adapted by me).

[28] See also material shortly before and shortly after the 6.4s: 6.373–6.374 and 6.52–6.522.

It seems to me that, when we take the Kantian strain of the 6.4s duly into account, we can see an isomorphism, albeit a rough one, between, on the one hand, Kant's contrast between thoughts which have content and thoughts which lack content and, on the other hand, Wittgenstein's contrast between propositions which have sense and certain nonsensical pseudo-propositions.[29]

	Left-hand side	*Right-hand side*
Kant	Thoughts with content	Thoughts without content
Wittgenstein	Propositions with sense	Certain nonsensical pseudo-propositions

And, granted this isomorphism, we can say that, for both Kant and Wittgenstein, transcendental idealism lies on the right-hand side; furthermore, that it serves in helping us to *see* that it lies on the right-hand side.[30] But obviously, if that were all it served to do, then we need have nothing more to do with it. To use an analogy that I have used before (Moore 2003: 184), it would then be like a plinth whose sole purpose was to support a sign reading 'Mind the plinth'. One reason why it is of greater use than that is that it also serves, for both Kant and Wittgenstein, in helping us to see what else is on the right-hand side. But its greatest service, I submit, again for both Kant and Wittgenstein, is in helping us to do justice to what is on the right-hand side, or, more strictly, in helping us to do justice to the forces at work when we produce what is on the right-hand side.

Neither Kant nor Wittgenstein thinks that our rational engagement with things—that part of our engagement with things which is made possible by the fact that we are rational, thinking beings—is exhausted by whatever finds expression in items on the left-hand side. It obviously *includes* whatever finds expression in items on the left-hand side, which for both Kant and Wittgenstein has as its paradigm discursive knowledge, the kind of knowledge that is embodied in natural science.[31] But, for Kant, it also includes each of the following:

[29] As far as the left-hand side is concerned, cf. 3.13, the sentence in parentheses. Note: I caution that this isomorphism is rough for all sorts of reasons. One is that analytic thoughts appear on both sides of Kant's contrast (A52–55/B76–79, A151–152/B190–191, and A258–259/B314–315), whereas analytic propositions appear on neither side of Wittgenstein's (4.461–4.4611 and 6.1–6.11). Another is that the right-hand side of Kant's contrast involves the exercise of concepts, whereas the right-hand side of Wittgenstein's contrast involves pure and utter nonsense (see further below, with references). A final caveat: what Wittgenstein means by 'thoughts' are simply propositions with sense (4).

[30] In saying that transcendental idealism lies on the right-hand side for Kant, I have in mind its appeal to things in themselves: see Bxxv–xxvii and A369. Someone might object that transcendental idealism is a doctrine that Kant thinks can be known. And to be sure, Kant does allow for—nay, he insists upon—the possibility of transcendental knowledge, in such a way as to render our grasp of transcendental idealism a case in point (A11–12/B25). This connects with a distinction that he draws between the 'transcendental' and the 'transcendent' (A296/B352–353). But my own view, which I cannot now elaborate, is that there is an inconsistency in Kant. I am extremely grateful to Mark Sacks for discussions that prompted me to add this footnote.

[31] Cf., in the case of Kant, A792/B820; and, in the case of Wittgenstein, 4.11 and 6.53.

- faith;
- hope;
- the practical use of pure reason;
- aesthetic judgement;

and, overlapping some of these, what he sometimes calls

- practical knowledge.[32]

(Kant takes it to be a fundamental mark of our finitude that our rational engagement with things fractures in this way.[33]) That we *recognize* ourselves as engaging with things in these various ways itself involves our engaging with things in a way that is not just a matter of discursive knowledge.[34] And in articulating what we recognize, we are led to transcendental idealism. For we contrast what finds expression in items on the left-hand side with our other modes of rational engagement with things, and we come to regard empirical reality—that to which all items on the left-hand side must answer—as correspondingly restricted. What it is restricted to, Kant insists, are objects of sensible intuition.[35] But, *in* so insisting, he sees himself as allowing for due acknowledgement of our other modes of engagement with things, which find expression, or at least partial expression, in thoughts that are not about objects of sensible intuition, but about value, about freedom, about God, and the like; items on the right-hand side.[36]

I think we can see the same structure in Wittgenstein. Admittedly, there are profound differences in how the structure is instantiated. Thoughts without content, for Kant, are nonetheless thoughts. They involve the genuine exercise of concepts (B146). And they are true or false (A820–831/B848–859). By contrast, nonsense, for Wittgenstein, is just nonsense. It comprises words to which no meanings have been assigned (5.4733 and 6.53). Thus, while Kant can countenance that which finds bona fide (if partial) expression in items on the right-hand side, Wittgenstein can at most countenance that which finds *apparent* or *attempted* expression in items on the right-hand side—what we might call, echoing the fact that items on the right-hand side are for Wittgenstein pseudo-propositions, pseudo-expression. Wittgenstein can *at most* countenance that. The point, however, is that he does. Like Kant, he recognizes modes of rational engagement with things other than that which finds expression in

[32] For references to practical knowledge (translated as 'practical cognition'—a translation of which I approve, though the use of 'practical knowledge', which can also be justified, is better suited to my current purposes), see Kant (1996: 5:103 and 2000: 5:195).

[33] See e.g. B135, B138–139, and B145. Cf. René Descartes, who writes, '[God's] understanding and willing does not happen, as in our case, by means of operations that are in a certain sense distinct from one another; we must rather suppose that there is always a single identical and perfectly simple act by means of which he simultaneously understands, wills, and accomplishes everything' (1985, Pt I, §23).

[34] There is a connection here with what Kant calls 'the sole fact of pure reason': see Kant (1996: 5:31 and 2000: 5:468).

[35] See again B147–148.

[36] See e.g. Kant (1996: 5:141).

items on the left-hand side. That is, he recognizes modes of rational engagement with things other than thought. I discussed one example in §I:[37]

- knowledge of objects,

which is also in effect understanding of names. There are other kinds of understanding too, including

- understanding of propositions (4.022, 4.024, and 4.1212),

and, quite differently, the state that we are supposed to get into as a result of reading the *Tractatus*:

- understanding of Wittgenstein (6.54).[38]

Then there are:

- the practice of philosophy (4.11–4.115);
- logical inference (5.13–5.133 and 6.12–6.1201);
- the practice of mathematics (6.2–6.211 and 6.233–6.234);

and, most significantly of all, all that comes under the head of

- evaluation,

including feeling the world as a whole, exercising the will, and being happy or unhappy (6.43 and 6.45).[39] *None* of these consists in having thoughts. *All* of them find pseudo-expression in items on the right-hand side.[40]

Sullivan casts Kant in the role of enemy throughout the 6s. I cannot see this. I think that in the 6.4s we see something profoundly Kantian. To be sure, Wittgenstein does want to dismantle a Kantian conception of the a priori, and he spends much of the early part of the 6s doing just that. But this is because he wants us to appreciate that what we

[37] There is a sense in which §III is a generalization of §I.

[38] For discussion, see Moore (2003, §§V–VIII).

[39] The fundamental distinction between fact and value that is implicit here is also the primary theme of Wittgenstein's 'A Lecture on Ethics'. He there gives two further notable examples of what I have in mind: wonder at the existence of the world; and the feeling of absolute safety (1965: 8). And he later remarks that 'the verbal expression which we give to these experiences is nonsense!' (ibid). He concludes with some highly pertinent reflections on how, in ethics, we 'run against the boundaries of language,' adding, '[Ethics] is a document of a tendency in the human mind which I personally cannot help respecting deeply and I would not for my life ridicule it' (1965: 11–12). Wittgenstein's *Notebooks* also contains highly pertinent reflections: see e.g. NB 76–89. (Particularly noteworthy is his comment at NB 80: 'The thinking subject is surely mere illusion. But the willing subject exists.') For helpful commentary on all of this material, see: Stenius (1960, esp. ch. XI); Janik and Toulmin (1973, esp. ch.6); Hacker (1986, esp. ch. IV); Diamond (2000), which, on p. 171, includes ideas that are very much in line with what I am arguing, though it is reasonably clear from what she says elsewhere that there is much in what I am arguing to which she would take exception; and Wiggins (2004).

[40] This is perhaps most tendentious in the case of mathematical practice. I am thinking of the fact that mathematical practice consists in manipulating items on the right-hand side in a way that can look for all the world like expressing thoughts (6.2 and 6.2341).

produce when we endorse transcendental idealism, of which the Kantian conception of the a priori is an integral part, is utter nonsense. What *leads* us to endorse it, on the other hand, continues to command Wittgenstein's deepest respect, and this is what we see most clearly in the 6.4s.[41] '*The book's point*,' he famously wrote in a letter to Ficker, '*is an ethical one*' (Wittgenstein 1967: 143, his emphasis). And part of that point is to uphold a fundamental separation of fact and value. It is in trying to come to terms with this separation that we construe the world as the totality of facts, that is to say as the totality of what can be thought and said, *to the exclusion of* value (6.41), that is to say to the exclusion of what can be affected by acts of will (6.43). There are, for Wittgenstein, genuine insights that lead us to cast the limits of the world as limitations in this way; genuine insights that lead us to endorse this version of transcendental idealism. They are inexpressible insights into what it is to think, into what it is to exercise the will, and into what separates these.

In conclusion, then, while I am still not entirely confident that I have got the dispute between Sullivan and myself into proper focus, I hope at least to have given some indication of why I want to answer my title question in the way I do, a way that marks a relatively clean division between us. And just to be clear about what my answer is: *was* the author of the *Tractatus* a transcendental idealist? *Not* in any straightforward sense—he thought that transcendental idealism was a tissue of nonsense—but, in the sense in which he held the propositions of logic to be true, or objects to be simple, yes he was.

References

Conant, James (1991) 'The Search for Logically Alien Thought: Descartes, Kant, Frege, and the *Tractatus*', *Philosophical Topics* 20, 115–80.

——(2000) 'Elucidation and Nonsense in Frege and Early Wittgenstein', in Crary and Read (2000), 174–217.

Crary, Alice and Read, Rupert, eds (2000) *The New Wittgenstein*. London: Routledge.

Descartes, René (1985) *Principles of Philosophy*, in *The Philosophical Writings*, vol. 1, trans. J. Cottingham, R. Stoothoff, and D. Murdoch. Cambridge: Cambridge University Press.

Diamond, Cora (2000) 'Ethics, Imagination and the Method of Wittgenstein's *Tractatus*', in Crary and Read (2000), 149–73.

Floyd, Juliet (2000) 'Wittgenstein, Mathematics and Philosophy', in Crary and Read (2000), 232–61.

Hacker, P. M. S. (1986) *Insight and Illusion: Themes in the Philosophy of Wittgenstein*, revised edition. Oxford: Oxford University Press.

Janik, Allan and Toulmin, Stephen (1973) *Wittgenstein's Vienna*. New York: Simon and Schuster.

Kant, Immanuel (1996) *Critique of Practical Reason*, in Immanuel Kant, *Practical Philosophy*, trans. and ed. M. J. Gregor. Cambridge: Cambridge University Press.

[41] A quotation from Iris Murdoch is very pertinent here. She writes, 'God does not and cannot exist. But what led us to conceive of him does exist' (1993: 508).

——(1998) *Critique of Pure Reason*, trans. and ed. P. Guyer and A. W. Wood. Cambridge: Cambridge University Press.

——(2000) *Critique of the Power of Judgment*, ed. P. Guyer, trans. P. Guyer and E. Matthews. Cambridge: Cambridge University Press.

McDowell, John (1996) *Mind and World*, 2nd edition. Cambridge, Mass.: Harvard University Press.

McGuiness, Brian (1990) *Wittgenstein: A Life—Young Ludwig (1889–1921)*. Harmondsworth: Penguin.

Moore, A. W. (1997) *Points of View*. Oxford: Oxford University Press.

——(2003) 'Ineffability and Nonsense', *Proceedings of the Aristotelian Society*, suppl. vol. 77, 169–93.

——(2006) 'The Bounds of Sense', *Philosophical Topics* 34, 327–44.

——(2007) 'Wittgenstein and Transcendental Idealism', in G. Kahane, E. Kanterian, and O. Kuusela, eds, *Wittgenstein and His Interpreters: Essays in Memory of Gordon Baker*. Oxford: Basil Blackwell.

Murdoch, Iris (1993) *Metaphysics as a Guide to Morals*. Harmondsworth: Penguin.

Stenius, Erik (1960) *Wittgenstein's Tractatus: A Critical Exposition of its Main Lines of Thought*. Oxford: Basil Blackwell.

Sullivan, Peter M. (1996) 'The "Truth" in Solipsism, and Wittgenstein's Rejection of the A Priori', *European Journal of Philosophy* 4, 195–219.

——(2002) 'On Trying to be Resolute: A Response to Kremer on the *Tractatus*', *European Journal of Philosophy* 10, 43–78.

——(2003) 'Ineffability and Nonsense', *Proceedings of the Aristotelian Society*, suppl. vol. 77, 195–223.

Travis, Charles (2006) *Thought's Footing*. Oxford: Oxford University Press.

Wiggins, David (2004) 'Wittgenstein on Ethics and the Riddle of Life', *Philosophy* 79, 363–91.

Williams, Bernard (2006) 'What Might Philosophy Become?', in his *Philosophy as a Humanistic Discipline*, ed. A. W. Moore. Princeton: Princeton University Press.

Wittgenstein, Ludwig (1961) *Tractatus Logico-Philosophicus* (TLP), trans. D. F. Pears and B. F. McGuiness. London: Routledge and Kegan Paul.

——(1965) 'A Lecture on Ethics', *The Philosophical Review* 74, 3–16.

——(1967) 'Letter to Ficker', in Paul Engelmann, *Letters from Ludwig Wittgenstein, with a Memoir*, ed. B. F. McGuiness and trans. L. Furtmüller. Oxford: Basil Blackwell.

——(1974a) *Philosophical Investigations* (PI), trans. G. E. M. Anscombe. Oxford: Basil Blackwell.

——(1974b) *On Certainty* (OC), ed. G. E. M. Anscombe and G. H. von Wright, trans. D. Paul and G. E. M. Anscombe. Oxford: Basil Blackwell.

——(1974c) *Philosophical Grammar* (PG), ed. R. Rhees, trans. A. Kenny. Oxford: Basil Blackwell.

——(1978) *Remarks on the Foundations of Mathematics* (RFM), 3rd edition, ed. G. H. von Wright, R. Rhees, and G. E. M. Anscombe, trans. G. E. M. Anscombe. Oxford: Basil Blackwell.

——(1979) *Notebooks: 1914–1916* (NB), 2nd edition, ed. G. H. von Wright and G. E. M. Anscombe, trans. G. E. M. Anscombe. Oxford: Basil Blackwell.

11

Idealism in Wittgenstein: a further reply to Moore

Peter Sullivan

1 Introduction

My title is explained by the long delay between Moore's first presentation of his paper and this response to it. In the interim another paper prompted by Moore's has appeared (Sullivan 2011), to which Moore (2011) has in turn generously replied. The response to Moore offered in that other paper is extended but notably oblique: in the spirit of moving beyond old disputes, I preferred there to shift our discussion to a new context, that set by Wittgenstein's later thought about the authority of rules and what it is to follow them. This response aims to be shorter but more direct.

Moore's chapter has as its title a question that we agree to be fundamental in estimating the tenor and value of Wittgenstein's early work, 'Was the author of the *Tractatus* a transcendental idealist?' Moore answers Yes to this question, while I have said No. My reasons for saying No are in fact very well portrayed in Moore's chapter (pp. 248, 250).[1] He suggests that, so far as those considerations go, there can be complete agreement between us (p. 250). He goes on to suggest, though, that those considerations belong to too narrow a view, first of Wittgenstein's book, and then in consequence of the issue between us. I welcome the first of these suggestions, and I suspect there is something right about the second. In §2 I will indicate what this might be, by contrasting broader and narrower understandings of Moore's title question. In the rest of the chapter I will try to explain why I persist in favouring the narrower view.

2 Broader and narrower understandings of the issue

As it is explained by Moore, in *Points of View*, transcendental idealism holds, first, that some aspect of the form or essential nature of the world is dependent on mind, the mind of the representing subject. But this first commitment seems only to characterize

[1] Unaccompanied page references are to Moore's chapter in this volume.

the broader species of idealism in general, and so to distinguish the transcendental variety Moore adds the 'rider' that the dependence in question is 'transcendent', that it cannot itself be represented or expressed (Moore 1997: 116). Whilst I am sympathetic to the framework of thought in which this definition is offered, I do not want to commit myself to it. For one thing, it makes even the characterization of the broad topic of our discussion rest on a sharper line between what can and cannot be represented than I know how to draw. For another, it creates the need for some questionable exegesis, designed to deny to the supposedly distinct category of the 'transcendental' its borderline status between the 'immanent' and the 'transcendent', even to make out that Kant was a transcendental idealist by this definition (cf. p. 251, n. 30).

Not having any better definition to offer, I have no choice but to let examples be the go-cart of judgement here,[2] and obviously Kant's is the example that counts for most. Pushed to an account, then, I might first say, vaguely and open-endedly, that a position will count as a version of transcendental idealism when it is shaped and motivated in ways like Kant's; or secondly, and somewhat more specifically, when it figures in the way his does as an essential plank in a transcendental explanation, one designed to legitimate representations by explaining a priori how it is that they constitute a mode of knowledge of objects (A56–7/B80–1). Since I've given up on definitions I have to leave the choice between the more open-ended and the more specific of these accounts as a matter of preference. But it is not a choice without consequences. This is particularly so if Kant has more than one motivation for his idealism, as he surely does. In that case something might fit the first account, by sharing *some* motivation with Kant, but not the second, because the shared motivation is not one captured in the more specific account, which derives from the Introduction to the Transcendental Logic of the first *Critique* and is informed by the concerns of the Transcendental Deduction. It now seems to me that at least part of the explanation for the differences between Moore and me about Wittgenstein lies simply in different choices on this matter, and so at bottom just on the different breadth of our respective visions. When I deny that the author of the *Tractatus* is a transcendental idealist, I have in mind the narrower account, focused on issues surrounding the Deduction. When Moore affirms that he was, he has in mind, in addition, motivations for Kant's idealism that don't surface there. The last two sections of this response, §§6–7, will be concerned with this difference.

First, though, I should explain how the terms in which I have set the issue here relate to the terms in which it is discussed in Moore's chapter and in my interim response. Those discussions and previous ones make play with a distinction between 'limits' and 'limitations'. The notion of a 'limit', as it is intended here, is a non-contrastive, non-excluding one: limits have no 'other side'; there is nothing 'beyond' them.

[2] Compare Kant (1998: A134/B173–4). Further references to this work will be given in the text in the easily recognizable 'A/B' format.

'Limitation', on the other hand, is a contrastive, excluding notion: limitations are boundaries separating what qualifies for inclusion within them from what does not. The previous discussion takes the construal of limits as limitations to be symptomatic of idealism, and then asks what place that construal has in Wittgenstein's thought. Is it only ever something that he aimed to expose as a confusion? Or is there some way in which, or some level at which, he is himself committed to it?

The connection with idealism as it is here much more conventionally described derives from the way Kant aimed to legitimate the structuring concepts of a discursive understanding. He held that the harmony between representations and reality that is needed for them to amount to a mode of knowledge could be secured only by recognizing a role for our representations in constituting the reality to which they answer. The limits of reality, or of the world about which our ways of representing are thereby qualified as ways of knowing, are set or determined by the intrinsic contours of our faculty of representation. Because of its role in an explanatory legitimation, this 'setting' or 'determining' cannot but be understood contrastively. And to do that is to turn the limits that are thus set into limitations.

Now, what I have just sketched, in indicating how a Kantian idealism will manifest itself in the construal of limits as limitations, is only a *one-way* implication. It plainly gives no ground for supposing that *every* way of construing limits as limitations will be symptomatic of idealism. For instance, my interim response mentions Descartes as an example of someone who adopted that construal without a hint of idealism (2011: 172). More relevantly here, though, it does not follow that every such construal that carries a Kantian echo will be symptomatic of idealism. The connection with idealism derives from the project of transcendental explanation, and whilst that is of course thought of by Kant as a theme that runs through and dictates the form of all three of his *Critiques*, not every echo of a connected theme need be a connected echo.

3 The outline shape of a response

That straightforward observation gives the general point around which I would aim to organize a detailed response to Moore's chapter. In it he points to various episodes in which, and various ways in which, the construal of limits as limitations figures in Wittgenstein's text. His discussion in the first section of the chapter gives prominence to how Wittgenstein thinks of our knowledge of, or acquaintance with, objects, which Moore takes to be a form of inexpressible understanding, and to how at 3.221 Wittgenstein deliberately conjures up the sense of an aspiration to express, or to put into words, what we then understand.

> Objects I can only *name*. Signs stand for them. I can only speak *of* them. I cannot *express* them. A proposition can only say *how* a thing is, not *what* it is. (3.221)

My interim paper made a start on responding to what Moore says about this example, in what I think might almost be called a 'resolute' manner. I (2011: 173) suggested that this is one case where Wittgenstein conjures up an idea in order to display its disintegration, and that the railing against the limitations of propositional thought that we first hear in this passage gives way to a realization that there is no more than an analytic triviality to rail against. The misperception of this triviality as a limitation or restriction is certainly something we are meant briefly to share, but in Wittgenstein's resolution of this episode there is, as I see it, no lingering commitment *either* to the misperception (as Moore would clearly agree) *or* (as I think he wouldn't) to the thought that some genuine insight motivates it.

But this is just one kind of example, and others that Moore mentions cannot be dealt with in the same way. When Wittgenstein says that 'The sense of the world must lie outside the world', that '*In* [the world] there is no value', and that what makes the world of value 'must lie outside the world' (6.41), it is not remotely attractive to suggest that he is encouraging a vivid imaginative engagement with a confusion so as more effectively to display its dissolution. 'Ethics is transcendental' (6.421)—the remark stands unqualified, and it lasts. And of course there are many cases in between. 'Logic', too, 'is transcendental' (6.13), although—despite Wittgenstein's clumsy first expression of the thought in 1916 (NB 77–9)—not at all in the same way as ethics. Logic, after all, pervades the world, and its limits are those of the world.

That logic is transcendental and that ethics is transcendental are of course both echoes of Kant, and what they echo are, in Kant, parts of a connected theme. But according to my general observation this is not enough to make them connected echoes. Surely Moore is right to hold that Wittgenstein's exclusion of ethics from the world, leaving everything in the world as accidental—as a totality of mere meaningless facts, not a realm of significance or values—cannot but be heard as some kind of limitation. So, if the construal of the limits of the world as limitations were sufficient for idealism, his case would be proved. But as I said, the implication that connects these two is one-way—the other way. Such a construal is no more than a symptom of idealism; and, as with most rashes, the symptom can have other causes.

4 A distrust of general patterns

Towards the end of his essay Moore displays a pattern discernible in the thought of both Kant and Wittgenstein and representing a rough isomorphism between them (p. 251). For both Kant and Wittgenstein, Moore contends, transcendental idealism as a doctrine occupies the same location within this pattern. The commitment of both to transcendental idealism emerges, though, not so much in the location they assign to it within the pattern, as in their endorsement of the pattern itself.[3] In §§6–7 below I will ask whether this pattern does more to suppress important differences than to highlight similarities. In this section I will question a second pattern which, I think, partly

[3] Stenius (1960: ch. 11) advances a very similar claim.

explains Moore's confidence in the first. This is a pattern in 'how we get to transcendental idealism' (p. 247) rather than in the position reached.

The movement of thought that constitutes this second pattern begins one step on from the initial misconstrual of limits as limitations, with the recognition that such a construal is incoherent. In trying to effect the correction this recognition calls for, Moore claims, we are liable to land ourselves back in the initial mistake.

> [I]n recognizing [the initial] incoherence, we have an urge to forbid any reference to the possibilities that limits exclude, in such a way that we ourselves make reference to the possibilities that limits exclude, and hence in such a way that we ourselves construe limits as limitations. (p. 247)

Suppose that the initial misconstrual involved taking to be meaningful some unobviously meaningless formulation, that we then imagined to characterize an excluded possibility. Recognizing the incoherence of the initial misconstrual will then involve 'see[ing] through the appearance of sense in [that] piece of nonsense'. We will in consequence 'have an urge to repress the appearance by redeploying the nonsense, using some such formula as "It does not make sense to say that . . . "'. And if we then go on to explain why it does not, 'we are liable to indulge in yet more nonsense of the same general sort'.

> This is how we get to transcendental idealism: a kind of nonsense born of the urge to combat nonsense of that very kind; a disease for which it itself purports to be the cure. (p. 247)

And that, Moore holds, is just what we find in the *Tractatus*,

I agree, of course, with the last observation here. Nonsense redeployed with the aim of exposing nonsense of that very kind is something we find throughout the *Tractatus*. But I do not agree with the suggestion that finding *that* much is finding anything very useful about what, in any particular episode, is going on. In particular, if we counter an initial misunderstanding in the general kind of way described here, I do not think it follows that, in any serious sense, 'we ourselves construe limits as limitations'. In any serious sense, 'to construe X as Y' ought to mean more than merely to use a form of words that might invite the understanding of X as Y (that, we might perhaps say, is 'to portray X as Y'); it ought to imply some share in, or at least sympathy with, that understanding. And that one's thinking exemplifies the general pattern Moore describes here does not, I think, establish anything like so much.

Nor, it seems, does Moore himself think so. He begins this phase of his discussion by remarking that 'We easily hear the claim that aunts have to be female not as *denying* the possibility of non-female aunts but as *presupposing* that possibility—and then excluding it' (p. 246). If we easily hear the claim in the second way, we presumably can also hear it in the first way. And why should we not continue so to hear it if the claim is put in some such form as 'It does not make sense to say that . . . ',[4] or again, if it occurs in

[4] As Moore (2011: 192) observes, the presence or absence of quotation marks is at best a fallible guide to the intended focus of a claim.

explanation of why it does not make sense to say that? If someone introduces himself to me as a long-lost cousin, saying 'My father is your aunt', and if, improbably, I can be bothered to explain what has gone wrong, then I show patience with his misunderstanding, but no share in it.

To portray limits as limitations, in the sense just recommended, is with Wittgenstein a characteristic idiom. At times we can see him deliberately accentuating the portrayal. One example comes in his recommendations (LO 27) for a far-from literal translation of the first sentence of 4.023:

> The proposition determines reality to this extent, that one only needs to say 'Yes' or 'No' to it to make it agree with reality.

The German text already inveigled us with the notion of a 'third' or 'middle' possibility to be excluded. But it did not so clearly suggest a view according to which the work of stating truth is accomplished through two sub-tasks: first, so fixing things that only two possibilities remain; second, picking the right one of those. In making this suggestion explicit Wittgenstein causes us to pause over what will now surely seem the more difficult of these sub-tasks,[5] and opens the way to an idealist conception of it, hinted at in the final sentence of 4.023: 'The proposition constructs a world with the help of a logical scaffolding'. We can describe the world—the idea naturally arises—only because to frame a proposition, to think, is already to determine reality in accordance with the logical form of thought; or again, thinking can so neatly divide logical space only because the space is shaped to be so divided.

But while it is easy to see Wittgenstein doing this, it is much less easy to be clear just *what* he is doing. One possibility, in the case in hand, is that the suggestion he has now made prominent represents his own view. David Pears supposes so, at least in part, in developing a view according to which the 'total demand' that a proposition makes of the world divides into its 'sense-conditions' and its 'truth-conditions' (1987: 70–1), though he resists an idealist conception of how it is contrived that 'sense-conditions' are satisfied. Stenius, by constructing his account of pictorial representation in the context of the notion of a subjectively organized 'world field' (1960: 20), swallows the suggestion whole. But a quite different possibility is that, for Wittgenstein, it is only by contrast with the conception invited here that his subsequent draining from the Law of Excluded Middle of all of the apodictic force of a law, to leave only an empty, failed indicative, can have its intended philosophical point. (To be a law-giver to Nature would be a grand thing; to be a dispenser of tautologies less so.) To put this second possibility more simply, Wittgenstein anticipates that his book will be understood only by someone who has had similar thoughts (Preface). Amongst those, and in a fashion familiar to anyone who has worked in the history of philosophy, are thoughts the author has explored and found wanting. If we are to understand him, when he later

[5] In the analogy Wittgenstein gives to Odgen, of fixing a shelf to the wall with two nails, it is clear that the real difficulty is in hammering home the first nail (LO 27).

remarks that 'There is only logical necessity' (6.37), we need to have had such other thoughts about what the scope and source of necessity might be—including those invited at 4.023.

Those are just two possibilities for what Wittgenstein might be doing on this one occasion of portraying a limit as a limitation. They are surely not exhaustive, and are not offered as models simply transferable to other such occasions. Moore would, I'm sure, agree with this stance; because in illustrating it I have, in effect, been illustrating his own description of the *Tractatus* as

> a text in which ideas are variously developed and suppressed, and temptations are variously indulged and dispelled, all within a context that allows for the eventual recognition that some of the words being put to critical use in these tasks are being put to use without meaning. (p. 241)

In return, I agree with Moore that transcendental idealism is the most important idea to be so treated, and the temptation Wittgenstein was most concerned to understand. It is in that sense that I take transcendental idealism to be 'there in the text'—not just in the 5.6s, or the 6.4s, but all over the text. But to appear 'in the text' in this way need not be to appear as a construal to which its author is even provisionally attached. In this respect the appearance of idealism at 4.023 is perhaps more typical. Room is made for the thought of it to arise, and an issue is knowingly reshaped to give prominence to one of the sources of its attractiveness, but we do not have even an 'apparent affirmation' of it. Fixating on the general pattern described at the beginning of this section is, I think, liable to obscure the complex variety in Wittgenstein's handling of this idea, variety that Moore rightly emphasizes in the passage just quoted.

5 Acquiescence

In his *Points of View* Moore succinctly characterizes the transcendental idealism that he and I agree was of central concern to Wittgenstein. It is

> the idea that the unity of reality is determined by its being held together at the transcendent level. This holding together is part of the transcendent process whereby we contribute a conceptual structure to reality, allowing for things to be some way or other. (1997: 125)

This passage occurs in summarizing Kant's view of how the unity of the represented world is forged through the application of the categories, this application being itself required to secure the unity of the representing subject. But whilst Kant is Moore's topic here, the possibility of a parallel understanding of Wittgenstein is clearly prefigured.

> The conceptual components of any judgement are configured in such a way as to say, '*This is how things are.*' And this most fundamental of ideas, the idea of things being a certain way, which so primordially characterizes the form of reality, is [thereby] brought within the ambit of Kant's idealism. (Moore 1997: 124)

This understanding is then made explicit in the following chapter.

[Wittgenstein's] preoccupation is with a unity in which everything is held together, an abstract, logical unity which contributes nothing to what the world is like but constrains what it could be like. It is the unity that attends every possibility. It is also the unity of self-consciousness.... The world's unity *is* the possibility of its being represented from a single point of view.... So the world is *my* world. (1997: 150–1)

In §I of his chapter, where he first summarizes the differences between us, Moore says that 'Kant...remains very much in the picture for Sullivan, but only as the enemy' (p. 242). Taken from context, and juxtaposed, as here, with Moore's summary statements of his own understanding of the parallel views of Kant and Wittgenstein, this remark could be misleading. For there is much in Moore's summary that I endorse. Wittgenstein is indeed 'preoccupied with a unity in which everything is held together'. He shares Kant's ambition to secure[6] simultaneously the unity of the subject (or, what comes to the same thing, the unity of language) and the unity of the world. The universal form of judgement—the general form of proposition, '*This is how things are*'—is for Wittgenstein the encapsulation of both. To fulfil its role it must indeed be 'an abstract logical unity which contributes nothing to what the world is like'. But if it is genuinely to fulfil this role—if it is to encapsulate the unity of the world *simpliciter* (rather than of a reality subjectively indexed, albeit invisibly or transcendently indexed), and of language *simpliciter* (rather than of a language somehow parochially confined)—then neither can it '*constrain* what the world could be like' (emphasis added). The logical form that unifies the world must impose no substantive structure on it, no a priori order in which the subject can discover a reflection of its uniqueness. It is in connection with this last point that I see Wittgenstein addressing Kant as a target, or 'the enemy'. But this point of disagreement, however fundamental, occurs in a context of alliance.

In the papers Moore refers to my broad intention has been to explain how Wittgenstein works to evaporate the idealism out of Kant's transcendental endeavour, to leave a kind of purified realism. The position he aims at is one that I would like to call, so as to register both its location within the transcendental tradition and its departure from Kant, 'transcendental realism'. But of course that label is already spoken for.

Towards the end of §II of his chapter Moore summarizes this broad view better than I have done.

[Wittgenstein] wants us to acquiesce in viewing the limits of language as precisely that: limits, not limitations.... Language is the totality of propositions; the world is the totality of facts; and the limits of these totalities coincide...But they coincide not because the limits of either are

[6] The word 'secure' is chosen as being neutral between Kant's and Wittgenstein's different conceptions of the task. For Kant it is a matter of constitutively explaining unity; for Wittgenstein it is a matter of neutralizing threats of disunity.

limitations somehow set by the other, nor yet because the limits of both are limitations somehow set by the subject, but precisely because they are the *limits* of two formal totalities that are to be conceived in terms of each other. (p. 248)

He expresses the hope that there is nothing in this summary to which I would take exception. Though it is no doubt ungracious to say so, there is: the word 'acquiesce'. The *Shorter* says: '*acquiesce*, v.i. 1. remain at rest or in quiet subjection to; rest satisfied. 2. agree, esp. tacitly; raise no objections.' As is vividly illustrated by Quine's notoriously sporadic 'acquiescence in the home language', acquiescence always reserves the right to take itself back. However much one says, 'Absolute limits are not limitations', if that is a matter of acquiescence—something to which one 'raises no objections', and accepts in 'quiet subjection' to the recognized incoherence of construing limits *as* limitations—then, I think, one still retains more sympathy with transcendental idealism than Wittgenstein intends we should. This sympathy might lead someone, for instance, to seek to construct another and more defensible outlet for the insights he takes to motivate transcendental idealism—perhaps in the fashion of Moore's own project in *Points of View*. Though it does not provide the precise statement of the difference between us that Moore is looking for, this difference over wording does convey something of its character.

6 Overdetermination

On the sources of idealism in Kant's theoretical philosophy, and on Wittgenstein's diagnosis of and response to them, Moore and I are, nonetheless, and on Moore's account, centrally in agreement. In summary of this he writes,

Kant arrives at his transcendental idealism by taking as a datum the existence of a priori knowledge that is not merely of limits, but of limitations; not analytic, but synthetic (Introduction, §§IV and V). Wittgenstein, in direct opposition to this, insists that 'there is no a priori order of things' (5.634; cf. 2.225, 3.04–3.05, 6.31, 6.3211). That is, there is no a priori knowledge of limitations. And this is how, for Wittgenstein, transcendental idealism is to be resisted. On this Sullivan and I are in agreement. (p. 250)

The paragraph containing this passage ends with a point of suspension. In the following paragraph, immediately after the point, Moore asks, '*Why* does Kant see limitations where Wittgenstein sees only limits?' He answers that Kant *needs* a contrast of the kind that limitations effect, a substantive restriction on knowledge and hence on the reality to which fully contentful thought is answerable, 'in order to make room for faith' (B xxx). Under that heading falls a concern with whatever is ultimately of value. Such matters are, for Kant, 'beyond the reach of discursive knowledge'. For Wittgenstein, similarly, they are beyond the reach of language. Wittgenstein's insistence on this intrinsic limitation of language is, Moore suggests, transcendental idealism's most blatant appearance in the *Tractatus* (p. 250).

I do not see how these two paragraphs cohere. My complaint is of course not that Moore does not spell out here how these two motivations towards transcendental idealism—the concern to vindicate substantive a priori knowledge, and the concern to make room for faith—are connected in Kant: he has done that elsewhere (Moore 2003). Nor is it that Moore does not spell out how the echoes of these connected Kantian concerns are themselves connected in Wittgenstein's argument. In their contributions to this volume Genia Schönbaumsfeld and Hanne Appelqvist both suggest—albeit with important differences—that in his attitude to ethics and to everything 'higher', expressed in the 6.4s, Wittgenstein is straightforwardly following through on his logical commitments. But, as Peter Hacker long ago observed, such a tight, logical connection into the overall argument of the book might be too much integration to hope for. It would be enough if the two lines of thought, even if originally independent in motivation, could be 'interwoven', and understood as converging on a common stance, each thereby offering some reinforcement to the other (Hacker 1986: 105). My complaint is rather that a Kantian understanding of Wittgenstein's remarks on value, when combined with the interpretation of his response to the theoretical sources of idealism on which Moore and I are agreed, rules out even the modest integration Hacker envisages. Knowledge must be restricted to make room for faith only if there is *some* sense in which knowledge and faith are claimants to the same territory. On Kant's account they are: what faith stipulates, and knowledge attains, are truths. Hence his practical philosophy *needs*, as Moore says, the contrast his theoretical philosophy effects. In the domain of theoretical philosophy Wittgenstein's response to Kant is to deny that any restrictive or contrastive limitation is needed. Only so can the self of solipsism shrink to an *extensionless* point, leaving the *whole* of reality coordinate with it. So, if Wittgenstein's ethics were to require such a contrast, it would not be loosely interwoven with his logical thought: it would be in direct collision with it. I think it follows that whatever room Wittgenstein wanted to make for faith cannot have been *anything like* the room Kant intended.

7 Isomorphism

In other words, the isomorphism between the views of Kant and Wittgenstein that Moore displays diagrammatically in the final section of his chapter cannot be right. Or more carefully, it cannot serve Moore's purposes. Outside of mathematics an isomorphism is simply *some* shared structure or pattern, and I of course don't doubt that the pattern Moore points to is there. He immediately cautions that the isomorphism is a rough one, that there are 'profound differences in how the structure is instantiated' (p. 252). That is no ground for complaint. As always, foregrounding similarities involves backgrounding differences. Some of the differences Moore then mentions as neglected by his pattern—such as the distinction Wittgenstein draws between nonsensicality and senselessness—are, in some contexts, of the first importance for understanding the *Tractatus* correctly. But still, that is no ground for complaint. There is no saying which similarities or differences are important independently of the issue in hand, no saying how rough an isomorphism can be while still being illuminating.

If the issue were simply whether, in the 6.4s, Wittgenstein casts the limits of the world as limitations, limitations that exclude something of enormous significance, and something which—or perhaps better, something sensitivity to which—commands his deepest respect, then *none* of the roughness in Moore's isomorphism would matter. His point would be proved. But as we noted at the start (§2), such a construal of limits is only ever a symptom of the matter that divides Moore and me.

I suggested at the end of the previous section that the issue in hand—at this point—is whether Wittgenstein thought of what is excluded in the 6.4s in 'anything like' Kant's way. In a final description of 'how we are led to transcendental idealism' Moore offers a way of making that somewhat less vague.

> For we contrast what finds expression in items on the left-hand side[7] with our other modes of rational engagement with things, and we come to regard empirical reality—that to which all items on the left-hand side must answer—as correspondingly restricted. (p. 252)

The phrase 'correspondingly restricted' is doing double duty here. Compare: 'If my neighbour extends his garden, mine will be correspondingly restricted (use 1); when new plans are lodged with the Land Registry, the red hatched area representing my garden will be correspondingly restricted (use 2).' The first use refers to two sides of one boundary, the second to 'the same' side of two coordinated boundaries. Moore explicitly uses the phrase in the second way: we come to regard reality as restricted in coordination with thought. But the *ground* for this use can only come from an appeal to the first. It is only if we *already* think of the contrast mentioned at the start of the passage on the model of one between two sides of a fence that there will be any such thing as a corresponding adjustment to the plan. This casts doubt, I think, on the passage's claim to describe 'how we get to transcendental idealism': it seems that the transition it describes, towards conceiving reality as subjectively constrained, can have any grip on us only if we are already operating within a framework that views thought as subject to a limitation—as reaching up to some fence but no further.

However that may be, the immediate helpfulness of the passage is as pointing to a necessary condition for Wittgenstein's exclusion of value from the world to be 'anything like' Kant's, which is to say, for it to be indicative of transcendental idealism. There must, namely, be *some* domain across which it marks a boundary with two sides. I already remarked in the previous section, and Moore himself is quick to stress, that the most obvious candidate domains are ruled out. For Wittgenstein, in contrast to Kant, all truth, all thought, and all meaning lies, in Moore's pattern, 'on the left-hand side'. Is there any viable alternative?

In his chapter Moore offers instead

[7] On the 'left-hand side' are contentful thoughts (Kant) and significant propositions (Wittgenstein).

our rational engagement with things—that part of our engagement with things which is made possible by the fact that we are rational, thinking beings. (p. 251)

The explanation given here seems to exclude, for instance, denting the chair one sits in, something a similarly shaped stone might do.—Unless perhaps denting the chair was one's aim in sitting, an intentional action? At any rate, at least some brutely causal bumpings into things will not count as rational engagement with them. Is this enough to yield a positive conception of the domain to be divided? I think it would be natural to complain here that Moore's 'made possible by' is too weak or heterogeneous a relation to yield what we need, that something might be possible only *for* a thinker yet still be nothing like a thought. But perhaps such a response is unhelpful. The question is whether Moore's proposal offers an alternative standard of likeness, and so is not to be answered by those we already have.

Suppose instead, then, that we approach the question from the other side, citing some things that we know will be included in the putative domain. Often in the *Tractatus*, and prominently in the 6.4s, Wittgenstein contrasts thoughts with *feelings*. Feelings, at least of the kind Wittgenstein has in mind, will clearly be part of our rational engagement with things. And certain very particular feelings are connected intimately by Wittgenstein with finding sense in the world (6.41), and finding anything to be of value (6.45). Knowing only this much about the putative domain we could then ask: Is thought limited by feeling? Do they butt up against each other, so that recognizing one is recognizing a corresponding restriction in the other? Did Wittgenstein conceive of them in that way? Well, we could ask these questions, but it's far from clear how we might go about answering them.

It is far from clear, I suggest, because we have lately (in the previous two paragraphs) tried to guide the discussion by what was acknowledged to be only a *necessary* condition for the drawing of a contrast to be indicative of transcendental idealism. There must be, we said, some domain that is divided. But when a domain of cherubs and pincushions is divided into the Angelic and the non-Angelic, idealism is not in the offing. To make headway, and to give more tractable form to the issues just raised, we have to draw more substantially on how the idea of a restricted domain figures in transcendental idealism, and in particular on how it contributes to the project of transcendental explanation.

Idealism, as I agreed at the outset with Moore, is a doctrine concerning something's form or essential nature. A domain to which it relates will be one united, or putatively united, by some common form. Paradigmatically[8] there will be principles claiming to articulate this uniting form, and thereby claiming authority or validity across the entire domain. But this claim, for whatever reason, is problematic, lacking in any apparent

[8] The paradigm is set in the sections entitled 'Conclusions' in the Transcendental Aesthetic (A26/B42–A30/B45, A32/B49–A36/B53).

grounding. The idea of a restriction on the domain enters as supplying that grounding: it is at once a validation of and a limitation on the claim it grounds.

The issue in hand is whether Wittgenstein's approach to ethical value has any share in this paradigm. In our previous fumbling approach to this issue we asked, in effect, whether thought and feeling putatively constitute a unified domain, or whether there are principles problematically claiming dominion across them both. The fuller appearance of the paradigm allows us to see that these questions were poorly framed. To ask them would be rather like asking whether appearances and things in themselves are side-by-side in space. It would, that is, be a matter of using terms that can only belong to the Kantian resolution of a problem to question whether there was ever a problem in the first place (and then very probably concluding—misled by the inept form of our question—that there wasn't). The right question should instead use terms in which the problem might present itself to us *before* its transcendental resolution. Part of it would then run:

(1) Do science and ethics at least putatively constitute a single domain of truths, governed by unifying principles?

And then a further part, needed to reconnect to the interpretive issue, would ask:

(2) Did they ever present themselves as such to Wittgenstein?

There is evidence enough in ethical theory—perhaps even in the *existence* of ethical theory—that the answer to question (1) is Yes. This implies than an approach to ethics of the kind envisaged here, a resolution broadly in accord with the Kantian paradigm of the problems caused by a positive answer to question (1), is at least a possible one—possible, that is, for *someone*. The further question, whether this approach might actually be Wittgenstein's, could be addressed in two ways. Textually, we would need to ask whether the paradigm helps us to make *detailed* sense of the 6.4s and their *Notebooks* antecedents—not just of isolated pronouncements, such as that 'there can be no ethical propositions' (6.42), which of course can be made to fit, but of the arguments, or suggestions of arguments, that lead to those pronouncements. And biographically, we might look to Wittgenstein's own engagement with value for signs that it sometimes took a form inviting a Kantian resolution.

For what it's worth, the biographical approach seems to me to speak unambiguously for answering No to question (2). It seems plain, that is, that value *never* presented itself to Wittgenstein in quasi-propositional form. From his earliest encounters with Russell, and so long before his philosophy took a Kantian turn, Wittgenstein regarded the inclination to treat of value or its appreciation in theoretical form, not as an inevitable illusion born of the proper function of reason, but as a sign either of superficial charlatanism or deep corruption. It is of course true that in *some* sense Wittgenstein did deny knowledge to make room for faith. He denied to language, reason, or discursive understanding any rights whatever in the region faith should occupy. Numerous documents, memoirs, and anecdotes confirm that. In doing so, I think, they often give us a better sense of the character of that denial than does Moore's rough isomorphism.

Consider, for instance, this passage from McGuinness, writing about Wittgenstein's relations with the Vienna Circle:

> Still, Wittgenstein spoke to them about 'their' topics . . . but he also spoke about themes of his own and in his own way, harking back to the more general lessons of the *Tractatus*. He told Feigl, for example, that it was necessary to get rid of metaphysics in order to make room for *den Ernst des Lebens*, the seriousness of life. Deep commitment was required, not rational justification, and whoever had the former . . . would not worry his head about the existence of the world or the ultimate substance of it. Nor indeed were the problems internal to science of real importance. *Kümmert euch nicht um den Lärm im Haus der Wissenschafter!* ('Don't concern yourself about the noise in the scientists' house'—as if it were 'the neighbours'). (McGuinness 2002: 191)

Here, with its mention of the need to clear away metaphysics to make room for the seriousness of life, things start out as well as they could for Moore's Kantian comparison. The continuation, elevating commitment over justification, is still broadly congenial. But then, for Moore's proposal, things take a definite turn for the worse. Perhaps we cannot rely on Feigl's recollection to capture every nuance of Wittgenstein's account, but even so the resolution portrayed is clearly Tolstoy's rather than Kant's. Someone who achieves commitment and faith no longer 'worries his head' about the metaphysical problems that used to torment him. Like Levin, he just rhythmically swings the scythe.

'For what it's worth', I said. The biographical approach to question (2) is suggestive, but only the textual approach could be decisive. I do not feel competent to tackle that task.[9] Fortunately, though, I don't have to. For the question whether Wittgenstein's remarks do in fact embody a Kantian resolution of the problematic of value is pre-empted by the question whether they *could* do so. And from what we have already seen in this section, I think it follows that they could not.

We asked whether Wittgenstein's locating value 'outside the world' (6.41), and beyond the reach of language, implies a corresponding restriction on language and the world it represents, a limitation of the kind indicative of transcendental idealism. And we saw that we could not make that question determinate—that is, that we could not make it bear determinately on Moore's title question—without drawing substantially on the project of transcendental explanation to make plain what *kind* of limitation that is. We saw, in other words, that one cannot separate the question, whether Wittgenstein's 'practical philosophy' displays a commitment to transcendental idealism, from the question whether his 'theoretical philosophy' provides the basis for satisfying that commitment. On the second of these questions, Moore says, 'Sullivan and I are in

[9] Michael Potter has undertaken it, in his (forthcoming) book *Wittgenstein, 1916*. As a result of many conversations I have no doubt drawn heavily on his conclusions.

agreement': it does not. So far as I can see, that leaves only one possible answer—the same answer—to the first.

References

Hacker, P. M. S. (1986) *Insight and Illusion: Themes in the Philosophy of Wittgenstein*, revised edition. Oxford: Oxford University Press.

Kant, Immanuel (1998) *Critique of Pure Reason*, trans. and ed. P. Guyer and A. W. Wood. Cambridge: Cambridge University Press.

McGuinness, Brian (2002) *Approaches to Wittgenstein*. London and New York: Routledge.

Moore, A. W. (1997) *Points of View*. Oxford: Oxford University Press.

——(2003) *Noble in Reason, Infinite in Faculty: Themes and Variations in Kant's Moral and Religious Philosophy*. London: Routledge.

——(2011) 'A Response to Sullivan', in R. Read and M. A. Lavery, eds, *Beyond the Tractatus Wars*. London: Routledge.

Pears, David (1987) *The False Prison*, vol. 1. Oxford: Clarendon Press.

Stenius, Erik (1960) *Wittgenstein's Tractatus: A Critical Exposition of its Main Lines of Thought*. Oxford: Basil Blackwell.

Sullivan, Peter (2011) 'Synthesizing without Concepts', in R. Read and M. A. Lavery, eds, *Beyond the Tractatus Wars*. London: Routledge.

Wittgenstein, Ludwig (1922) *Logisch–Philosophische Abhandlung/Tractatus Logico–Philosophicus* (TLP). Trans. C. K. Ogden, London: Kegan Paul Ltd, 2nd edition 1933. Trans. D. F. Pears and B. F. McGuinness, London: Routledge and Kegan Paul Ltd, 1961, 2nd edition 1974.

——(1973) *Letters to C. K. Ogden* (LO), ed. G. H. von Wright. Oxford: Blackwell.

——(1979) *Notebooks 1914–1916* (NB), 2nd edition, ed. G. H. von Wright and G. E. M. Anscombe, trans. G. E. M. Anscombe. Oxford: Blackwell.

Index

acquaintance, principle of 8–10, 146–51, 179–81, 184–6, 190n., 191–6, 201–2, 217–19, 228n.
acquiescence 248, 262–4
aesthetics 40ff., 72
Ahonen, H. 50n.
Anscombe, E. 35–6, 84, 141n., 204n.
anti-realism 9, 143, 161–7
Appelqvist, H. 2–4, 265
Armstrong, D. 171n., 174n., 175n., 178n.

Baker, G. 99n.
Barcan formula 175n.
Barrett, C. 55
Bazzocchi, L. 13, 26–7
beauty 44–8, 50–4, 57
Berkeley, G. 10, 170–8, 185–7, 197, 203–5, 214–15, 220–6, 232–3
bipolarity 52, 106n., 239
bivalence 161–6
Boltzmann, L. 3
bracket-expressions 129–32, 136
Brentano, F. 77
buddism 49n.
Byrne, A. 224n.

Carnap, R. 100, 131, 156n.
Cheung, L. 111n., 121n.
Child, W. 2, 8–9
Christianity 49, 62–3, 70
  eastern 49n.
  western 49n.
compositionality 97, 100–1, 106–7, 110
Conant, J. 59n., 60, 64, 82n., 86n., 205n., 210n., 211n., 213n., 233n., 241n.
concept *horse* problem 4–5, 76ff.
conceptual relativism 220–2, 226–31
conceptual scheme 170, 220–7, 230n.
consequentialism 64
Craig, E. 150n.
Crary, A. 143, 161n.
Creegan, C. 59n.

Davidson, D. 10, 85, 106n., 108, 114, 121n., 122, 165, 170, 220–33
Deans, R. 213n.
definite descriptions 82n., 84–85, 91, 189
Descartes, R. 252n., 258
de Silentio, J. 63, 69–73
Diamond, C. 8–9, 143–53, 157–8, 161–6, 199n., 205n., 211n., 213n., 233n., 241n., 253n.
disinterestedness 46–8
Dostoevsky, F. 49n., 59, 62
Drury, M. 59
Dummett, M. 6, 99n., 109–10, 143–4, 161–7

elementary sentences, *see* propositions, elementary
elucidation 72, 86n., 92–3, 156
Engelmann, P. 4, 30–5, 37, 61–4, 68–70
epistemology 150–1
ethics 4, 22, 38, 40ff., 58ff., 253n., 259, 265–8
Evans, C. 59n.
excluded middle, law of 261
exclusivity thesis 77, 83
expressivism 50n.

faith 55n., 61–3, 68–71, 250–2, 264–5, 268–9
Fara, D. 85n.
Ferreira, J. 59n.
Fletcher, W.M. 15
Floyd, J. 133n., 140n., 240n.
Fodor, J. 223n.
Fogelin, R. 7, 116–17, 131
formalism 1, 45, 50
formal relation 141
form-series 133, 136–41
Frege, G. 1–8, 30–3, 76–94, 97–122, 125–36, 161, 165, 179, 188–93, 214n., 218, 229

Geach, P. 5, 7, 76, 85–88, 93, 131–2
generality 6–9, 91, 98, 125–41, 145–6, 149n., 178, 187–8, 196–205, 212n., 214n., 215–17, 223, 227–9
  formal 126, 136n., 139
  multiple 132
Geschkowski, A. 13, 19–24
Glock, H. J. 41–4, 48n.
God 30, 32, 41, 49–51, 55, 62–72, 170, 250–2, 254n.
Goldfarb, W. 151n., 221n.
Grasshoff, G. 33, 35
Grillparzer, F. 50n.
Groay, H. 30, 32

Hacker, P. 41–3, 60n., 99n., 101n., 161n., 204n., 253n., 265
Haecker, T. 61
Hanslick, E. 50n.
Hay, W. H. 176n.
Hegel, G. 65–6, 71
Heidegger, M. 214n.
Hertz, H. 3
Hilbert, D. 93
Hintikka, J. 45, 193n.
Hintikka, M. 45, 193n.
holism 215–20, 222n., 223n., 229–32
Hübner, A. 61
Hume, D. 11
Hutto, D. 60n.
Husserl, E. 105
Hylton, P. 99n., 100, 121n., 129n.

idealism 10–11, 170–233, 256–8
  master argument for 10, 170–8, 185–7, 220, 226n., 232
  see also transcendental, idealism
ineffability 60, 200, 205n.
infinity 220n.
intuitionism 1
irresolute reading 2, 4, 143, 157–8, 212n., 240n.
Ishiguro, H. 42n., 149n., 151n.
isomorphism 42, 99n., 104n., 251, 259, 265–9

Janik, A. 45, 50, 253n.
Johnson, W. E. 15

Kang, J. 13, 22–3, 27
Kannisto, H. 41–4
Kant, I. 3–4, 10, 38, 40–57, 64–71, 207n., 231, 239–44, 249–54, 257–9, 262–9
Kerry, B. 77–9
Keynes, J. M. 1, 33
Kierkegaard, S. 3–4, 59–74, 214n.
Kierkegaardian despair 62–4
King, J. 105, 110n.
Klement, K. 105, 107n., 110n.
Kraus, K. 3, 247n.
Kremer, M. 133n., 205n., 212n.
Kripke, S. 161n.

Lampert, T. 33, 35
Levine, J. 2, 8–10
Lippitt, J. 60n.
Locke, J. 11, 176
logical category distinctions 76, 89–94, 241–2
  see also concept *horse* problem
logical constants 6, 97–100, 105–22, 135
logical form 41–3, 50, 141, 165, 184, 193–4, 216, 261–3
logical possibility 183, 246
logical space 116, 126, 193–9, 208–13, 216–25, 230–2, 261
logicism 11
Loos, A. 3
Luckhardt, C. G. 54n.

Mackie, J. L. 171–2, 174n., 177–8, 203–4, 214, 232
Malcolm, N. 59
mathematics 1, 35, 38, 88, 265
McDowell, J. 229n., 233, 249n.
McGinn, M. 60n., 212n.
McGuinness, B. 4, 13, 18–22, 28–31, 37, 41–2, 151n., 163n., 240n., 269
meaning
  picture theory of 52, 100, 104, 113–16, 122, 212n.
  truth-conditional theory of 100, 106, 108, 121
  use theory of 146–51
Meinong, A. 165
Meyendorff, J. 49n.
Milne, P. 2, 6–7, 110n.
mind, philosophy of 2, 143
Monk, R. 41, 61, 72
Moore, A. W. 2, 10–11, 44, 94n., 170n., 212n., 239n., 240n., 243n., 244, 246n., 251, 253n., 256–70
Moore, G. E. 14–17, 29, 37, 106n., 120, 149n., 196n., 215n.
Morrel, O. 16, 59
Mounce, H. 204n.

Nagel, T. 10, 170, 220–33
necessity 42, 54–5
  logical 53–5, 64, 71, 262
Newman, M. 35
new reading, *see* resolute reading
Nientied, M. 59n.
Nietzsche, F. 62
nonsense 2, 11, 41, 60, 83, 86n., 87, 157, 203, 211, 212n., 214, 242, 247–54, 260
  philosophers' 85
  substantial 60
N-operator 5–6, 23, 127, 131, 140–1, 195, 201

Ogden, C. K. 35–6, 190n.
Ostwald, W. 35–6
overdetermination 264

Parsons, T. 79, 83
Pears, D. 42, 156n., 193n., 200n., 207n., 261
Phillips, D. Z. 59n.
Pinsent, D. 15, 18, 29
Platts, M. 106n., 122
Potter, M. 2–3, 13, 135n., 137n., 269n.
practical reason 41

private language argument 8, 143–67, 199n.
Priest, G. 170n., 174n.
Prior, A. N. 170–8, 181, 185, 197, 220–1, 226n., 228n., 234n.
Proops, I. 2, 4–5, 60n., 98n., 101, 103n., 118–21, 212n.
propositional content 171, 189, 194n., 221–4, 229–32
propositional function 6–7, 91, 129–41, 147, 151n., 152–3, 163
propositions
  elementary 5–6, 8–9, 38, 102–4, 110–11, 114–18, 126–41, 146–7, 151n., 152–3, 163, 194–203, 206, 216–20
  molecular 6–7, 98
pseudo-propositions 239–40, 248n., 251–2
punctuation marks 6, 97, 122
Putnam, H. 226n.

quantification 6–7, 85, 125–41, 145–6, 165
  existential 228
  first-order 133, 138
  higher-order 125, 129, 133–9
  monadic 133
  second-order 133–9
quietism 143
Quine, W. O. 106n., 264

Ramsey, F. 5, 7–8, 34–6, 131, 196n.
Read, S. 108n., 212n.
realism 9, 143, 152, 161–7, 207n., 225n., 226–31, 243n.
  Dummettian 144, 162–3
  global 161, 164–6
  pure 209, 263
  see also transcendental, realism
religion 22, 38, 59–60, 70n., 72
resolute reading 2–4, 143–4, 153–5, 157–61, 212n., 240n., 258
Ricketts, T. 2, 6–8, 92, 103n., 118–19, 210n.
Rhees, R. 41, 56
Rogers, B. 133n.
Rorty, R. 223n., 226n.
Russell, B. 1–10, 13–38, 59, 80n., 82–4, 91–4, 97–8, 111, 118–22, 125–39, 144–9, 157, 161–7, 170–1, 178–220, 227–8, 230n., 232–3, 268

Sachlage 27, 35–6
Sachverhalt 27, 35, 115, 127
Sacks, M. 251n.
saying / showing distinction 76, 93, 154–7, 204–6, 210–14, 225n.
Schlick, M. 1, 70, 156, 208n.
Schönbaumsfeld, G. 2, 4, 265
Schopenhauer, A. 3–4, 11n., 38, 41–4, 48n., 49n.
Schubert, F. 45
sense-data 144n., 148, 155–7, 165n., 181, 190n., 191–3, 216n.
sentence-function, *see* propositional function
Shier, D. 117n.
Silesius, A. 59
singular term 81, 84–6, 91, 125, 165–7
solipsism 9–11, 159n., 170ff., 265
Spengler, O. 3
Sraffa, P. 3
St. Augustine 59
Stenius, E. 41, 49n., 97n., 253n., 259n., 261
Stern, D. 160n.
Stokhof, M. 41, 49n., 56
Stroud, B. 226n.
subject
  metaphysical 42n., 43–4, 47–9, 57
  transcendental 43
subjective purposiveness 50
subjective universality 52
suicide 55–6, 62
Sullivan, P. 2, 10–11, 60n., 81n., 91n., 127n., 137n., 149n., 175n., 205n., 207n., 212n., 219n., 240–54, 256

Tarski, A. 108, 121–2
taste 4, 45–54
tatsachen 36
Tilghman, B. 50
Tolstoy, L. 3, 49, 50n., 59, 62n., 269
Toulmin, S. 45, 50, 253n.
traditional reading, *see* irresolute reading
transcendent 49, 251n., 257, 262
transcendental
  deduction 41, 257
  idealism 2, 41–4, 239ff., 256ff.
  knowledge 251n.
  realism 263
Travis, C. 243n.

value 4, 40–54, 64–72, 250–4, 259, 265–9
van Heijenoort, J. 80
verificationism 151, 223, 225n.
Vienna Circle 1, 70, 208n., 214n., 269
von Ficker, L. 4, 26, 34, 61, 72–3, 254
von Wright, G. H. 37–59

Waismann, F. 60
Wehmeir, K. F. 133n.
Weiner, J. 79n., 82n., 89n.
Weininger, O. 3, 55
Whitehead, A. 34, 98, 129n., 138, 188n., 196
Wiggins, D. 253n.
Williams, B. 176n., 226n., 247n.
Williamson, T. 223n., 229n.
Wittgenstein, H. 4, 18–29, 61
Wrinch, D. 34–5
Wuchterl, K. 61